UNCONVENTIONAL COMPUTING, ARTS, PHILOSOPHY

Volume 2

WSPC Book Series in Unconventional Computing

Print ISSN: 2737-5218
Online ISSN: 2737-520X

Published

Vol. 2 *Unconventional Computing, Arts, Philosophy*
edited by Andrew Adamatzky

Vol. 1 *Handbook of Unconventional Computing*
(In 2 Volumes)
Volume 1: Theory
Volume 2: Implementation
edited by Andrew Adamatzky

https://doi.org/10.1142/9789811257155_fmatter

Preface

Unconventional computing is about breaking boundaries in thinking, acting, and computing. Typical topics of this non-typical field include, but are not limited to, physics of computation, non-classical logics, new complexity measures, novel hardware, mechanical, chemical, and quantum computing. Unconventional computing encourages a new style of thinking while practical applications are obtained from uncovering and exploiting principles and mechanisms of information processing in, and functional properties of, physical, chemical, and living systems; in particular, efficient algorithms are developed, (almost) optimal architectures are designed, and working prototypes of future computing devices are manufactured.

The term "unconventional computing" has no exact definition. Proceeding by inclusiveness we could say that the following research topics are most commonly, but not necessarily, classified as 'unconventional': physics of computation (e.g., conservative logic, thermodynamics of computation, reversible computing, quantum computing, collision-based computing with solitons, optical logic); chemical computing (e.g., implementation of logical functions in chemical systems, image processing and pattern recognition in reaction-diffusion chemical systems, and networks of chemical reactors); bio-molecular computing (e.g., conformation based, information processing in molecular arrays, molecular memory); cellular automata as models of massively parallel computing complexity (e.g., computational complexity of non-standard computer architectures; theory of amorphous computing; artificial chemistry); non-classical logics (e.g., logical systems derived from space-time behaviour of natural systems, logical reasoning in physical, chemical, and biological systems); smart actuators (e.g., molecular machines incorporating information processing, intelligent arrays of actuators); novel hardware systems (e.g., cellular automata VLSIs, functional neural chips); mechanical computing (e.g., micromechanical encryption, computing in nanomachines, physical limits to mechanical computation).

There are two discipline-wise paths to unconventional computing. First, you are initially trained as a mathematician or computer scientist, then you rebel and start pushing the limits of conventional science, and eventually find yourself outside the well-established tracks. Second, more common, you are trained as a chemist, biologist, physicist, then you get involved in computation and get eager to understand the meaning of information and computation in natural systems, and subsequently start realizing computing devices in novel substrates. Unconventional computing remains at the forefront of emergent technologies. This forefront is continuously changing its shape and meaning. Only static fields can be defined formally. Unconventional computing is dynamic and fluid and therefore the field defines formal definitions.

The book brought together scientists, philosophers, artists, and engineers and encouraged them to express their vision of unconventional computing in a lively informal manner. A range of ideas presented is impressively wide. A reader from any walk of life and with level of education will find topics discussed in the book exciting and inspiring.

Andrew Adamatzky
Unconventional Computing Laboratory,
UWE Bristol,
United Kingdom

WSPC BOOK SERIES IN UNCONVENTIONAL COMPUTING *Volume 2*

UNCONVENTIONAL COMPUTING, ARTS, PHILOSOPHY

Editor **Andrew Adamatzky**

University of the West of England, UK

World Scientific

NEW JERSEY • LONDON • SINGAPORE • BEIJING • SHANGHAI • HONG KONG • TAIPEI • CHENNAI • TOKYO

Published by

World Scientific Publishing Co. Pte. Ltd.
5 Toh Tuck Link, Singapore 596224
USA office: 27 Warren Street, Suite 401-402, Hackensack, NJ 07601
UK office: 57 Shelton Street, Covent Garden, London WC2H 9HE

Library of Congress Cataloging-in-Publication Data
Names: Adamatzky, Andrew, editor.
Title: Unconventional computing, arts, philosophy / editor,
Andrew Adamatzky, University of the West of England, UK.
Description: New Jersey : World Scientific, [2023] | Series: WSPC book series in
unconventional computing, 2737-520X ; 2 | Includes bibliographical references and index.
Identifiers: LCCN 2022013719 | ISBN 9789811257148 (hardcover) |
ISBN 9789811257155 (ebook for institutions) | ISBN 9789811257162 (ebook for individuals)
Subjects: LCSH: Computer science--Technological innovations. |
Computer simulation--Philosophy. | Natural computation. | Artificial intelligence.
Classification: LCC QA76.24 .U53 2023 | DDC 003/.3--dc23/eng/20220527
LC record available at https://lccn.loc.gov/2022013719

British Library Cataloguing-in-Publication Data
A catalogue record for this book is available from the British Library.

For any available supplementary material, please visit
https://www.worldscientific.com/worldscibooks/10.1142/12870#t=suppl

Desk Editors: Nandha Kumar/Steven Patt

Typeset by Stallion Press
Email: enquiries@stallionpress.com

https://doi.org/10.1142/9789811257155_fmatter

About the Editor

Andrew Adamatzky is Professor of Unconventional Computing and Director of the Unconventional Computing Laboratory, University of the West of England Bristol, UK. He conducts research on molecular computing, reaction-diffusion computing, collision-based computing, cellular automata, slime mold computing, massive parallel computation, applied mathematics, complexity, nature-inspired optimization, collective intelligence and robotics, bionics, computational psychology, nonlinear science, novel hardware, and future and emergent computation. He has authored seven books, mostly notably *Reaction-Diffusion Computing*, *Dynamics of Crow Minds*, and *Physarum Machines*, and has edited 22 books in computing, most notably *Collision-Based Computing*, *Game of Life Cellular Automata*, *Memristor Networks*, and *Handbook in Unconventional Computing*. He has also produced a series of influential artworks published in the atlas "Silence of Slime Mould". He is Founding Editor-in-Chief of *Journal of Cellular Automata* and *Journal of Unconventional Computing* and Editor-in-Chief of *International Journal of Parallel, Emergent, Distributed Systems* and *Parallel Processing Letters.*

https://doi.org/10.1142/9789811257155_fmatter

Contents

https://doi.org/10.1142/9789811257155_0001

Chapter 1

Labyrinth: From Literature to Algorithms

José Félix Costa[*,‡] and Paula Gouveia[†,§]

[*] *Instituto Superior Técnico, Department of Mathematics, University of Lisbon, Portugal*

[†] *Instituto de Telecomunicações, Portugal*

[‡] *jose.felix.costa@tecnico.ulisboa.pt*

[§] *paula.gouveia@tecnico.ulisboa.pt*

In this chapter, we confront the labyrinth's reverie with the scientific solution of the labyrinthine tour given in the 19th century, a solution that was, however, ignored by the literary world of the 19th and 20th centuries. Present in literature since Antiquity, the concept of the labyrinth has been worrying philosophers, theologians, and mathematicians for more than 2,500 years. From the Daedalus multicursal labyrinth to Chartres Cathedral unicursal labyrinth, from Dante's *La Divina Commedia*, Gil Vicente's *Auto da Alma*, and *Queste del Saint Graal* (13th century) to the labyrinth of Jorge Luis Borges's *La Biblioteca de Babel* and Umberto Eco's *Il Nome Della Rosa*, individual growth processes settled down that unfold while one walks through the intricacies of the various labyrinths facing the self, the real Minotaur of Knossos. Nowadays labyrinths turn out to provide the subject of musicals such as David Bowie's *Labyrinth*, the series *Maze Runner*, and the movie *Harry Potter and the Goblet of Fire*, inter alia. However, we realize that the labyrinth we are familiar with has a simple, although ignored, mathematical solution in a new language in Mathematics unexpectedly introduced ad hoc by Leonhard Euler in the 18th century. How does a problem that has a simple mathematical solution continue, in the form of an insolvable problem, decorating the world of art, literature, and movies? It is just that crossing the labyrinth has become a metaphor for mental processes and dialectical learning, already described in Plato's *Euthydemus*, but which still does not have an adequate language in contemporary science. The contradiction is highlighted by Gaston Bachelard when he says that "... walking in the dark woods or in the tenebrous cave, getting lost, being lost are typical situations that provide innumerable images and metaphors in the clearest activity of the spirit, although [...] actual experiences of this kind are at the end of very rare accounts. [...] We are afraid of losing ourselves, without ever being lost."

1. Origins of the Labyrinth Problem

> The Labyrinth in modern literature is an image of multiple significance, but it was not always so. Indeed, the twentieth century can justifiably be called the 'age of the labyrinth' in that it tends to see labyrinths where there are none.[1]
>
> André Peyronie

Being or feeling lost in a labyrinth is a common condition in Neo-Gothic and Postmodern literature of the 19th and 20th centuries. Gaston Bachelard comments on the reverie of being lost: "Following a long narrow passage, or finding yourself at a crossroads, it determines two somewhat complementary anxieties. We can even free ourselves from one for the other. Let us enter this narrow path, at least we will not hesitate any longer. Let's get back to the crossroads, at least we won't be dragged any more. But the nightmare of the labyrinth encompasses these two anxieties and the dreamer experiences a strange hesitation: he hesitates in the middle of the single path."

From a long list of literary works addressing the theme of the labyrinth or expressing the maze's reverie, we highlight the following: *Queste del Saint Graal* (13th century), Dante's *Divina Commedia* (1307–1321), Shakespeare's *A Midsummer Night's Dream* (1594), Charlotte Brontë's *Jane Eyre* (1847), Wilkie Collins's *The Woman in White* (1859), Jorge Luis Borges's, *La Biblioteca de Babel* (1944), Lawrence Durrell's *The Dark Labyrinth* (1947), Umberto Eco's *Il Nome Della Rosa* (1980), and J.K. Rowling's *Harry Potter and the Goblet of Fire* (2000). Coincidently, mathematical statements about traversing mazes start appearing in the 19th century and beginning of the 20th century, by the hand of authors such as Lucas,[2] Ball,[3] Ahrens,[4] and Dudeney,[5] where many discussions and exercises on labyrinthine paths can be found. Let us thus summarize the story of

Fig. 1. (left) Knossos coin and (right) Argentine coin honoring Jorge Luis Borges.
Source: (left): File:001-knossos-01.jpg, *Creative Commons license* (*Creative Commons license* is *Creative Commons Attribution-Share Alike 2.5 Generic*, or *Creative Commons Attribution-Share Alike 2.0 Generic*, or *Creative Commons Attribution-Share Alike 3.0 Unported*, or *Creative Commons Attribution-Share Alike 4.0 International.*); (right): File:Jorge_Luis_Borges_Coin_1899-1999_-_Flickr_-_The Alieness_GiselaGiardino %C2%B%C2%B3.jpg, *Creative Commons license.*

the labyrinth which is detailed and profusely illustrated in the books by Matthews,[6] de Freitas,[7] and Hackworth.[8]

Born of an unnatural relationship between Parsiphaë and the white bull offered by Neptune for sacrifice to her husband, King Minos of Crete, the Minotaur had to be hidden from the eyes of the world in a labyrinth purpose-built by Daedalus, the architect. However, Minos's putative labyrinth engraved on the Knossos coin in Figure 1 does not seem very convincing, as it shows a unicursal path spiraling toward the center. There are no crossroads, no free will... The ancient labyrinths seem to consist of only one tortuous path culminating in a sanctuary. Surely the labyrinth design engraved on the Knossos coin is more of a clue to get out of the labyrinth than the architecture of the Minotaur labyrinth. Surely, the labyrinth of Knossos would be more like the multicursal labyrinth of the Gortyna caves in Crete, as seen in Figure 2, and has been described unambiguously in Ovid's *Metamorphoses*:

> The scandal concerning his family grew, and the queen's [Parsiphaë] unnatural adultery was evident from the birth of a strange hybrid monster. Minos resolved to remove this shame, the Minotaur, from his house, and hide it away in a labyrinth with blind passageways. Daedalus, celebrated for his skill in architecture,

Fig. 2. (left) Cave of Gortyna in Crete: possibly, in one interpretation, the real Cretan labyrinth (namely in Lawrence Durrell's *The Dark Labyrinth*) and (right) a replica of the ambiguous conventional maze of Daedalus.
Source: (left): https://freerepublic.com/focus/chat/2364359/posts; (right): File: 2015-09-18_Cockington_Green_Gardens_-_Maze.jpg, *Creative Commons license.*

> laid out the design, and confused the clues to direction, and led the eye into a tortuous maze, by the windings of alternating paths. No differently from the way in which the watery Maeander deludes the sight, flowing backwards and forwards in its changeable course, through the meadows of Phrygia, facing the running waves advancing to meet it, now directing its uncertain waters towards its source, now towards the open sea: so Daedalus made the endless pathways of the maze, and was scarcely able to recover the entrance himself: the building was as deceptive as that.
>
> Ovid. *Metamorphoses*, Book VIII: 155–168,
> A.S. Kline's translation.

A similar description appears in Virgil's *Aeneid* to illustrate the complexity of the games and military exercises performed by the men of Aeneas in honor of his father Anchises, upon their landing in Sicily:

> As once in Crete, the lofty mountain-isle,
> that-fabled labyrinthine gallery
> wound on through lightless walls, with thousand paths
> which baffled every clue, and led astray
> in unreturning mazes dark and blind:
> so did the sons of Troy their courses weave
> in mimic flights and battles fought for play,

Fig. 3. Cretan labyrinth in art: (left) *Theseus and Minotaur* by Antonio Canova, (center) Oscar Estruga's *Pasiphaë and the Bull*, Vilanova i la Geltrú, and (right) Dante guided by Virgil meets the Minotaur raging at the edge of the Seventh Circle of Hell, where the violent dwell, by Dürer.
Source: (left): File:Etienne-Jules_Ramey_Th%C3%A9s%C3%A9e_et_le_Minotaure_Tuileries_Paris.jpg#/media/File:Jardin_des_Tuileries_IMG_1843.JPG, *Creative Commons license*; (right): File:DVinfernoMinotaurOnCliff_m.jpg, public domain; (center): https://puntodincontro.mx/articoli2014/storiadiparoleitaliane 04022014-sp.htm.

like dolphins tumbling in the liquid waves,
along the African or Carpathian seas.

Virgil. *Aeneid*, Book V: 591–509,
Theodore C. Williams's translation, 1910.

Regardless of our inclination to enjoy such a cross-cutting and interdisciplinary theme as labyrinths, the truth is that the concept and its history undermine art as illustrated in Figure 3.

Leaving aside the possible relationship of the labyrinth with dance and military exercises (as in Virgil's *Aeneid*), we see, however, that the image of the labyrinth is used by poets throughout history in the description of intricate movements, such as the planetary orbits that in the models of Ptolemy, Copernicus, and Tycho Brahe were the result of the composition of a large number of elementary or less elementary circular motions, as Milton puts in evidence:

That day, as other solem dayes, they spent
In song and dance about the sacred Hill,
Mystical dance, which yonder starrie Spheare
Of Planets and of fixt in all her Wheeles

Resembles nearest, mazes intricate,
Eccentric, intervolv'd, yet regular
Then most, when most irregular they seem:
And in thir motions harmonie Divine
So smooths her charming tones, that Gods own ear
Listens delighted.

Milton. *Paradise Lost.*, 618–627.

Whether the labyrinth of Knossos was unicursal or multicursal (according to Plutarch as the intrincate architecture of the palace itself), does not change the fact that both concepts have a history until our days, although only the multicursal labyrinth was in fact of mathematical and, later in our days, computational interest, modeling diverse networks: social, conceptual, computer, etc. On the one hand, the classical multicursal labyrinth has a central place hiding a danger, either the Minotaur or, more recently, the second volume of Aristotle's *Poetics* (on comedy), a dangerous and forbidden book to the Benedictine order (in Umberto Eco's *The Name of the Rose*). On the other hand, the unicursal labyrinth was relevant to the moral imperative of Christianity: "Do not deviate from the path of righteousness if you want to reach the heavenly Jerusalem." In the adoption of the symbol by clergy and Christian writers, the old idea of multiplicity, whether of the paths or of the deities, was such an unacceptable concept that the labyrinth has been turned into its opposite. From an image of complexity, the labyrinth became a single path (see Figure 4). Christ takes the place of Minotaur in the labyrinths of 12th-century Italy and 13th-century northern France. Wright[9] argues that many of the labyrinths of churches, when interpreted as places of "pilgrimage to Jerusalem" in the Middle Ages, are fabricated in the period of the Neo-Gothic. St. Augustine compares the multicursal labyrinth to paganism. Christ provides the true path (unicursal), free from error, an idea that Dante resorts to in *The Divine Comedy.* We can see the unicursal labyrinth as a multicursal labyrinth where, in each crossroads, our guide, or guardian angel, shows us the way forward. We find this idea expressed in Gil Vicente's *Auto da Alma,* where the soul dragged in the labyrinth of life is guided by the angel:

Fig. 4. (left) Notre Dame d'Amiens Cathedral, (center) Notre Dame de Chartres Cathedral, 1215–1221, and (right) Grace Cathedral, San Francisco.

Source: (left): File:AmienCathedralLabyrinth.JPG, *Creative Commons license*; (right): File:Grace_Cathedral_interior_with_labyrinth.jpg, *Creative Commons license*; (center): https://roddreher.substack.com/p/the-thin-chartres-blue-line.

Angel. O who thus hinders thee?
On, on!
How loiterest thou on glory's path
So slowly!
O God, sole consolation!
Now is there none
Who of that victory honor hath
That is most holy.
Soul, already dost thou tire
Sinking so soon beneath thy burden?
Nay, soul, take heart!
Ah, with what a glowing fire
Of desire
Cam'st thou couldst thou see what guerdon
Were then thy part.
Forward, forward let us go:
Be of good cheer, O soul made holy
By this thy strife.

Devil. But what is all this coil and woe?
Why to and fro
Flutterest thou in haste and folly?
Nay, live thy life.
For very piteous is thy plight,
Poor, barefoot, ruined utterly,
In bitterness,
Carrying nothing to delight
As thine by right,
And all thy life is thus to thee
A thing senseless.
But don this dress, thy arm goes there,
Put it through now, even thus, now stay
Awhile. What grace,
What finery! I do declare
It pleases me. Now walk away
A little space.

Devil. Enjoy, enjoy the goods of Earth,
And great estates seek to possess
And worldly treasures.
Who to the hills, exiled from mirth,
Thus sends thee forth?
Who speaks to thee of foolishness
Instead of pleasures?
This life is all a pleasaunce fair,
Soft, debonair,
Look for no other paradise:
Who bids thee seek, with false advice,
Refuge elsewhere?

Gil Vicente. *The Soul's Journey,* 30–32, 32–34, and 23–24,
translation and notes by Aubrey F. G. Bell, 1920.

Referring to the building itself, the labyrinth can be a hedge garden (in this case it is common to find a centric labyrinth), a network of interconnected caves (like Gortyna in Crete, acentric), or simply a centric or acentric imaginary building. Since the width of the passages and meanders do not matter, a labyrinth can be identified with a connected graph where the vertices are the junctions or crossroads and the passages or meanders are the edges. The entrance and the exit to and from the labyrinth are distinguished vertices. The most classical challenge is to reach its center such as in J.K. Rowling's *Harry Potter and the Goblet of Fire*:

> That's right! – said Bagman. – A maze. The third task's really very straightforward. The Triwizard Cup will be placed in the centre of the maze. The first champion to touch it will receive full marks.
> We seemply 'ave to get through the maze? – said Fleur.
> There will be obstacles, – said Bagman happily, bouncing on the balls of his feet. Hagrid is providing a number of creatures... then there will be spells that must be broken... all that sort of thing, you know. Now, the champions who are leading on points will get a head start into the maze. – Bagman grinned at Harry and Cedric. – Then Mr. Krum will enter... then Miss Delacour. But you'll all be in with a fighting chance, depending how well you get past the obstacles. Should be fun, eh?
> The towering hedges cast black shadows across the path, and, whether because they were so tall and thick or because they had been enchanted, the sound of the surrounding crowd was silenced the moment they entered the maze. Harry felt almost as though he were underwater again. He pulled out his wand, muttered, – Lumos –, and heard Cedric do the same just behind him. After about fifty yards, they reached a fork. They looked at each other. – See you –, Harry said, and he took the left one, while Cedric took the right.

In an acentric labyrinth, the purpose of finding the center is not justified, but rather ensuring that one successfully leaves the labyrinth. Mathematics and Computer Science have added a broader purpose, which is to enter and exit the labyrinth after being sure that all of its galleries and crossroads have been visited. This is the purpose of our chapter, bridging together the concept of the

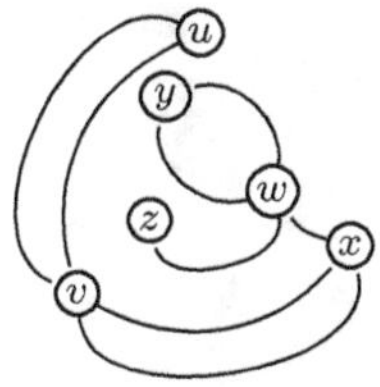

Fig. 5. Toy labyrinth.

literary labyrinth with the algorithms to escape from labyrinths. As a toy example, we will consider throughout the following sections the labyrinth which underlying graph is displayed in Figure 5.

This is the organization of the chapter. First, we discuss the language introduced by the mathematician Euler to solve a similar problem and the question of the labyrinth itself. Then we inquire if one can successfully traverse the labyrinth at random. Obtaining a negative answer, we rest on the possibility that the subject has access to the architect's plan of the building. Not at random and in absence of the plan, we explore Ariadne's thread in several ways to see how far we can go exploring the meanders of the labyrinth. Then we discuss William's training on labyrinths in Eco's novel, and we conclude by introducing a further quite unknown algorithm to escape from labyrinths.

2. From Ehler to Euler

In the 18th century, there were seven bridges over the Pregel River in the city of Königsberg (now Kaliningrad, Russia) that connected two islands, either to each other or to the river banks (see Figure 6). The inhabitants of Königsberg were delighted with a problem that has become famous: could it be possible to cross all bridges at once without passing the same bridge twice? The Königsberg bridge problem was solved by Leonhard Euler in 1736. It is assumed that Euler became aware of the problem through a letter addressed to him by the mayor of Danzig (Gdansk), Carl Leonhard Ehler, with whom he corresponded in the period 1735–1742: "You would do me a great favor, and our friend Kühn (a friend mathematician), most learned Sir, if you would send us the solution, which you certainly

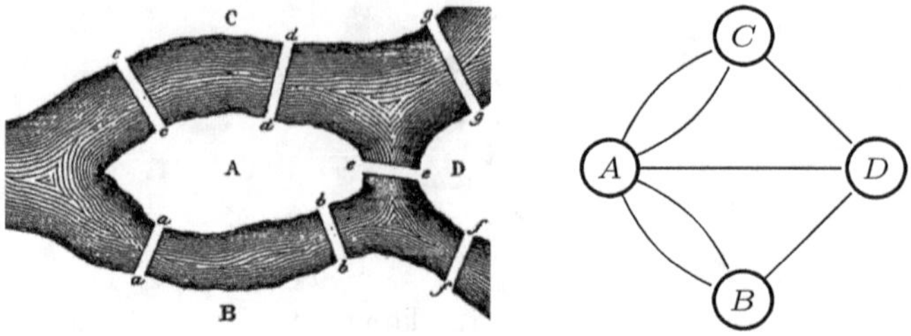

Fig. 6. Königsberg bridges problem. The concrete problem (on the left[2]) is solved after being interpreted by a model graph (on the right).

know well, to the problem of the Königsberg bridges, together with proof of its correctness. It will become a remarkable example of the calculus of position revealing your great genius. I enclose an outline of such bridges." Four days after, in a letter addressed to the Italian mathematician Giovanni Jacobo Marinoni, in March 1736, Leonhard Euler wrote "This is a trivial problem, but it seems to me worthy of attention, because not even what we know about geometry or the art of counting is enough to solve it," and, in the answer addressed to Ehler, in April of the same year, it can be read "This solution seems not have a relationship with Mathematics; I don't understand why you expect a mathematician to find it and not anyone else, because the solution is based on pure reasoning and its discovery does not depend on any mathematical principle." In Figure 6, we can see a sketch of the Pregel river and its bridges, as well as the graph that models the situation: the vertices are regions (margins B and C and islands A and D) and the edges represent the bridges.

Looking at the graph in Figure 5, we find edges or arcs converging to vertices. In the labyrinth, the vertices constitute the junctions or crossroads and the edges, the passages or meanders. We say that each edge of the graph is incident on its extremities.[10,11] We also say that the graph is connected if there is a path between any two of its vertices.[12] Moreover, we say that a certain edge is a bridge if, once removed, the resulting graph is no longer connected.

Definition 1. A *Eulerian graph* is a graph that has a *closed Eulerian trail*, i.e., a closed walk that runs along all the edges and vertices of the graph not repeating any edge. An *open Eulerian trail* is an open walk that runs along all the edges and vertices of the graph not repeating any edge. A *traversable graph* is a graph that contains an open Eulerian trail.[12]

The graph in Figure 6 is not Eulerian. There is not even an open Eulerian trail that contains all the edges of the graph and does not repeat any of them. We can give, indeed, as Euler did, a general answer to Königsberg bridge-type problems.

Theorem 1 (Euler's Theorem). *A graph is Eulerian if and only if it is connected and every vertex has an even number of incident edges.*[12]

To find a closed Eulerian trail in an Eulerian graph, we can apply the so-called Fleury algorithm (19th century) described in Figure 7. It can be proved that Fleury's algorithm always produces an Eulerian closed trail:

Theorem 2. *The trail built using Fleury's algorithm on a Eulerian graph is a closed Eulerian trail, that is all the edges are traversed (and only once) during the construction and one returns to the initial vertex.*

Fleury's algorithm can also be used to obtain an open Eulerian trail in a traversable graph. It follows from Theorem 1 that a graph is traversable if and only if it is connected and has exactly two vertices

FLEURY'S ALGORITHM, 1883

Start at an arbitrary crossroads and whenever reaching a crossroads with fresh passages (not yet traversed, including the starting crossroads), look at the plan of the labyrinth and apply one of the following rules:

- (Rule 1) Choose randomly one passage incident at that crossroads that is not a bridge; remove the passage from the plan and add it to the trail in construction;
- (Rule 2) If such a passage does not exist, then add the bridge to the trail and remove it from the plan.

Fig. 7. When adding the passages to a trail it is assumed that the crossroads in which they are incident are also added.

with an odd number of incident edges (so-called odd vertices). Hence, given a traversable graph, we get a Eulerian graph just by adding a new edge connecting the odd vertices. If we first apply Fleury's algorithm to the new graph and then remove the new edge from the closed trail obtained executing the algorithm, we get an open Eulerian trail in the traversable graph. In fact, in practice, the new edge can be omitted, starting the algorithm at one odd vertex. The other odd vertex is necessarily the last visited vertex. Figure 8 depicts such a run of Fleury's algorithm over the toy labyrinth in Figure 5. In step (c) we cannot traverse passage xw because it is a bridge. If we choose this passage, we are still able to get a trail ending in z, but not an Eulerian trail. Passage xu, for instance, will not be traversed, unless we cross xw more than once.

Note that Fleury's algorithm assumes that the plan of the labyrinth is given to the maze runner... so that the runner can avoid bridges before traversing them at the right time. It is not the case of labyrinth runners in general, e.g., in Lawrence Durrell's *The Dark Labyrinth* (1947), a Postmodern attempt to recreate the real labyrinth of Minos, together with its full potential of psychological effects.

3. Rushing at Random Through this Inextricable Labyrinth

The first attempt to get out from a maze is... to rush at random in panic! Jules Verne in his book *Journey to the Centre of the Earth* (translated with an introduction and notes by William Butcher, 1992) reports on the affliction of the runner:

> Lost in the labyrinth, whose multiple meanderings criss-crossed in all directions, I could no longer try an impossible flight. I had to die from the most terrifying of deaths and – strangely enough – it came into my mind that if one day my fossilised body was found again, encountering it seventy miles into the bowels of the Earth would raise serious scientific questions.
>
> I lost my head. I raised my arms in front of me, trying to feel my way in the most painful fashion. I started fleeing, rushing at random through this inextricable labyrinth, going down all the

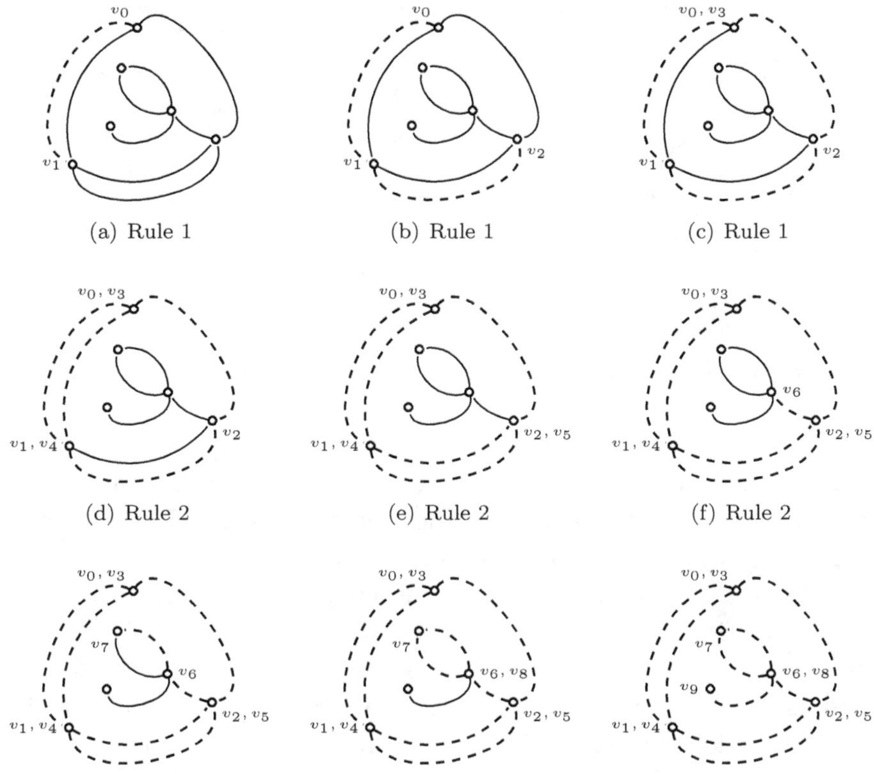

(a) Rule 1 (b) Rule 1 (c) Rule 1

(d) Rule 2 (e) Rule 2 (f) Rule 2

(g) Rule 1 (h) Rule 2 (i) Rule 2

Fig. 8. Toy labyrinth: emulation of Fleury's algorithm.

> time, running through the Earth's crust like an inhabitant of the underground faults, calling, shouting, screaming, soon bruised on the rock projections, falling and getting up covered with blood, trying to drink the blood flooding over my face, but constantly waiting for some wall of rock to come and offer an obstacle for my head to break on.
> Where did this mad running take me? I shall never know. After several hours, undoubtedly at the end of my strength, I fell like an inert mass along the wall and lost all awareness of existence.

Let us assume that all labyrinths, when modeled as graphs, are Eulerian (and have three or more vertices). Is there any vertex of the graph from which the closed Eulerian trail can be built randomly?

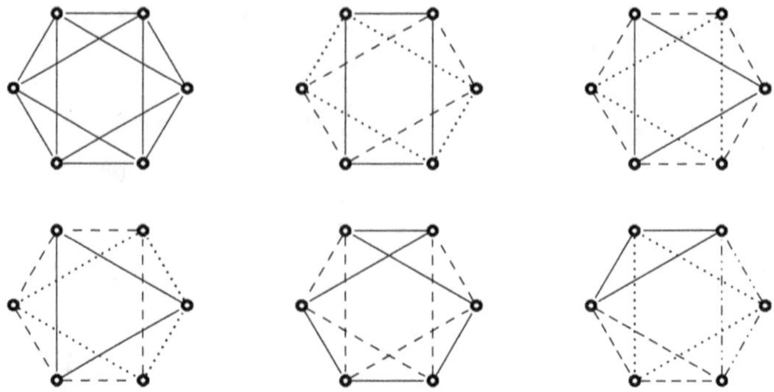

Fig. 9. Decompositions in cycles of the graph in the top left corner.

If that is true for the entry crossroads of a labyrinth, then one has not to worry about being lost. The algorithm would be: run across the meanders and crossroads, but do not repeat passages! In this section the reader will find the answer to that question.

We start by characterizing Eulerian graphs in terms of cycle decompositions. The notion of cycle will be understood here as a closed path that does not repeat edges or vertices, except for the first and the last vertices. For example, Figure 9 shows all possible decompositions in cycles of the graph in the top left corner. A graph is said to be decomposable into cycles when the cycles, taken two by two, have no edges in common (of course they have vertices in common) and the original graph can be restored by adding all the edges and vertices of all the cycles of the decomposition.

The following theorem relates the existence of a closed Eulerian trail in a labyrinth with the decomposition of the labyrinth into cycles:

Theorem 3 (Veblen's Theorem). *A connected graph (with three or more vertices) is Eulerian if and only if it is decomposable into cycles.*

We can introduce the concept of *randomly Eulerian labyrinth from a crossroads* as a labyrinth that can be traversed at random, without repeating passages (thus not duplicating passages), starting from that crossroads and returning to it. Equivalently, a labyrinth is

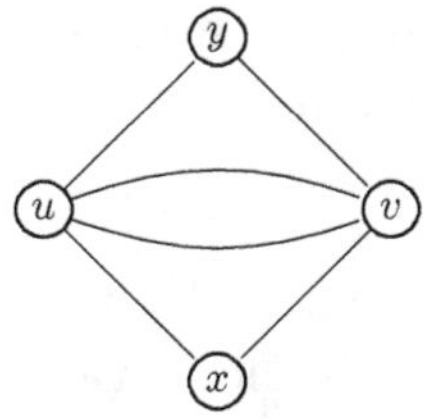

Fig. 10.

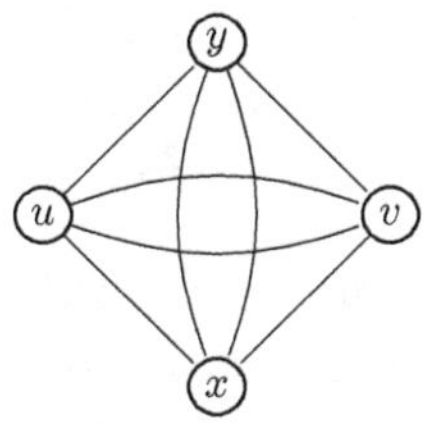

Fig. 11.

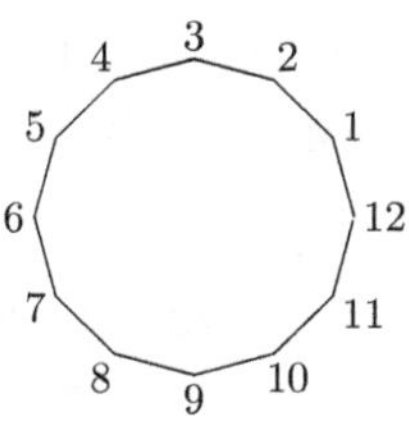

Fig. 12.

said to be randomly Eulerian starting from a crossroads if, starting from that crossroads, one can randomly choose a trail that closes an Eulerian trail at that crossroads. The labyrinth with underlying graph in Figure 10 is randomly Eulerian from the crossroads u and v. The labyrinth with underlying graph in Figure 11 is not randomly Eulerian from any crossroads. Finally, a labyrinth is said to be *randomly Eulerian* if it is randomly Eulerian from all its crossroads. A cycle labyrinth is randomly Eulerian from all its crossroads, as shown in the example in Figure 12. Ore's Theorem characterizes such vertices.

Theorem 4 (Øystein Ore's Theorem). *A graph is randomly Eulerian from a vertex if and only if all cycles of the graph pass through that vertex.*[13]

It follows from this statement that a graph is randomly Eulerian from a vertex if and only if the new graph obtained by removing this vertex as well as all its incident edges has no cycles. It also follows that a graph is randomly Eulerian if and only if it is itself a cycle (Figure 12). Thus, if the underlying graph of a labyrinth is randomly Eulerian, then the labyrinth is acentric and unicursal. Let us call the number of incident edges to a vertex (the number of passages convergent in a crossroads) the degree of this vertex.

Theorem 5. *If a graph is randomly Eulerian starting from a vertex, then this vertex has the maximum degree.*[13]

However, an Eulerian graph is not necessarily randomly Eulerian from a vertex of maximum degree, as shown in the examples in Figures 13–15. Thus, we cannot rest in finding the vertex of maximum degree. Moreover, as a consequence of Theorems 3 and 4 we have the following statement, recalling that we are considering only Eulerian graphs herein:

Theorem 6 (Fujie and Zhang's Theorem). *A graph is randomly Eulerian from none, one, two, or all of its vertices.*[13]

As a consequence of Theorem 6, we conclude that it is highly unlikely that the underlying graph of a labyrinth is random Eulerian from the entry vertex — the case of one vertex — or it is random Eulerian from the entry and the center vertices (if the graph of a

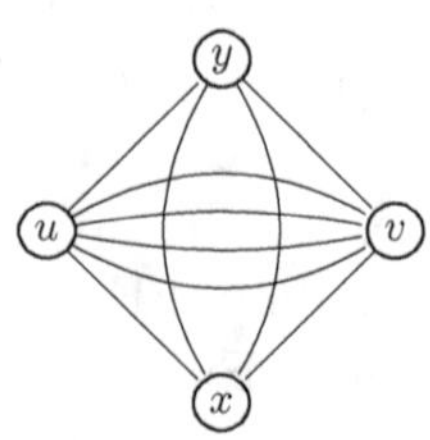

Fig. 13.

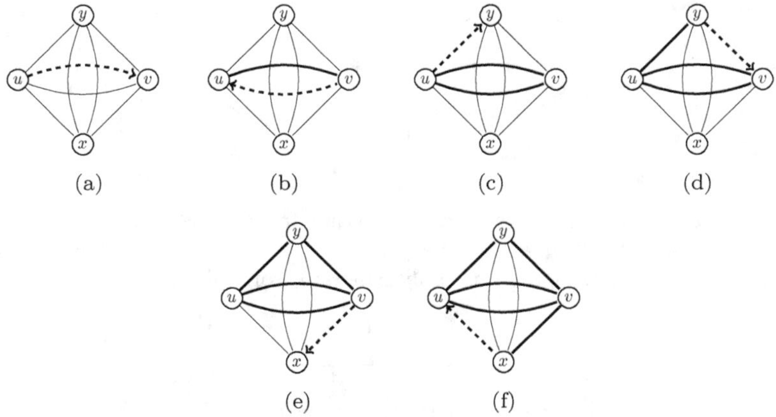

Fig. 14.

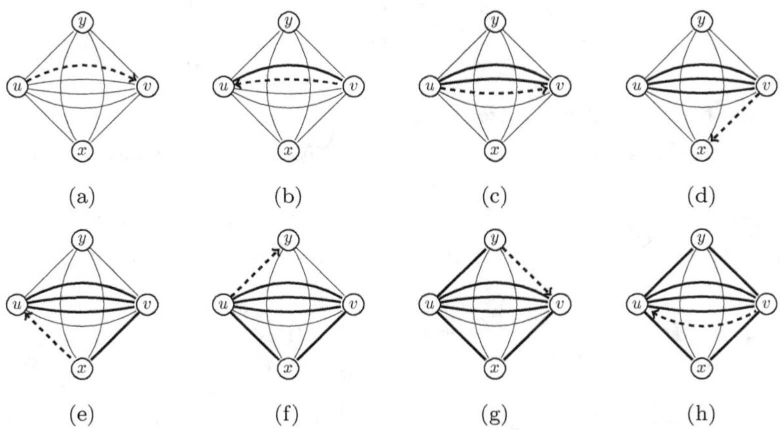

Fig. 15.

centric labyrinth) — the case of two vertices — or it is random Eulerian from all its vertices. To escape from the labyrinth one has to find an algorithm... that is not entirely random! We were already aware of this since in Fleury's algorithm we cannot arbitrarily cross bridges. However, the fact that Fleury's algorithm needs the plan of the labyrinth does not disprove that another algorithm exists that would do without the plan. Now we know that most probably none

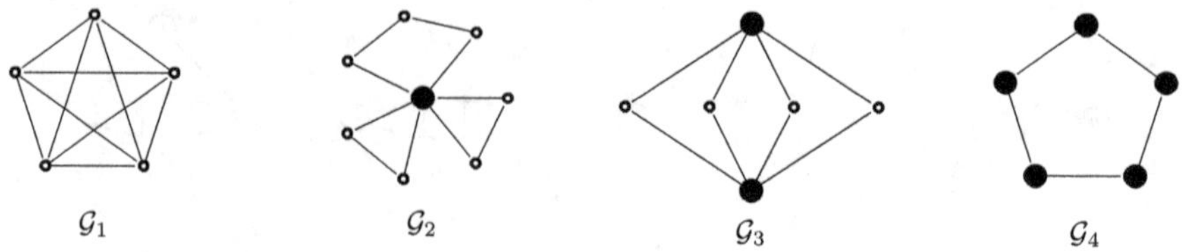

Fig. 16. The graph $\mathcal{G}_1$ is randomly Eulerian from no vertex, $\mathcal{G}_2$ is randomly Eulerian from only one vertex, the graph $\mathcal{G}_3$ is randomly Eulerian from two vertices, and the graph $\mathcal{G}_4$ is randomly Eulerian from all its vertices.

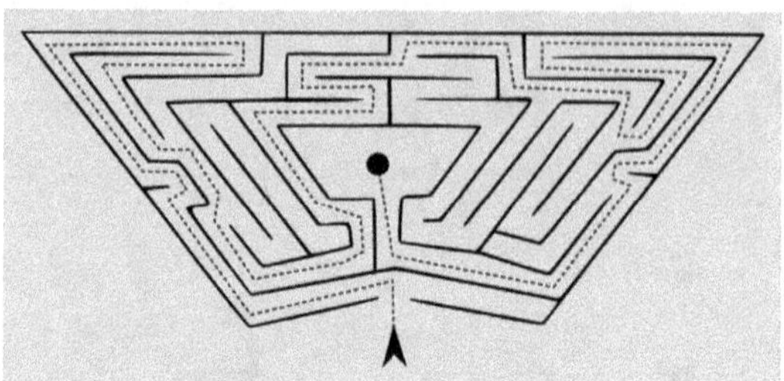

Fig. 17. Hampton Court maze (UK).
Source: File:Hampton_court_maze_sol.svg, *Creative Commons license.*

will do, unless we are allowed to repeat passages! And this is, it seems, the reason why we feel lost... Goodbye maze runners!...

4. *L*, *R*, *R*, *L*, *L*, *L*, *L* — The Need for the Architect's Design

Euler's Theorem applies to the architecture of mazes as we will describe in what follows. As already mentioned, in the maze, the corridors correspond to the edges and the crossroads correspond to the vertices of a graph. The entrance and the center of a centric labyrinth like Hampton Court's (see Figure 17) are in general odd crossroads. If these are the only odd junctions, then we already know that the maze is traversable from the entry to the center and vice-versa. Regardless of how many odd junctions there are in the labyrinth, there is always a trail that connects the initial junction — the entrance to the labyrinth — to the final junction — e.g., the center of the labyrinth — which, like any trail, can be traversed without repeating passages. However, in practice, either we will be lost, or we will not be visiting all the crossroads, or we will not be traversing all the passages of the labyrinth. Euler's theorem itself provides a solution: we must duplicate each passage of the labyrinth, which denotes forward and backward connections between two labyrinth crossroads, thus obtaining an Eulerian graph. The following theorem was stated by Euler himself and reduces to Theorem 1 in the following way:

Theorem 7. *Every (finite) connected graph can be considered a closed walk where every edge is traversed exactly twice.*

How? We duplicate all the edges of the starting graph, thus producing a new graph where every vertex is necessarily even. The new graph is Eulerian, so that there is a closed trail in which all the edges are covered exactly once. Identifying the twin edges of this trail, we obtain a new trail, this time in the original graph, where each edge is traversed exactly twice. It follows from Theorem 7 that it is possible to traverse any connected graph in such a way that each edge is not visited more than twice. However, to do this, we need the plan of the labyrinth... Thus, without the plan, to get out of a labyrinth, it is necessary to mark the crossroads in a suitable way, as the labyrinth is traversed. The mathematicians Trémaux and Tarry, in the 19th century, who were labyrinth fanatics, developed specific techniques to escape the never ending circuits inside labyrinths. Armed with these rules, any explorer can escape from any maze! Note that the same rules can be applied in three-dimensional labyrinths, for example to get out of the catacombs. Meanwhile, to be sure that we are successful in exiting from Hampton Court's labyrinth, we have to keep looking at the plan and avoid the bridges. A formula L, R, R, L, L, L, L given by the guide will do, telling us to turn left (L) or right (R) at each junction, but it does not solve the problem with the labyrinth... when we know nothing about its plan. The algorithmic attempt "do not repeat passages" does not work, as we lose our way as soon as we inadvertently cross a bridge before the proper time.

Armed with Ariadne's thread we will not in general succeed traversing all the passages of a labyrinth. The idea of Ariadne's thread is to enter the labyrinth armed with a ball of yarn and unroll it until we find an already visited junction, and then rewind the ball walking backwards until we find a crossroads of a fresh passage ready to be crossed. And it continues... until we are back to the entry of the labyrinth. Although we can always return to the beginning, it is not guaranteed that we will either reach the center in a centric labyrinth or we will really find a Minotaur in an acentric labyrinth. Thus, the legend of Theseus's arrival to Crete, described by Plutarch (45–129 AD) and many others, is prior to the discovery of the first algorithm to get out from a labyrinth:

> When he [Theseus] arrived at Crete, as most of the ancient historians as well as poets tell us, having a clue of thread given him by Ariadne [daughter of Minos], who had fallen in love with him, and being instructed by her how to use it so as to conduct him through the windings of the labyrinth, he escaped out of it and slew the Minotaur, and sailed back, taking along with him Ariadne and the young Athenian captives.
>
> Plutarch. *Lifes.*[14]

Perhaps the oldest rule, next to Ariadne's thread, for getting out of a high-walled maze or a hedge maze, is the rule of thumb. Once entering such a maze, the subject must walk with his left (or right) hand in contact with the building wall. This rule, illustrated in Figure 18, always allows you to escape the maze of corridors, meanders, and crossroads. However, the rule does not allow you to visit all the galleries of all labyrinths. In particular, you may not find the center (in the case of centric labyrinths). The rule only allows you to conquer compact labyrinths, that is, labyrinths in which meanders and crossroads are made of a single block. For example, the hand's rule is not suitable for the labyrinth of the Getty Museum (Los Angeles, Robert Irwin, 1997) as shown in Figure 18 (right). We find the hand's rule encoded in Dante's *The Divine Comedy.* On his descent to hell, guided by the Roman poet Virgil, who had described

Fig. 18. (left) Right hand's rule and (right) Getty Museum's labyrinth.
Source: (left) File:Maze01-02.png, *Creative Commons license*; (right) File: Flickr_-_brewbooks_-_The_Central_Garden_Getty_Center_(2).jpg, *Creative Commons license.*

the katabasis of Aeneas (in the Aeneid), Dante describes his negative vortex always turning to the left:

> [10.133–135] Following that, his steps turned to the left,
> leaving the wall and moving toward the middle
> along a path that strikes into a valley
>
> [18.19–21] This was the place in which we found ourselves
> when Geryon had put us down; the poet
> held to the left, and I walked at his back.
>
> [23.67–69] A tiring mantle for eternity!
> We turned again, as always, to the left,
> along with them, intent on their sad weeping;
>
> *The Divine Comedy*, Digital Dante Edition, Colombia University.

5. Wiener's Algorithm

Christian Wiener was the first to deal with the ancient labyrinthine walk problem as a mathematical problem in his paper titled "On a problem in the geometry of position."[15] Wiener's algorithm constitutes a method for traversing graphs called today *depth search.* We first describe Wiener's algorithm and then illustrate its application to the labyrinth of Figure 5.

WIENER'S ALGORITHM, 1873
Starting walking in an arbitrary crossroads, one traverses passage after passage, crossroads after crossroads, at random, numbering new crossroads and keeping the sequence of these numbers as one traverses the labyrinth, until no longer possible, or because you find a dead end or because you reach a crossroads already visited (Rule 1). Having found such a crossroads or a dead end, one goes back along the same walk until one finds a crossroads where a passage hits that has not yet been traversed (Rule 2). We now repeat the procedure from this crossroads, traversing the new passage, and proceeding as before, until we arrive at the initial crossroads and realize that all the passages leaving the initial crossroads have been traversed at least twice.

Wienner's algorithm is not efficient since the passages must be, in general, traversed quite often. It would be very hard to traverse a labyrinth either small or large (see Figure 19) by means of this Wiener's method. As we will see in the sections to follow, a walk covering all passages of the labyrinth exists, in which no passage is crossed more than twice.

Fig. 19. (left) Longleat maze and (right) Parc del Laberint (Barcelona), where some scenes of the movie *Perfume: The Story of a Murderer* have taken place. *Source*: (left) File:Longleat-maze.jpg, *Creative Commons license*; (right) File: Parc_del_Laberint_(Barcelona)_-_1.jpg, *Creative Commons license.*

Theorem 8 (Wiener's Theorem). *The depth search algorithm applied to a labyrinth (with a underlying connected graph) ends at the crossroads where it starts after all the passages have been traversed.*

It follows from the former theorem that, if the labyrinth is centric, sooner or later during the execution of the algorithm, the center will be reached, although the algorithm is designed to traverse all the passages and meanders of the labyrinth at least once. The sequence of operations shown in Figure 20 illustrates the in-depth search of a labyrinth, following Wiener's algorithm. In the end, after having exhausted the passages of the labyrinth, one finds himself again at the initial crossroads z. Lost in a labyrinth, to succeed with Wiener's algorithm we have to mark the way as we proceed back and forth along the passages. One way of doing this consists of attaching numbers to the crossroads and to the exits from the passages, keeping on a paper a list of the sequence of crossroads already visited. In detail: (a) the crossroads are numbered as we pass through them, not forgetting to register the number of the crossroads from which we came at the exit of each passage, distinguishing different passages coming from the same crossroads, (b) we carry paper and pen with us and keep track of the crossroads we pass and the passages used to get there, always complying with Wiener's rules, and (c) as soon as we revisit the starting crossroads and all the passages leaving it have already been traversed at least twice, we are sure that we have covered all the passages. Then we notice that the sequence of crossroads crossed is a palindrome.

6. Trémaux's Walk

The major breakthrough in the labyrinthine walk is due to the telegraph engineer Trémaux (c. 1886), who, as Lucas[2] says, was graciously communicated to him by Trémaux himself. We can understand the algorithm as follows. We enter the labyrinth equipped with a ball of yarn that is large enough to unwind and walk randomly through the meanders until we find a crossroads already visited. We were careful enough to mark the beginning and the end of each passage, for example using a pebble, to remind us later that we have

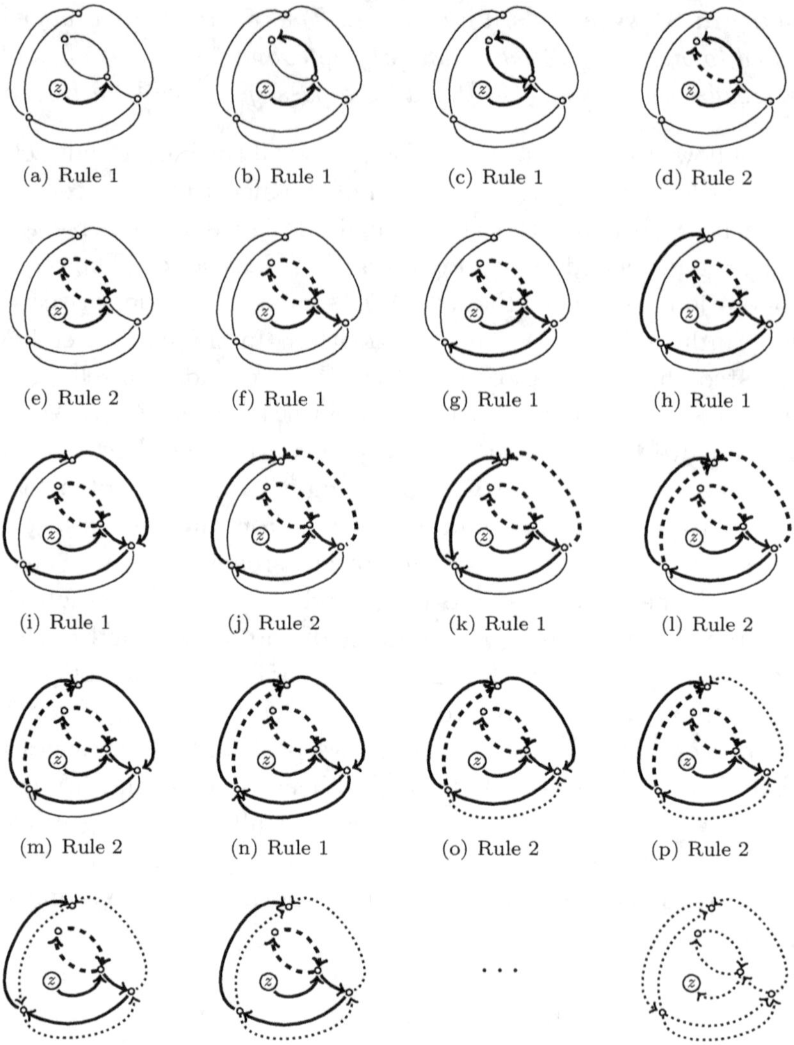

(a) Rule 1 (b) Rule 1 (c) Rule 1 (d) Rule 2

(e) Rule 2 (f) Rule 1 (g) Rule 1 (h) Rule 1

(i) Rule 1 (j) Rule 2 (k) Rule 1 (l) Rule 2

(m) Rule 2 (n) Rule 1 (o) Rule 2 (p) Rule 2

(q) Rule 2 (r) Rule 2

Fig. 20. Toy labyrinth: Emulation of Wiener's algorithm.

already been there once. When we arrive at a crossroads already visited, through a passage we have just traversed for the first time, we go back to the previous crossroads, winding the thread and leaving one more pebble at the beginning and the end of that passage.

When, through a passage traveled twice, we reach a crossroads that has already been visited, we mark the exit with one more pebble and we continue unwinding the thread through a passage that has not yet been traveled or, if such a passage does not exist, through a passage traveled only once (thus marked with just one pebble at both crossroads) and we add one more pebble to its beginning and end. In the particular case of reaching a dead end, we just go back, rewinding the thread and leaving two pebbles at the entry. When we reach a crossroads where we can neither go back nor walk forward, because all the passages have been traversed twice, then (eureka) we have necessarily arrived (!) at the starting point with the ball of yarn all rolled up just like when we started. The ball of yarn is not really needed, but gives us an idea of the reversibility of the algorithm. That is: the reverse of a Trémaux walk is also a Trémaux walk.

TRÉMAUX'S ALGORITHM

(Rule 1) Whenever you reach an unvisited crossroads, follow any untraveled passage;

(Rule 2) Whenever, by a passage not yet traversed, you reach a crossroads already visited or a dead end, retreat along the same passage to the preceding crossroads;

(Rule 3) Whenever by a passage already traversed you reach a crossroads (already visited), follow a passage that has not yet been traversed, or if it does not exist, follow a passage that has already been traversed just once.

Note that the directions in which the edges were traversed are not assumed to be known. This means that the marks we leave on each passage between two crossroads only indicate the number of times we have passed there. This can be done using 0, 1, or 2 pebbles. The algorithm guarantees that we will return to the entrance of the maze after having traversed each of the passages exactly twice. It also assures us that we have unwound Ariadne's thread and rewound it. Thus, at the end all the labyrinth corridors are marked with two pebbles at both extremities.

Theorem 9 (Trémaux's Theorem). *Starting a walk at a certain crossroads of a labyrinth (with an underlying connected graph), advancing persistently, as long as it is possible to comply with the Trémaux rules at each crossroads, one returns to the initial crossroads, after having traversed each passage of the labyrinth exactly twice, once in each direction.*

The first full proof of Trémaux's Theorem is due to König.[16] Trémaux, as it seems, did not publish his proof. Lucas's proof[2] (pp. 47–51) is — as he says — a slight modification of Trémaux's proof. König comments on Lucas's proof saying that it is not complete, because, first of all, the case where the edges are identical (a case that can actually happen) is left out of consideration and, secondly, an essential assertion at the end of the proof is taken care of by the short remark "on démontrera de même." In Ahrens,[4] both of these critical points are dealt with in more detail, but his proof also contains an essential gap. Let it be noted here that neither Lucas nor Ahrens mentions that the edges are traversed twice (in opposite directions) as a consequence of Trémaux' rules. In Figure 21, we illustrate the application of Trémaux reasoning to our toy labyrinth.

7. William's Adoption of Tarry's Solution

In this section, we will analyze Tarry's algorithm,[22] referred to by Umberto Eco's *The Name of the Rose*:

> To find the way out of a labyrinth, – William recited – there is only one means. At every new junction, never seen before, the path we have taken will be marked with three signs. If, because of previous signs on some of the paths of the junction, you see that the junction has already been visited, you will make only one mark on the path you have taken. If all the apertures have already been marked, then you must retrace your steps. But if one or two apertures of the junction are still without signs, you will choose any one, making two signs on it. Proceeding through an aperture that bears only one sign, you will make two more, so that now the aperture bears three. All the parts of the labyrinth must have been visited if, arriving at a junction, you never take a passage with three signs, unless none of the other passages is now without signs.

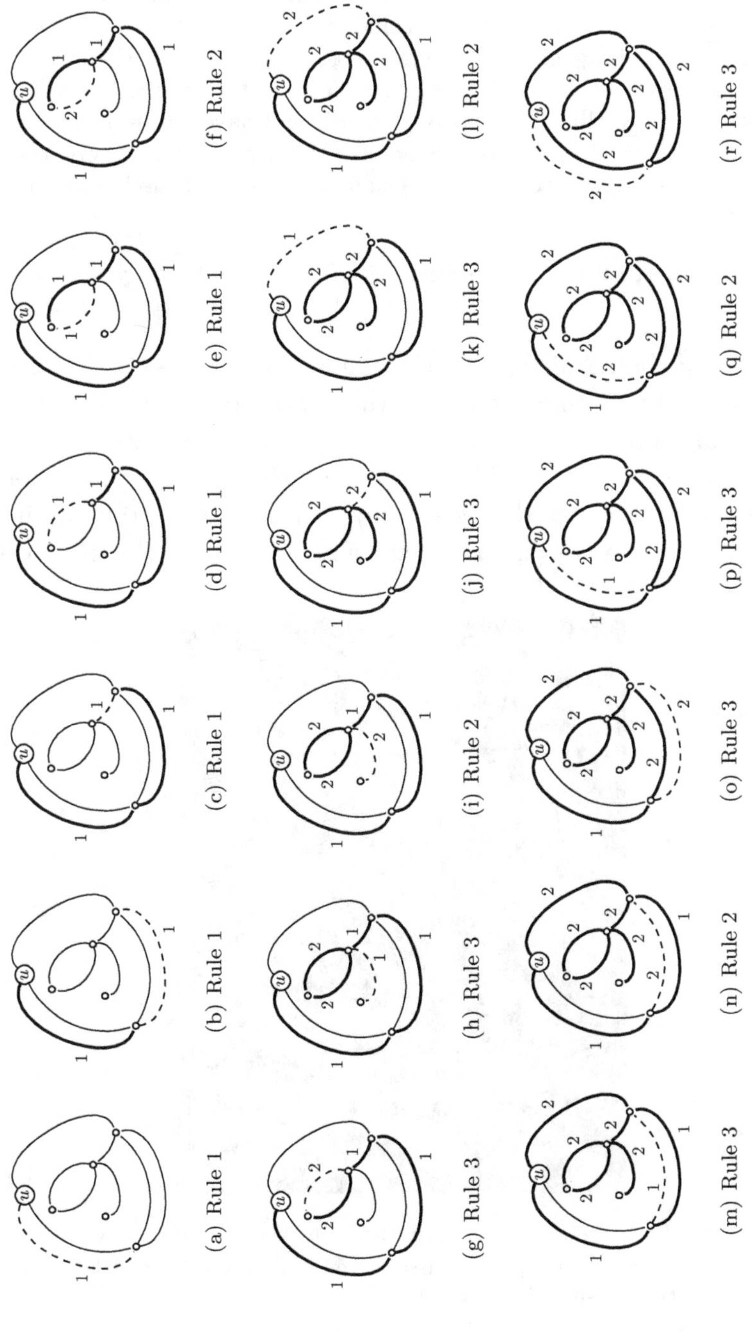

Fig. 21. Toy labyrinth: Emulation of Trémaux's algorithm.

How do you know that? Are you an expert on labyrinths?
No, I am citing an ancient text I once read.
And by observing this rule you get out?
Almost never, as far as I know. But we will try it, all the same. And besides, within the next day or so I will have lenses and time to devote myself more to the books. It may be that where the succession of scrolls confuses us, the arrangement of the books will give us a rule.

Umberto Eco. *The Name of the Rose*, translated by William Weaver, 1984.

The Benedictine library in Eco's novel, illustrated in Figure 22, is centric as the Minotaur labyrinth. Umberto Eco returns to the classical labyrinth as opposed, e.g., to the former accentric labyrinth of Jorge Luis Borges in *The Library of Babel.* Jorge Luis Borges imagined that the world was confined to a gigantic (albeit finite) library made up of a three-dimensional mesh of hexagonal rooms

Fig. 22. The three-dimensional labyrinth of the Benedictine library. *Source*: https://www.dicopathe.com/merveilleuses-bibliotheques-5eme-partie-quelques-bibliotheques-imaginaires-et-utopiques/.

(see Figure 24), like a beehive. Each room has bookcases and the books consist of 400 ordinary pages of randomly arranged letters and punctuation marks. In some books not a word can be read. There are others in which one or another word is legible. But other books feature completely understandable chapters, and some tell the story of our own future. Lost in the Library of Babel, there are blind pilgrims in search of answers to the problems facing humanity, particularly walking in search of the catalog of catalogs.

The homage of Umberto Eco in *The Name of the Rose* to Jorge Luis Borges is clear and studied by several authors[8,17]: both the venerable Benedictine librarian, Jorge de Burgos, and Jorge Luis Borges are blind; Jorge de Burgos and Jorge Luis Borges both really live in libraries; the Minotaur of the Benedictine library, the Benedictine's most dangerous book (or the Benedictine Minotaur) is Aristotle's *Poetics*, volume II (a really lost document) on comedy; and Babel's most wanted book is the catalog of catalogs — the most important challenge of humanity — Jorge de Burgos is the Man of the Book; like Daedalus's labyrinth, the Benedictine labyrinth is centric (although Babel's library is acentric), the mirror at the central door of the Benedictine library is the mirror in the vestibule of the library of Babel; Benedictine's labyrinth is anachronic and Babel's labyrinth, postmodern. As another curiosity, we see that in each floor the Benedictine labyrinth follows the pattern of Rheims Cathedral, as the reader may notice in Figure 23, and the hexagonal contour at each corner follows from Borges Babel's library (see Figure 24).

Let us see how Tarry's algorithm succeeds the labyrintine walk:

TARRY'S ALGORITHM, 1895

Whenever you reach a vertex v that has not yet been visited by the uv edge, follow the vz edge that you like (Rule 1); you should never choose, now in the opposite direction, the edge through which you first arrived at a vertex v, except as a last resort (Rule 2; this condition does not apply to the initial vertex).

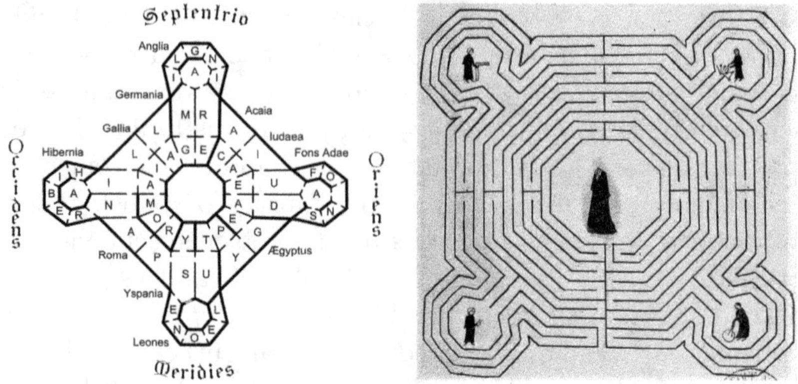

Fig. 23. (left) The Benedictine library and (right) Rheims Cathedral labyrinth. *Source*: (left): File:Labyrinthus_Aedificium.svg#file, *Creative Commons license*; (right): File:Jacques_Cellier_-_Cath%C3%A9drale_de_Reims_-_Labyrinthe_(1583-1587).jpg, *Creative Commons license.*

Fig. 24. Jorge Luis Borges's Labyrinth of Babel, 1980.
Source: https://aubreymcfato.com/tag/borges/.

For every arrival at a crossroads P it ought to be known for every passage KP whether or not the passage was already traversed toward K, and it ought always to be possible to determine the edge through which a crossroads was reached for the first time.

Theorem 10 (Tarry's Theorem). *Starting a walk at a certain crossroads of a labyrinth (with an underlying connected graph), advancing persistently, as long as it is possible to comply with Tarry's* *rules in each visited crossroads, one returns to the entrance after having traversed each meander exactly twice, once in each direction.*

The sequence of figures presented in Figure 25 illustrates the application of Tarry's algorithm to find a walk in the toy labyrinth of Figure 5 that visits all its crossroads and traverses its meanders exactly twice. The plotted path *is random* as long as it complies with Tarry's rules. The first entry in each of the crossroads other than the initial one is marked by E at the closest end of the respective arc.

An interesting fact worth mentioning is that any a Trémaux walk is at the same time a Tarry walk. If Tarry's rules were modified so that, instead of requiring that no edge is traversed twice in the same direction, it was only required that no edge is traversed three times, then the labyrinth problem with this new rule would not be solved. The proofs of these facts can be found in König's.[16]

As in the case of Trémaux's algorithm, Tarry's algorithm requires marking: (a) it is necessary that at every visited crossroads (except the initial one) one finds a passage marked in a special way (an E in Figure 25) so that one knows by which passage one has arrived there in the first place and (b) at the beginning and end of each passage already traversed we must find an indication of the directions already taken.

Different authors have found different marking systems. In his novels that make history or even history of science, such as in *Il Nome della Rosa* or *L'Isola del Giorno Prima*, Umberto Eco often closely follows works or treatises of the time, such as those of Giordano Bruno, Galileo, Kepler, etc. In his novel *Il Nome della Rosa*, Eco takes the information about Labyrinths from Matthews's treatise.[6] It is interesting to note that William's explanation to Adso in *The Name of the Rose* follows very closely the marking system described by Mattheus in[6] for Tarry's algorithm as we illustrate in Figure 26.

The idea of Tarry's algorithm marking system is poorly explained by Matthew and, consequently, poorly explained by Eco. At the entrance of a passage we leave two pebbles, and at its exit one pebble if the crossroads has already been visited and three pebbles if we are at the crossroads for the first time. So, arriving at a crossroads, we

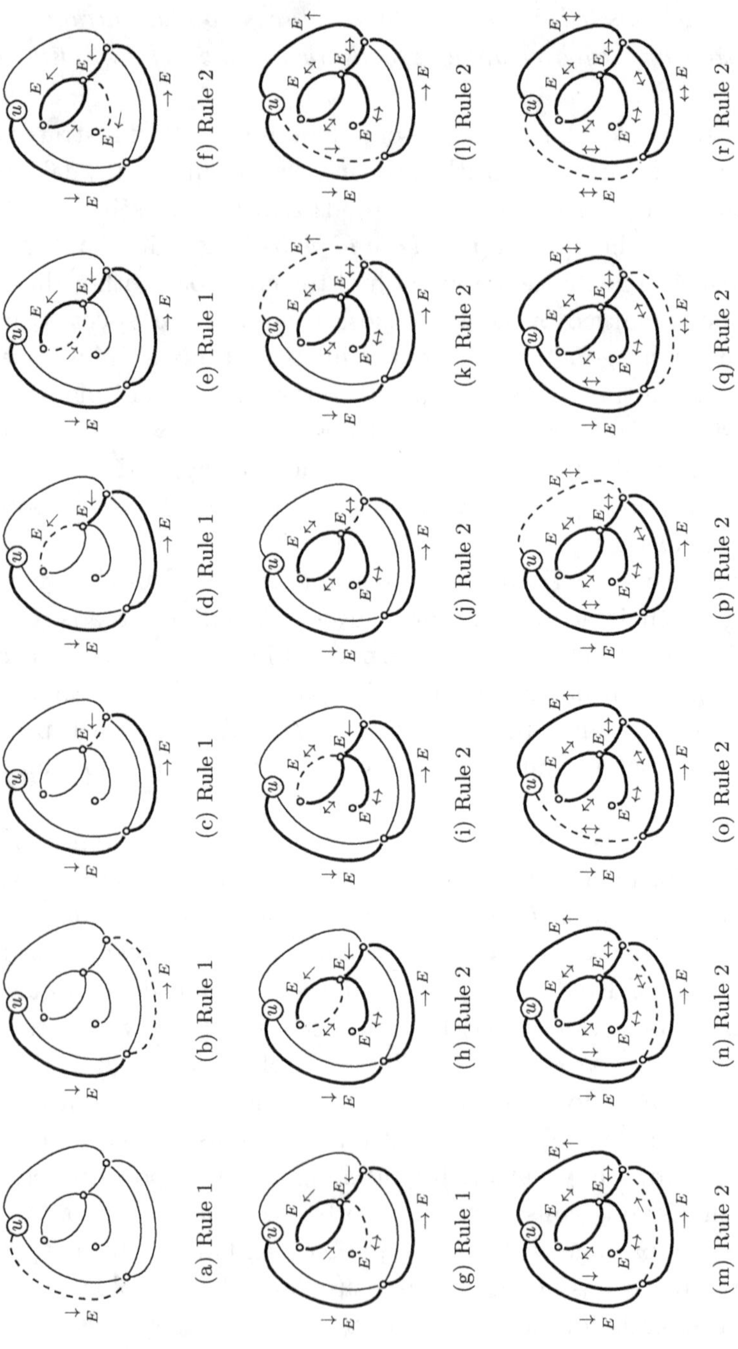

Fig. 25. Toy labyrinth: Emulation of Tarry's algorithm.

At every new junction, never seen before, the path we have taken will be marked with three signs. If, because of previous signs on some of the paths of the junction, you see that the junction has already been visited, you will make only one mark on the path you have taken. If all the apertures have already been marked, then you must retrace your steps. But if one or two apertures of the junction are still without signs, you will choose any one, making two signs on it. Proceeding through an aperture that bears only one sign, you will make two more, so that now the aperture bears three. All the parts of the labyrinth must have been visited if, arriving at a junction, you never take a passage with three signs, unless none of the other passages is now without signs.	On arriving at a node which, by the absence of marks, you know you have not already visited, mark the path by which you have just arrived by three marks; if you see by marks on other paths that you already been to that node, mark the arrival path with one mark only. If now there are no unmarked paths at this node, it means that you have explored this particular branch-system and must retrace your steps by the path by which you have arrived. If, however, there are one or more unmarked paths leading from the node, select one of them, and, as you enter, mark it with two marks. We can now make certain of visiting every part of the maze if we make it a rule that, on arrival at a node, we shall never take a path with three marks unless there are no paths unmarked or with one mark only.
Umberto Eco[8]	W. Matthews[6]

Fig. 26. Eco vs. Matthews.

avoid the passage marked with the three pebbles and follow along any other passage either not yet marked or marked with just one pebble, leaving more pebbles, two in the entrance of a non-marked passage or one more pebble at the entrance of a passage already marked with one pebble. Any passage with a beginning and end marked with two pebbles cannot thus be crossed again in any direction. Arriving at a crossroads and noting that all possible passages are marked with two or three pebbles, we follow the passage marked with the three pebbles. Note that it is always possible to turn back at a crossroads if traversing a passage for the first time we arrive at a crossroads already visited. In this case, we do not forget to mark the exit with two pebbles, one for the arriving and another for turning back.

Reading Eco's novel, like in other crime novels, the reader not only faces the centric Benedictine labyrinth, but he or she also falls in the acentric labyrinth of deduction to find the murderer. Plato in *Euthydemus* is the first, as we know, to notice the similarity between the physical labyrinth and the labyrinth of deduction in

Physical and psychic trap	Intellectual trap
Theseus and the Minotaur **Mythos**	**Socrates** **Logos**
Ovid's Metamorphoses	Plato's Euthydemus
Building	Metaphor
Center	Periphery
Multicursal centric	Multicursal acentric
Error on the way	Error in reasoning

Fig. 27. Comparison between the physical labyrinth and the labyrinth of reasoning.

a proof system. The error in reasoning is compared to the error in following the labyrinthine walk:

> Socrates: Find! my dear sir, no indeed. And we cut a poor figure; we were like children after larks, always on the point of catching the art, which was always getting away from us. But why should I repeat the whole story? At last we came to the kingly art, and inquired whether that gave and caused happiness, and then we got into a labyrinth, and when we thought we were at the end, came out again at the beginning, having still to seek as much as ever.
>
> Plato. *Euthydemus*, translated by Benjamin Jowett, 2013.

Thus, the labyrinth of reasoning can be compared with the physical labyrinth in the way displayed in Figure 27.

7.1. *Proof methods*

The reader will wonder how all these theorems we presented in our chapter can be proved. The answer lies in the language of Graph Theory.[10,11] We choose Tarry's rules to illustrate the idea of a proof concerning the mathematics of labyrinths. Tarry's Theorem 10 is, in fact, much easier to prove than Trémaux's Theorem 9, but at the same time it is interesting enough to display the tools of thought in such proofs. The proof that follows is partially due to Tarry fully done by König.[16]

Let us assume that our labyrinth has crossroads u_1, $u_2, \ldots, u_m$. Let u_1 be the entrance to the labyrinth, and let u_1, u_2, $u_3, \ldots,$ in that order, be the sequence of the crossroads as they are being visited through the maze, removing repetitions and always respecting the rules of Tarry, until all possibilities were exhausted. Note that the sequence represents the order in which crossroads are visited for the first time. Let us start by proving that the crossroads u_1 is not only the first junction, but also the last junction visited in the walk through the labyrinth following Tarry's rules. Note that a crossroads is the last one when one can no longer leave it because all passages confluent to it have already been traversed in both directions. The last crossroads of the walk cannot be a junction u distinct from u_1. In fact, if u does not coincide with u_1, then for each passage uv traversed in the direction vu (entry in u), there is a possible passage uv' (which in the last case may be the passage uv) to be traversed in the direction uv' (exit from u). Thus, it cannot happen that one cannot get out of u because a situation has been reached in which all the passages confluent to it have already been covered in both directions. On the other hand, all passages u_1u (u not coincident with u_1) were traversed in the direction u_1u, otherwise the walk would not end in u_1. However, as for each passage u_1u traveled in the direction u_1u (exit from u_1), there must be a passage u_1u' traveled in the direction $u'u_1$ (entry in u_1), we conclude that all passages u_1u were traversed twice, once in each direction.

We now show that this same phenomenon happens with all the other crossroads of the traveled walk. Contrary to this conjecture, let u_n ($n > 1$) be the first crossroads of the sequence $u_1, u_2, u_3, \ldots$ that does not satisfy the requirement, that is the first crossroads such that, due to the necessary coincidence of the number of entries and exits from u_n, there are crossroads u and u' in the maze such that the passage u_nu was not traversed in the direction uu_n (entry in u_n) and the passage u_nu' was not traversed in the direction u_nu' (exit from u_n). By the hypothesis, the passage u_iu_n, for some $i < n$, was traversed twice, one in each direction, since u_n is the first crossroads of the sequence u_1, u_2, $u_3, \ldots$ that does not exhaust entries and exits. However, the passage $u_i u_n$ is the entry passage in u_n and therefore, due to Tarry's rules, it cannot have been traversed in the opposite direction since the passage $u_n u'$, with u' not coincident with u_i, was not traversed.

Finally, we show that all the crossroads of the labyrinth were visited. Suppose that v is an unvisited crossroads of the maze. As the underlying graph is connected, there is a path $\mathcal{P} = u_1, v_1, v_2, \ldots, v$ starting in u_1 and ending in v. Let v_i be the first crossroads of $\mathcal{P}$ that is not part of the sequence $u_1, u_2, u_3, \ldots$ Consequently, there is a passage $v_{i-1}v_i$ ($v_{i-1} = u_1$ in the case $i = 1$) that has not been traversed. However, this fact is not in agreement with the previous proof that, for every crossroads u_n of the sequence u_1, u_2, $u_3, \ldots$ every possible passage $u_n u$ has been traversed twice, once in each direction.

8. Maurice's Solution

Maurice, who had been a student at the École Polytechnique, introduced Lucas[2] to another very curious algorithm that allows us to escape from any maze. Again we impose the condition that we can only go through each of the passages twice with no need to specify directions or any special markings on crossroads' first arrivals (as imposed by Tarry's algorithm). The fundamental rule consists of a modification of Rule 2 of Trémaux's algorithm: whenever we arrive at a crossroads already visited through a passage not yet traversed, we can also proceed making a bridge between the exit of this passage — passage 1 — and the entrance of another passage not yet traversed — passage 2 — through a mark of the same color, exclusive to those two passages, 1 and 2, in that crossroads; then, when turning back along passage 2 we are forced to follow back passage 1 as well, that is, passages 1 and 2 are connected in a unique corridor. In Figure 28, we illustrate the application of Maurice's algorithm.

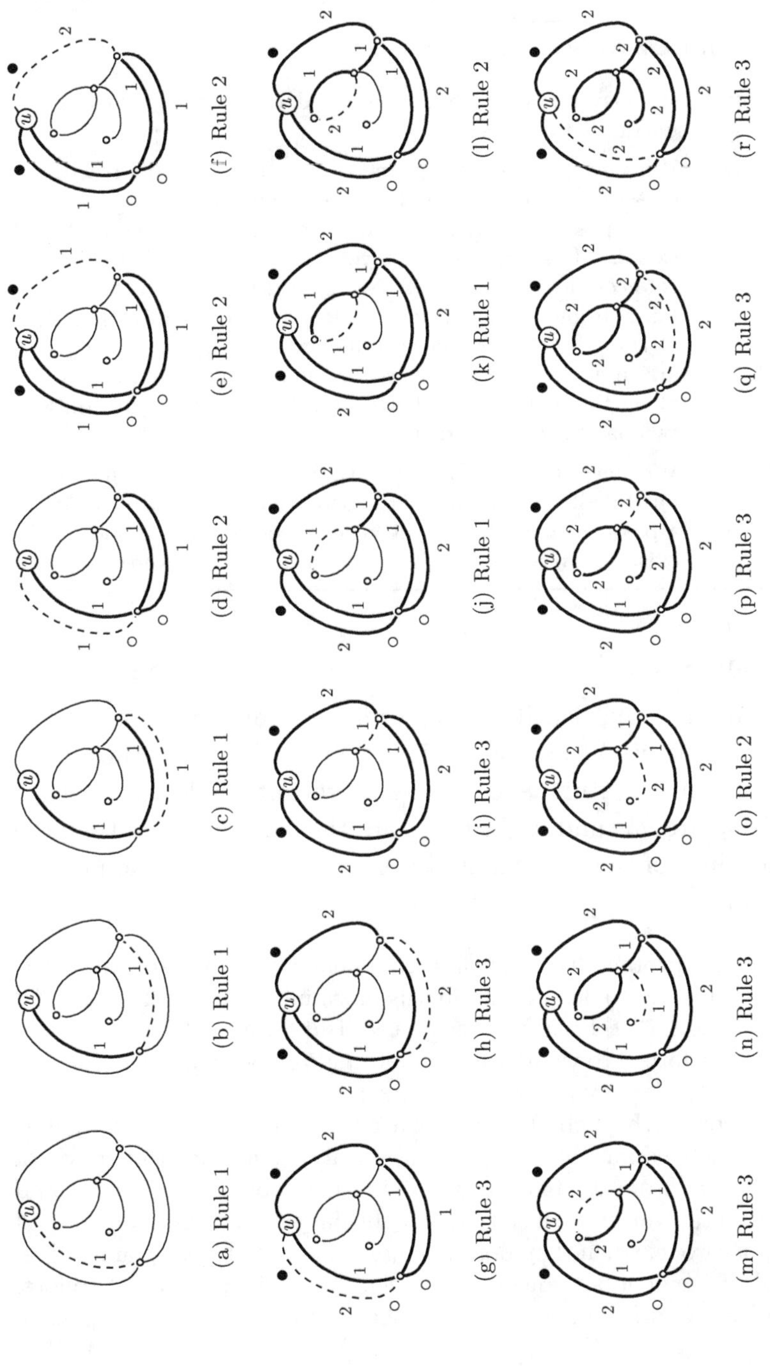

Fig. 28. Toy labyrinth: Emulation of Maurice's algorithm.

MAURICE'S ALGORITHM

(Rule 1) Whenever you reach an unvisited crossroads, follow any untraveled passage;

(Rule 2) Whenever, by a passage not yet traversed, you reach a crossroads already visited or a dead end, retreat along the same passage to the preceding crossroads or follow, if it exists, a fresh passage (not yet traversed); in this case the connection between these two passages should be marked (as a bridge or a single corridor) in such a way that if, at some point, turning back along the second passage you arrive again at that crossroads, you should continue back along the first; and that will be the only way to use these two passages a second time;

(Rule 3) Whenever by a passage already traversed (and not marked by Rule 2) you reach a crossroads (already visited), follow a passage that has not yet been traversed, or if it does not exist, follow a passage that has already been traversed once.

9. Final Remarks

Assuming that the algorithms to evade labyrinths we described above can indeed be used successfully, it is weird to find that the labyrinths occurring in literature and movies were not made to escape from them. Daydreams and nightmares are the same since Antiquity, e.g., the labyrinth of the city is half soluble as well described in Petronius's *Satyricon*:[19]

> And the protagonist anguishly comments: "What were we miserable wretches to do, shut up in this newfangled labyrinth. [Rome]"? [Satyricon, 73.1] Finally they succeed. But once out from the house, they are forced to wander unfamiliar and rocky paths. Groping and guided by the rare references retained, there they manage to retrace their path, through the unknown city, back to the pension. It is not yet freedom: the door, closed, it only opens with the arrival of a servant from Trimalchio who, without knocking, pushes it and goes into. But as soon as they left this labyrinth, they soon entered another one, when they boarded on Lica's ship. That is, human life, full of incidences, dangers, of difficulties, it is a maze. Therefore, as Nair Castro Soares observes, the protagonist of the Satyricon,

lost in a world where light was rare and traps and surprises lurked on every corner, complaining that "we didn't even have a torch of refuge, to open paths for the wandering poor."

In Labirinto e Minotauro de José Ribeiro Ferreira.[20]

In the same way, the labyrinth in David Bowie's musical or the dynamic labyrinth in the trilogy *The Maze Runner*, directed by Wes Ball, or even the labyrinth of consciousness in *Inception*, directed by Christopher Nolan, are not to be solved, they are only instruments of torture as Gaston Bachelard already comments in Ref. 21: "In short: the labyrinth is a primary suffering, a childhood suffering. It will be a birth trauma? It will be, on the contrary, as we believe, one of the traits clearest of a psychic archaism? The suffering always imagine the instruments of their torture." In this sense, the language of Mathematics of graph traversing algorithms is not enough to describe the problem or its possible solutions. That mathematical solutions did not yet seem to help poets and novelists at all is witnessed by the proliferation of poems and novels whose message is emphasized by Sophia De Mello Breyner Andresen who, despite everything, doesn't get lost in the labyrinth:

Imagine,
In Crete where the Minotaur reigns I went through the wave
Eyes open fully awake
Without drugs or a potion
Just the wine I drank and the solemnity of things
For I belong to the race of those who go through the labyrinth
Without ever losing the linen thread of the word

Acknowledgments

The research of José Félix Costa is supported by Fundação para a Ciência e Tecnologia, projeto FCT I.P.:UID/FIL/00678/2013. The work of Paula Gouveia is funded by FCT/MCTES through national funds and when applicable co-funded EU funds under the project UIDB/50008/2020.

References

1. A. Peyronie, *The Labyrinth.* In *Companion to Literary Myths, Heroes and Archetypes*, Pierre Brunel, editor. Routledge, London, 1992, 685–719.
2. É. Lucas, *Récréations Mathématiques, Volume 1.* Gauthier VIllars, Imprimeure-Libraire, 1882.
3. W. W. R. Ball and H. S. M. Coxeter, *Mathematical Recreations and Essays*, 13th edition. Dover Publications, Inc., New York, 1974, 1987.
4. W. Ahrens, *Mathematische Unterhaltungen und Spiele*, 13th edition, Leipzig; 1st edition, 1901; latest edition: vol. 1, 3rd edition, 1921; vol. 2, 2nd edition, Leipzig, 1918.
5. H. E. Dudeney, *Amusements in Mathematics.* Dover Publications, London, 1917, 1958, 1970.
6. W. H. Matthews, *Mazes & Labyrinths, Their History & Development.* Dover Publications, Inc., New York, 1922, 1970.
7. J. Lima De Freitas, *O Labirinto.* Coleção Biblioteca Arcádia/Arte, 1975. Arcádia, Lisboa, 1975.
8. M. Hackworth, *Solved by Walking, Paradox and Resolution in the Labyrinth.* CreateSpace Independent Publishing Platform, Printed in Great Britain by Amazon, 2012.
9. C. Wright, *The Maze and the Warrior: Symbols in Architecture, Theology and Music.* Harvard University Press, Cambridge, 2004.
10. G. Chartrand, *Introductory Graph Theory.* Dover Publications, New York, 1977, 1985.
11. R. J. Wilson, *Introduction to Graph Theory.* Prentice-Hall, Edinburgh Gate, Harlow, Essex, England, 1972, 1996.
12. G. Chartrand and A. T. White, Randomly traversable graphs. *Elements of Mathematics* 25:101–107, 1970.
13. F. Fujie and P. Zhang, *Covering Walks in Graphs.* Springer Briefs in Mathematics. Springer, New York, 2014.
14. Plutarch, *Plutarch Lifes.* The Dryden translation, edited with preface by Arthur Hough Clough. The Modern Library, New York, 2001.
15. C. Wiener, Ueber eine Aufgabe aus der Geometria situs. *Mathematische Annalen*, 6:29–30, 1873.
16. D. König, *Theory of Finite and Infinite Graphs.* Birkhäuser, Boston, 1990.
17. R. G. Mendoza, *Giordano Bruno's Prelude to Contemporary Cosmology.* Element, Shaftesbury, Dorset, 1995.
18. U. Eco, *The Name of the Rose*, translated by William Weaver. Harcourt, San Diego, 1984, 2009.
19. P. Arbiter, *The Satyricon.* In *The Project Gutenberg EBook of The Satyricon, Complete*, 2021.
20. J. R. Ferreira, *Labirinto e o Minotauro — Mito de Ontem e de Hoje.* Centro de Estudos Clássicos e Humanísticos da Universidade de Coimbra, Coimbra, Portugal, 2008.
21. G. Bachelard, *La Terre et les Rêveries de la Volonté.* J. Corti, Paris, 1948.
22. G. Tarry, Le Problème des Labyrinthes. *Nouvelles Annales de Mathématiques* 14(3):187–190, 1895.

https://doi.org/10.1142/9789811257155_0002

Chapter 2

Computational Non-universality: Philosophical and Artistic Perspectives

Selim G. Akl
School of Computing
Queen's University
Kingston, Ontario, Canada
akl@cs.queensu.ca

This chapter draws connections from the science of computation to philosophy and the visual arts. The motivation for this endeavor is computational non-universality, a fundamental theorem in theoretical computing established relatively recently. Two distinct mathematical proofs of this result are offered, one proof by counterexample, and one proof by contradiction. Both proofs show that simulation, upon which rests the principle of universality, is not always possible, thereby making the existence of a universal computer a myth. Both proofs are inspired by philosophy of science. Both are illustrated using an artist's conception.

1. Introduction

The science of computation has witnessed a tremendous success since its inception in the middle of the 20th century. Today, computers are crucial in every aspect of modern society, transforming communication, transportation, education, business, health care, and entertainment, to name just a few of the areas benefiting from information technology. The idea behind this extraordinary impact is a simple one, namely, *universality.* The *principle of universality* states that given sufficient time and memory space, any computation that can be performed by a general-purpose computer can, through *simulation,* be performed by any other general-purpose computer (regardless of any architectural differences between the simulating and simulated computers, or how efficient or inefficient the simulation is). This is accomplished by having the second computer *simulate* (that is, imitate exactly in order to obtain the same effect) every step executed by the first computer, using its own (hardware and software) resources. Thus, for example, an email program that runs on a laptop can be made to run equally well on a mobile phone. This applies to far more complex computations, from studying the subatomic realm to exploring the far reaches of our universe. It is indeed fair to say that simulation is the engine that would make universality possible. Consequently, according to the principle of universality, every general-purpose computer is universal, capable of carrying out through simulation any computation that is possible on any other computer. And yet, it seems natural to ask: is simulation *always* possible?

In reality, the principle of universality is but a conjecture, sometimes referred to as the Church–Turing thesis. This conjecture is impossible to prove in general because an all encompassing and agreed upon definition of what constitutes a computation does not exist. This is the case, despite the overwhelming number of instances providing evidence of its validity. An abundance of examples confirming a claim is not a proof, however. By contrast, it is actually possible to *disprove* the universality conjecture, and this is precisely what this chapter is about.

It was shown recently that, in fact, the principle of universality is false in general; it does not apply to all computations. The reason for this is that, as it turns out, simulation is the weakest link, the Achilles heel in the quest for universality in computation. Put simply, *simulation is not always possible.* It immediately follows that a universal computer cannot exist. This result, to which we refer as *computational non-universality* (or *non-universality in computation*) is established using two distinct approaches:

(1) A proof by counterexample, whereby an otherwise *computable* function, cannot be computed on a putative "universal" computer, that is, on a computer that is finite and fixed once and for all.

(2) A proof by contradiction, in which it is assumed that there indeed exists a "universal" computer, and this assumption is then shown to lead to an absurd situation whereby this computer embarks on an unending computation.

These two approaches to disproving the existence of a universal computer are reviewed in this chapter. The role of philosophy in interpreting these results is highlighted. The equally important contribution of the visual arts in illustrating the proofs is put in evidence.

2. Non-universality and Incompleteness

In this section, we draw an analogy between Gödel's Incompleteness Theorem in mathematical logic, and the impossibility of achieving a Universal Computer in computer science, that illustrates the similarities in the formal structure and philosophical implications of the two results. Specifically, Gödel proved that there exist formal systems of mathematics that are consistent but not complete. In the same way, we show that there does not exist a general-purpose computer that is universal in the sense of being able to simulate any computation executable on another computer.

2.1. *Proving non-universality in computation by counterexample*

Let $\mathcal{U}_1$ be a general-purpose computer. For the purpose of this proof, we suppose further that time is divided into discrete time units, and that $\mathcal{U}_1$ is capable of $V(t)$ elementary operations at time unit number t, where t is a positive integer, $t = 1, 2, 3, \ldots$. Here, an elementary *computational* operation may be any one of the following:

(1) Obtaining the value of a fixed-size variable from an external medium (for example, reading an input, measuring a physical quantity, and so on),
(2) Performing an arithmetic or logical operation on a fixed number of fixed-size variables (for example, adding two numbers, comparing two numbers, and so on), and

(3) Returning the value of a fixed-size variable to the outside world (for example, displaying an output, setting a physical quantity, and so on).

Each of these operations can be performed on every conceivable machine that is referred to as a *computer*. Together, they are used to define, in the most general possible way, what is meant by *to compute*: the acquisition, the transformation, and the production of information.

Now all computers today (whether theoretical or practical) have $V(t) = c$, where c is a constant (often a very large number, but still a constant). In order to make the non-universality result even stronger, in what follows we do not restrict $V(t)$ to be a constant. Thus, $V(t)$ is allowed to be an increasing function of time, such as $V(t) = t$, or $V(t) = 2^{2^t}$, and so on.

Finally, $\mathcal{U}_1$ is allowed to have an unlimited memory in which to store its program, as well as its input data, intermediate results, and outputs. Furthermore, no limit whatsoever is placed on the time taken by $\mathcal{U}_1$ to perform a computation.

Non-universality Theorem: $\mathcal{U}_1$ cannot be a universal computer.

Proof. Let us define a computation P_1 requiring $W(t)$ operations during time unit number t, for $t = 1, 2, 3, \ldots$. If these operations are not performed by the end of time unit t, the computation P_1 is said to have failed. Let $W(t) > V(t)$, for some t. Clearly, $\mathcal{U}_1$ cannot perform P_1. However, P_1 is successfully completed by another computer $\mathcal{U}_2$ capable of $W(t)$ operations during the tth time unit, for $t = 1, 2, 3, \ldots$. □

It is important to note that, by the definition of universality, $\mathcal{U}_1$, once its features have been specified is fixed and cannot change during the computation. Despite being allowed extraordinary powers (such as, for example, the ability to increase the number of operations it can do at every consecutive time unit), $\mathcal{U}_1$ still fails to perform P_1. The computer $\mathcal{U}_2$ on the other hand is especially tailored to carry out P_1 and succeeds in doing so. This establishes that P_1 is definitely computable. Yet surprisingly, $\mathcal{U}_1$ is unable to simulate the actions of

$\mathcal{U}_2$, notwithstanding the fact that no limit is placed on its memory or the time it is allowed to run. Would $\mathcal{U}_2$ be the new Universal Computer? Of course not, as we can easily define a computation P_2 requiring $X(t) > W(t)$ operations during time unit t that $\mathcal{U}_2$ cannot perform. A more powerful computer $\mathcal{U}_3$ can execute P_2, but is in turn defeated by a third computation P_3, and so on forever.

2.2. *A pictorial example*

As shown in Figure 1, N clocks are hanging on a wall, $N > 1$. The clocks are digital, each displaying the time as a quadruple of digits AB:CD; for example, 19:48. All the clocks are working, ticking away synchronously. At every tick, each clock displays a new, but random, quadruple AB:CD — a time perhaps different from the ones displayed by the other $N - 1$ clocks. No clock has a memory; therefore, when at the following tick a new time is generated and displayed, the previous quadruple is lost forever. The wall is long at will, allowing N to be big at will. The problem to be solved is

Fig. 1. Computing a function of N clocks.

the following: For an arbitrary number of clocks N, it is required to compute a function (for example, the average) of the N times displayed at a given moment T. A computer capable of exactly N operations per time unit, and claiming to be "universal," readily solves the problem. It does so by reading the times displayed by the N clocks, and computing a function of these values all in one time unit (before time unit $T + 1$ when the clocks update their displays). This computer, however, is thwarted if *even one* clock is added to the wall!

2.3. *An algorithmic counterexample to universality*

Consider the well-known quintessential computational problem of sorting a sequence of numbers stored in the memory of a computer. For a positive even integer n, where $n \geq 8$, let n distinct integers be stored in an array S with n locations $S[1], S[2], \ldots, S[n]$, one integer per location. Thus $S[j]$, for all $1 \leq j \leq n$, represents the integer currently stored in the jth location of S. In a variant of the standard sorting problem, it is required to sort the n integers in place into increasing order, such that:

(1) After step i of the sorting algorithm, for all $i \geq 1$, no three consecutive integers satisfy:

$$S[j] > S[j+1] > S[j+2], \tag{1}$$

for all $1 \leq j \leq n-2$.

(2) When the sort terminates, we have:

$$S[1] < S[2] < \cdots < S[n]. \tag{2}$$

An algorithm for a computer $\mathcal{M}$ capable of $n/2$ operations per time unit solves the aforementioned variant of the sorting problem handily in n steps, by means of predefined pairwise swaps applied to the input array S, during each of which $S[j]$ and $S[k]$ exchange positions (using an additional memory location for temporary storage). Thus, for example, the input array

7 6 5 4 3 2 1 0

would be sorted by the following sequence of comparison/swap operations (each pair of underlined numbers are compared to one another and swapped if necessary to put the smaller first):

$$\underline{7\ 6}\ \underline{5\ 4}\ \underline{3\ 2}\ \underline{1\ 0}$$

$$6\ \underline{7\ 4}\ \underline{5\ 2}\ \underline{3\ 0}\ 1$$

$$\underline{6\ 4}\ \underline{7\ 2}\ \underline{5\ 0}\ \underline{3\ 1}$$

$$4\ \underline{6\ 2}\ \underline{7\ 0}\ \underline{5\ 1}\ 3$$

$$\underline{4\ 2}\ \underline{6\ 0}\ \underline{7\ 1}\ \underline{5\ 3}$$

$$2\ \underline{4\ 0}\ \underline{6\ 1}\ \underline{7\ 3}\ 5$$

$$\underline{2\ 0}\ \underline{4\ 1}\ \underline{6\ 3}\ \underline{7\ 5}$$

$$0\ \underline{2\ 1}\ \underline{4\ 3}\ \underline{6\ 5}\ 7$$

$$0\ 1\ 2\ 3\ 4\ 5\ 6\ 7$$

However, an alleged "universal" computer capable of fewer than $(n/2)$ operations per time unit, cannot solve the problem consistently. Confronted with the input array shown above, it fails to satisfy the requirement that at no time three consecutive values are listed in decreasing order. Is $\mathcal{M}$ universal? Certainly not, for it too cannot solve the sorting problem when the input sequence has length greater than n.[1]

2.4. *A philosophical precursor to non-universality*

In 1931, the 25-year-old Austrian logician Kurt Gödel published his famous Incompleteness Theorem, arguably the most important result in the history of mathematical logic. The theorem established that there exist non-trivial formal systems of mathematics that, if consistent, cannot be complete.[2]

In order to make his point, Gödel chose the formal system of simple arithmetic, that is, the natural numbers with equality, addition, and multiplication. Denoting this system by $\mathcal{A}_1$, consider the following proposition G_1, expressible within $\mathcal{A}_1$:

$$G_1 =< \text{This statement cannot be proved within } \mathcal{A}_\infty > .$$

Stepping outside of $\mathcal{A}_1$, Gödel proved that G_1 cannot be proved within $\mathcal{A}_1$. Indeed, proving it true within $\mathcal{A}_1$ would mean that a false statement is true, while proving it false within $\mathcal{A}_1$ would mean that a true statement is false. Since G_1 cannot be proved within $\mathcal{A}_1$, it follows that G_1 is true. This means that $\mathcal{A}_1$ is incomplete as it contains a true statement that cannot be proved within $\mathcal{A}_1$.

To appreciate the significance of this result, consider adding the recalcitrant proposition G_1 to $\mathcal{A}_1$, thus obtaining a new system $\mathcal{A}_2$. Is the latter now complete? Surely not, for now we can create a proposition G_2 not provable within $\mathcal{A}_2$. We can prove G_2 in a new system $\mathcal{A}_3$, which in turn has its own problem proposition G_3 not provable within it, and so on forever. This result became known as Gödel's Incompleteness Theorem.

The infinite ascent of formal systems $\mathcal{A}_1$, $\mathcal{A}_2$, $\mathcal{A}_3$, ..., in the Incompleteness Theorem, is directly paralleled by the infinite ascent of aspiring "universal" computers $\mathcal{U}_1$, $\mathcal{U}_2$, $\mathcal{U}_3$, ..., in the Non-universality Theorem.

It is interesting to note in passing the way in which various philosophical movements took hold of the result as a validation of their agenda. Thus, for example, to the postmodernists, Gödel's Incompleteness Theorem implied that no firm foundation exists for any system of logic. The existentialists saw in it an end to rational and objective thought. Some mathematicians and an assortment of thinkers in various disciplines argued, on the strength of the theorem, that humans are superior to machines. One physicist even suggested that, thanks to Gödel's work, it is now obvious that the human brain is not a deterministic computer; rather, it is a quantum computer.

3. Non-universality and unending recursion

In this section, the philosophical concept of infinite regress is used to develop a proof of non-universality that is distinct from the proof by counterexample presented in Section 3. The proof itself is presented within the general framework of logic known as proof by contradiction. When invoking contradiction to prove a theorem, we begin by assuming that the claim in the theorem is false; we then show that this assumption leads logically to an absurdity (hence, the Latin name *reductio ad absurdum* for this style of proof in mathematics).

3.1. *Proving non-universality in computation by contradiction*

An alternative proof of the Non-universality Theorem is provided in what follows.[3] Let $\mathcal{U}_1$ be a general-purpose computer.

Non-universality Theorem. $\mathcal{U}_1$ cannot be a universal computer.

Proof. Let us assume, as is commonly the case in computer science, that there exists a "universal" computer capable of simulating any possible computation C, the latter being the result of another computer M executing a certain program on an input I. For brevity, in what follows we use M to represent both the computer as well as the program being simulated. □

In order to be specific, and without any loss of generality, let the assumed "universal" computer be $\mathcal{U}_1$. We write $\mathcal{U}_1(M, I)$ to express the fact that $\mathcal{U}_1$ takes M and I as input and simulates the computation C by performing the actions of M on I. Note that $\mathcal{U}_1$ is used here, by definition, as a *simulating computer.* Such a computer needs the description of another computer (M) in order to simulate *that* computer as it runs a program on its input (I). In other words, $\mathcal{U}_1$ does not act directly on an input (such as I).

In what follows, let $C = (M, I)$ be a terminating computation, meaning that M runs on I and halts in a finite number of computational steps. We write $\mathcal{U}_1[C]$ as a shorthand for $\mathcal{U}_1(M, I)$, the simulation of C by $\mathcal{U}_1$, the latter also a terminating computation.

It is evident, from the principle of universality, that the actions of $\mathcal{U}_1$ itself should be possible to simulate. The question is: "Who" is to simulate a computation performed by a universal computer? Specifically, how are the actions of $\mathcal{U}_1$ — as it simulates C, that is, $\mathcal{U}_1[C]$ — *themselves* to be simulated?

There are two options.

Option 1. $\mathcal{U}_1$ simulates itself. We write $\mathcal{U}_1[\mathcal{U}_1[C]]$ to indicate that $\mathcal{U}_1$ is simulating $\mathcal{U}_1[C]$. This means that $\mathcal{U}_1$ is executing the actions of itself ($\mathcal{U}_1$, the computer), on the simulation ($\mathcal{U}_1[C]$, the input); we write:

$$\mathcal{U}_1[\mathcal{U}_1[C]] = \mathcal{U}_1(\mathcal{U}_1, \mathcal{U}_1[C]).$$

The right-hand side of the above expression has three $\mathcal{U}_1$s: the first is the simulator, the second is the computer being simulated, and the third, $\mathcal{U}_1[C]$, is the input. We therefore have:

$$\begin{aligned}
\mathcal{U}_1[\mathcal{U}_1[C]] &= \mathcal{U}_1(\mathcal{U}_1, \mathcal{U}_1[C]) \\
&= \mathcal{U}_1(\mathcal{U}_1[\mathcal{U}_1[C]]) \\
&= \mathcal{U}_1(\mathcal{U}_1(\mathcal{U}_1, \mathcal{U}_1[C])) \\
&= \mathcal{U}_1(\mathcal{U}_1(\mathcal{U}_1[\mathcal{U}_1[C]])) \\
&= \mathcal{U}_1(\mathcal{U}_1(\mathcal{U}_1(\mathcal{U}_1, \mathcal{U}_1[C]))). \\
&\ \ \vdots
\end{aligned} \tag{3}$$

This leads to a *self-referential* infinite regress, with $\mathcal{U}_1$ simulating itself, simulating itself, simulating itself, ... The computation (3) just described is a non-terminating process (one could say that it is not even a computation, by definition, since it does not halt, but that is not important for our purpose). It follows that $\mathcal{U}_1$ has failed to simulate itself, while executing a *terminating* computation C.

Option 2. We can stipulate the existence of a more powerful universal computer $\mathcal{U}_2$ that simulates $\mathcal{U}_1[C]$. Again, this leads, in turn, to an *ascending* infinite regress, featuring a sequence of ever more powerful computers $\mathcal{U}_2$, $\mathcal{U}_3$, $\mathcal{U}_4, \ldots, \mathcal{U}_n$, $\mathcal{U}_{n+1}$, $\mathcal{U}_{n+2}, \ldots$, performing simulations C_1, C_2, C_3, ..., C_n, C_{n+1}, C_{n+2}, ... Formally,

$$\begin{aligned}
C_1 &= \mathcal{U}_2[\mathcal{U}_1[C]], \\
C_2 &= \mathcal{U}_3[\mathcal{U}_2[\mathcal{U}_1[C]]], \\
C_3 &= \mathcal{U}_4[\mathcal{U}_3[\mathcal{U}_2[\mathcal{U}_1[C]]]], \\
&\ \ \vdots \\
C_{n+1} &= \mathcal{U}_{n+2}[\mathcal{U}_{n+1} \ldots \mathcal{U}_4[\mathcal{U}_3[\mathcal{U}_2[\mathcal{U}_1[C]]]] \ldots], \\
&\ \ \vdots
\end{aligned} \tag{4}$$

and so on *ad infinitum*. In the unending sequence of computations described by (4), computer $\mathcal{U}_{n+1}$ is needed to simulate $\mathcal{U}_n$, for $n = 1, 2, 3, \ldots$, given that computer $\mathcal{U}_n$ cannot simulate itself (as shown in Option 1 for $\mathcal{U}_1$).

Both Option 1 and Option 2 are computationally absurd, leading to one conclusion: the assumption regarding the existence of a universal computer $\mathcal{U}_1$ is false. No computer is universal. □

3.2. *Absolute idealism and self-referential cartography*

As the sun was setting on the 19th century, idealist philosopher Josiah Royce proposed an interesting thought experiment. Imagine, he suggested, we could draw a detailed map of England. The map would be so precise as to contain every road, every river, every hill, every plain, and so on. A flat terrain will be chosen and the map inscribed on the surface of the English countryside. But then something unexpected would happen. Since the map is now part of the landscape, in order for it to be exact, it would necessarily have to contain a copy of itself. That copy would therefore inevitably contain a copy of itself as well, and the copy of the copy a copy, ...; the process will continue indefinitely. This self-referential map-within-a map construction is known as the map of England problem and is sometimes used in studies of consciousness.[4] A whimsical illustration of self-reference and the potential for infinite regress is shown in Figure 2.

4. Conclusion

On a hot August day in the year 1900, the illustrious German mathematician David Hilbert addressed the International Congress of Mathematicians assembled at the Université Sorbonne in Paris. Hilbert presented his colleagues with a list of problems on which, he believed, they should spend their time in the new century. Among these problems was the question of whether there exists a fixed set of true mathematical statements that can be used to prove *automatically* any new mathematical statement. Hilbert's objective was the formalization of mathematics.

Fig. 2. Self-reference and unending recursion.

The purpose of Hilbert's program in formalizing mathematics was twofold. The first goal was to contain infinity. Proofs for all true statements of a formal system were to be produced from a finite set of axioms. The second goal was to eliminate intuition from mathematics. By mechanizing proof generation, serendipity would no longer be part of the process of doing mathematics. Gödel's work demonstrated that, on the contrary, infinity is an integral part of mathematics and cannot be tamed. Mathematicians will always use their intuition to reason about the infinite.

Likewise, the Non-universality Theorem shows that no finite computer can be universal. A new machine will always be needed to cope with the next challenge. The resemblance between the Non-universality Theorem and the Incompleteness Theorem is captured by the drawing in Figure 3. For every computer thought to be universal, there exists a computational problem that it cannot solve, even if given unbounded time and space; a more powerful computer is required. Similarly, for every set of axioms thought to be complete,

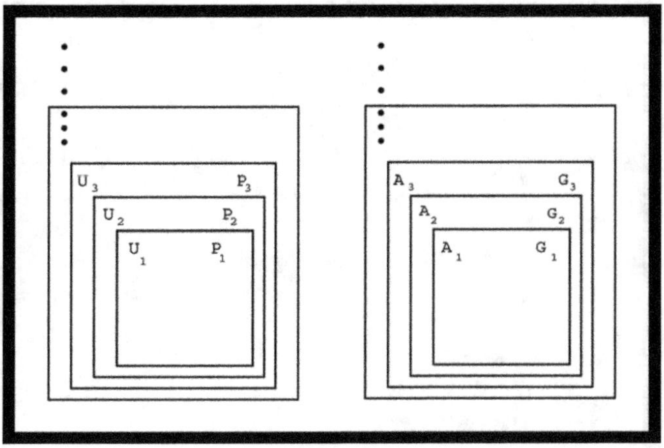

Fig. 3. Non-universality and incompleteness.

there exists a proposition that it cannot prove; an augmented set of axioms is required.

The idea of non-universality in computation is not a new one. Several proofs *by counterexample* of the non-existence of a universal computer have been presented since the early 21st century. They described different unconventional computational paradigms that falsify the idea of a universal computer. Examples of such paradigms include computations with: time varying variables, time varying computational complexity, rank varying computational complexity, interacting variables, uncertain time constraints, mathematical constraints, global variables, and so on. These computational problems imply that computation is a fundamental category of Nature, and as such it has no bounds. Its parameters are limitless. Time passes, inexorably, changing everything in its path. The constituents of our physical space constantly interact with one another, mutually affecting each other. As our world evolves, computations are taking place everywhere, all the time. The genie simply does not fit in the bottle.

By contrast with the proof by counterexample, the proof of non-universality in computation *by contradiction* is a more recent result. It was motivated when an obvious and logical question was asked,

Fig. 4. Self-simulation and the ensuing infinite regress.

apparently for the first time ever: If simulation is the bedrock of computing, and the supposed "universal" computer can simulate any computation, how can the actions of the "universal" computer *itself* be simulated? The answer led once again to the collapse of the notion of universality. This, in turn, leads here to the following interesting observation. The invention in 1899 of the self-referential map-of-England metaphor, bringing about an inevitable infinite regress along with it, predates by 121 years the discovery of a proof, by self-simulation, of computational non-universality. Miraculously, the former (old result) serves as a striking illustration of the latter (new result), as demonstrated in Figure 4.

We note in closing that non-universality in computation applies to all known computational models and existing conventional computers, both sequential and parallel, as well as applying to all future unconventional computers, including quantum computers, biomolecular computers, chemical computers, and so on. Like the Halting Problem in Computer Science, Incompleteness in Mathematics, and the Uncertainty Principle in Physics, Non-universality in Computation is a limiting theorem, an impossibility result.

Acknowledgment

I am grateful to my daughter Sophia for the beautiful drawings in this chapter.

References

1. S. G. Akl, Unconventional computational problems. In: *Encyclopedia of Complexity and Systems Science*, R. A. Meyers, editor. Springer, 2018, 631–639.
2. K. Gödel, Über formal unentscheidbare Sätze der Principia Mathematica und verwandter Systeme I. *Monatshefte für Methematik und Physik* 38:173–198, 1931.
3. S. G. Akl, A map of England, the simulator simulated, and nonuniversality in computation. *Int. J. Unconv. Comput.* 15(3):219–223, 2020.
4. J. Royce, *The World and The Individual, First Series: The Four Historical Conceptions of Being*. Dover Publications, 1959, 504–505.

https://doi.org/10.1142/9789811257155_0003

Chapter 3

Natural Computation of Cognition, from Single Cells Up

Gordana Dodig Crnkovic
Department of Computer Science and Engineering, Chalmers University of Technology, 41296 Gothenburg, Sweden
dodig@chalmers.se

For describing continuum of natural cognition, from basal cellular to human level, we use evolutionary info-computational framework, where natural/physical/morphological computation drives generation of increasingly adept cognitive systems. We build on novel developments in information physics, bioinformatics, information chemistry, computational neuroscience, complexity theory, self-organization, theory of evolution (new synthesis), information, and computation, a constructive interdisciplinary framework for understanding of cognition in the context of computing nature. Interactions between constituents at different levels of organization lead to complexification of agency and increased cognitive capacities of organisms. Recently, possibility of controlling basal cell cognition attracted a lot of interest for its possible applications in medicine, new computing technologies, as well as micro- and nano-robotics. Bio-cognition of cells connected into tissues/organs and organisms with the group (social) levels of distributed information processing provides insights into cognition mechanisms that can help us to better conceptualize human cognition with its evolution and support the development of new AI models and cognitive robots. It will also give us better understanding of the human as a natural being and our connectedness with the rest of nature.

1. Introduction

When talking about cognition, it is typically assumed to be a human capacity,[1] seldom including animal cognition and hardly ever cognition of plants[2–5] or even simpler living organisms, such as bacteria[6–14] or slime molds.[15] Even though cognition is traditionally considered as a result of human mind and thought, cognition in nature appears throughout biological systems[16–20] in different degrees, and it is important to understand its evolutionary development from the basal/basic/elementary cognition to the human level.[21–23] Models of cognition in this context have considerable explanatory value.

Traditionally, cognition is argued to conflict with computational models, where computation is understood as abstract disembodied mechanical symbol manipulation. It has been claimed that natural embodied cognition is irreconcilable with computational models, and that dynamical systems cannot be computational. Those claims have been demonstrated to be false, based on empirical observations, once it is understood that computation fundamentally always is embodied and that cognitive processes are information processes. If computation is understood as information processing in nature (natural computation/physical computation/unconventional computation/morphological computation), it is necessarily both embodied and dynamic.

This chapter presents computational framework of natural cognition based on info-communication (info-computation) in living agents. In this naturalistic approach, the underlying assumption is that cognition in nature is a manifestation of biological processes[24–27] that subsume chemical and physical processes — from single cells to humans.

A significant move from human-only cognition, toward a broader approach, can be found in the recent work of Piccinini,[28] who credits cognition to all organisms with nervous systems. Moreover, Piccinini argues that cognition is a result of computation, neurocomputation. Even though he goes a step beyond conventional understanding, Piccinini does not include all living organisms. However, there is rich empirical evidence that "cognitive operations we usually ascribe to brains — sensing, information processing, memory, valence, decision-making, learning, anticipation, problem solving, generalization, and goal directedness — are all observed in living forms that don't have brains or even neurons."[23] Thus, it is possible, and even necessary to generalize cognition a step further, to include all living forms, not only those with nervous systems.

Based on empirical and theoretical insights about cognition and its evolution and development in nature,[29,30] from basal/basic/primitive/elementary/cellular up to complex form of human cognition[21,23,25,27,31] modeled on natural information processing (natural computation), we can identify several generative mechanisms of cognition that deserve more attention.

The rest of the chapter is organized as follows. Section 2 addresses info-computational approach to cognition in biological systems, "thinking" fast and slow through sub-symbolic and symbolic computational processes and connection of cognition with evolution. Section 3 presents natural processes of info-computation as a basis of natural cognition, and morphogenesis as (morphological) computation. Section 4 discusses the importance of time aspects of cognition. Section 5 offers conclusions.

2. Info-computational Approach to Natural Cognition. Evolution of a "Sensitive Soul" Through Extended Evolutionary Synthesis

Knowledge and skills of a living agent are embodied in an info-computational natural process that generates cognition. This process produces intelligent behavior (problem-solving ability) of natural systems in complex environments. Traditionally, cognition has been studied as a result of brain activity in humans, where there was an opposition between three different types of brain activities supposed to underlie cognition: brain function as a result of symbol processing of Turing computation type,[32] distributed computation models,[33] and dynamic models of a brain as a meta-stable oscillating system.[34] It is important to realize that all three "modi operandi" in the brain can suitably be modeled as natural computations (natural information processing) on different levels of organization. Let us start by introducing the framework for naturalizing cognition, with two basic elements: natural (embodied) information, and natural (physical) computation.

2.1. *Naturalized cognition. Thinking fast and slow — sub-symbolic and symbolic computing*

The info-computational naturalist model of cognition is a hybrid connectionist-symbolic, biologically realistic conceptual framework aiming to integrate current knowledge from a variety of research fields, such as cognitive science, computational neuroscience, bioinformatics, computability, biology, and new evolutionary synthesis.[35] This approach is based on the view of hierarchical recursive structure of information processing in nature, which is especially important for living organisms, from cells, to tissues, organs, organisms, and their groups — all of them communicating at different levels of organization by exchanging specific types of information — physical (elementary particles, electromagnetic), chemical (electric, molecular), biological, and symbolic.

From the time when the first models of cognition had been proposed until now, a lot has changed in our understanding of cognition, embodiment, functioning of the brain, neurons, and neuronal networks. Within AI, the field of artificial neural networks with deep learning has made an impressive progress in modeling perception on the level of data/signal processing. In humans, two basic cognitive systems have been recognized, System 1 (reflexive, non-conscious, automatic, intuitive information processing, which is fast) and System 2 (reflective, conscious, reasoning, and decision-making, which is slow).[36,37] As Kahneman explains, System 1 and System 2 stand for informational processes that are functional abstractions, not the brain regions. As deep learning models are inspired by the human brain information processing, recent advances in understanding of natural cognitive systems can contribute both to better explanatory models and to future developments of constructive engineered cognitive models. Deep learning level corresponds to Kahneman's fast, intuitive System 1,[36] and current developments in AI are continuing toward even more ambitious goals of modeling System 2 symbolic reasoning.[38] Here we should add that Bengio's and Kahneman's interpretations of Systems 1 and 2 are not identical, which was evident from the discussion at the AAAI-2020 conference, Fireside Chat with Lecun, Hinton, Bengio, and Kahneman, https://vimeo.com/390814190. However, the details are not essential for our present exposition.

It has long been recognized that mechanisms of cognition based on natural computation are far more sophisticated than the machine-like classical computationalist models based on abstract symbol manipulation.[39] They conform to the view expressed by[40–42] that rule-based machines are not good enough models of natural cognition which appears in highly complex living organisms. Embodiment is the fundamental feature of cognition, which implies that valence, affect, feelings, and emotions must be taken into account as constitutive elements in the models of cognition,[27,30,43–45] and they impact both System 1 and System 2 information processing.

2.2. *Information processing in embodied and extended cognition*

Naturalized cognition as a systemic perspective means broadening the scope beyond neurocentrism[17,23,27,46,47] to networks of information-processing agents, down to the molecular level as computational basis of distributed cognition. The physical mechanism of natural computation is morphological computation[48–51] where morphology refers to form, shape, and structure which define interactions.[52]

As already discussed, cognitive models typically identify cognition with information processing in the brain. However, with the rise of embodied cognition, neurocentrism is being challenged by a systemic view of cognition, where the body shapes cognitive functions. Furthermore, as Cowley *et al.*[46] show, living beings also connect their bodies with artifacts, in a sense of extended cognition (Clark, 2008). The central importance of interaction we learn as well from Ginsburg and Jablonka,[53] view of the evolution of "the sensitive soul," providing the naturalized communication between an agent and its environment, which changes both.

2.3. *Evolutionary view of cognition in nature. Scaling from basal to complex*

If we want to learn how cognition functions in humans as the most complex living organisms, it is instructive to see how this ability developed through evolution, resulting in a variety of cognitive architectures of organisms from bacteria to humans.[21,27,53] In a naturalist approach of Maturana and Varela, cognition in any living organism is a result of embodied processes that make the organism alive,[24] or as Stewart[25] puts it, "Cognition = Life." Here life includes capability of growth and reproduction. All living systems are cell-based, from unicellular to complex ones, with cells organized in tissues and organs, where each cell possesses cognition. Groups of organisms like swarms and flocks exhibit social cognition. Cognitive capacities on different scales make living systems goal-directed, robust, and adaptive.

In biomimetic (nature-inspired) robotic systems, cognition is represented by the equivalents of living functions, implemented in a robot provided with sensors, actuators, and information processing units. This basal level of cognition can be of practical interest, as robots do not always need human-level abilities to perform their tasks.[54] Cognition of a different, non-human type can be adequate in biomimetic soft robots.[55] Levin *et al.*[16,21–23,56] describe a variety of mechanisms of cognition where robust adaptive information processing and behavior can be used to develop new computational techniques in biological and engineered systems. In the naturalized, evolutionary concept of cognition, the development goes from the simplest organizational form of a single cell, as "cellular mind," up to the brain as "the society of mind."[57] In this process, the body plays a vital role in shaping the mind.[48] Organisms learn about the world by means of information exchange/communication.[58–61] Reality for a cognitive agent is an informational structure[50] and biology computes. Processes of change in informational structures establish computational dynamics. This model of reality for an agent includes both the information about the agent itself and about the world as it appears for the agent via interactions with the environment. The proposed naturalist framework provides computational models for cognitive info communications which the author has been developing.[31,45,62–65] In the info-computational approach, the evolutionary process unfolds in living organisms, and it happens in the sense of extended evolutionary synthesis[53,66–68] as a result of interactions (communication) between natural agents, be it cells, their groups, or multicellular organisms.

Unlike the "Modern evolutionary synthesis," which supplemented Darwinism with added genetic determinism, in the computing nature approach, the emphasis is on the role of the interaction/communication with the environment for development and evolution. Origins of life can be found in the interactions of the first simplest pre-biotic chemical agents, leading to more complex forms such as viruses and furthermore first cells as bacteria, continuing up in complexity through the information structures' self-organization. Genes are important, but not solely responsible for the development of cells and their aggregates up to organisms and ecologies. As Ginsburg, Jablonka, and Witzany[41,53,67] describe, the interplay between the genetic code with the environment, through material embodiment, is crucial. Even Rovelli[69] argues for the central role of evolution as a mechanism that generates mental (intentionality, purpose, agency) from physical: "Meaning and Intentionality = Information + Evolution."

In the framework of info-computational nature, living organisms are cognitive agents, from single cells to humans.[45,70] Cognitive artifacts can also be seen as natural physical systems with various degrees of cognitive capacities.[19,52] Cognition is an open-ended process of self-organization where computation proceeds as signal processing at physical and chemical levels, while on the biological and cognitive levels it takes the form of symbol manipulation and language-based communication.[71,72]

Simultaneous developments of minds and bodies has been studied by Schröder[73] as natural information processes. In a cognitive agent, variety and its dynamics is tackled through the dual concept of selective and structural aspects of information. Biological evolution of a species is a dynamical processing of information. What for a cognitive agent appears as "the world" is an interface, a shared boundary across which the information is exchanged,[74] with perception based on data/information obtained from the senses. A similar idea of a computational boundary of a "self" is put forward by Ref. 22 who describes mechanisms driving biological agents toward multicellularity and scale-free cognition.

3. Natural Processes of Info-Computation — Computing as Morphogenesis — Morphological Computing

In our study of cognition as a natural phenomenon, we adopt Floridi's Informational structural realism[75,76] as an approach in which reality is "informational structure for an epistemic agent interacting with the universe by the exchange of data as constraining affordances." The dynamics of that informational structure is conceptualized through the idea of Computing nature — where the physical dynamics of informational nature is natural computation.[49,77–81] This provides a unified naturalist setting for studying structures and processes in both the animate and inanimate worlds.[26] The underlying fundamental property of information which makes it suitable as a basis of the naturalist approach is the Landauer principle of naturalization of information: "there can be no information without physical implementation."[82] Our approach is thus based on naturalized metaphysics. As Ladyman *et al.*[83] explain, naturalists construct science-based, structural-realist, computational ontology. Physicist Rovelli contributes to the naturalization program with his proposal to build the foundation of physics on relative information.[69] Here Shannon's relative information between two physical systems defines a purely physical notion of information, which can be used to "glue everything together."[64] That means to connect networks of information processing nodes in nature. Interactions between physical systems are exchanges of information that establish physical correlations between them, through Shannon's relative information (correlation). By combining physical correlations with Darwinian evolution, Rovelli builds a ground for emergence of meaning in nature.

Insights from informational chemistry help further in bridging the gap between physics and biology, with supramolecular chemistry that connects molecular recognition, molecular information processing, and self-organization.[84,85] Biology and cognitive sciences are already established as information-based and their processes have been modeled as natural computation.[86–89] Even the evolution of life has been modeled as a process of morphological info-computation (metamorphogenesis).[90] The whole sequence of information-based sciences — from physics to chemistry, biology (including evolution of species), and cognitive sciences (including social cognition) makes it possible to understand cognition as natural information processing/ natural computation.

3.1. *Natural/physical/morphological cognitive computation*

The concept of natural computation as presented by Refs. 77, 91 addresses information processing (both discrete and continuous) as spontaneously appearing in nature.[92] Models of natural computation/natural information processing differ from the Turing model of computation, that is symbol manipulation. From the point of view of organization of computational processes, natural computation is different from von Neumann computation. Natural computation models of biological organisms with their multi-level processes of computing and distributed information processing are capable of capturing the dynamic behavior of natural cognitive agents, including neuronal networks.[93] In the framework of info-computational nature, the fundamental mechanism is morphological computation, i.e., a process of information self-organization such as described by Haken.[94,95] The author has addressed this topic in Refs. 31, 45, 62, 63, 65, 91. Morphological computation in living nature is a network of morphological informational processes for cognitive agents, with cognition as layered morphological computation.

Recently, in robotics, specific use of the term "morphological computation" has been adopted to denote decentralized embodied control in robots. In the context of robotics, appropriate body morphology is saving information processing (computational) resources as well as enabling learning through self-structuring (self-organization) of information.[48,96,97] This is a macroscopic view of morphological computation that does not concern lower levels of organization, such as cellular, molecular, or quantum computation.

Natural computation appears on all levels of organization in nature, from physical, chemical, biological to cognitive. Of special interest for us are the levels with chemical and biological computation contributing to cognitive behavior such as presented by Ref. 98, showing biochemical basis of connectionism. On the level of cells and tissues, there are numerous computational approaches such as Refs. 20, 99, 100. Proposals have been made for synthetic analog computation of living cells with artificial epigenetic networks.[101] Extensive literature exists on specific neuronal computation,[28,102] as well as computational models of the brain.[103–108]

It is important to keep in mind the difference between new computational models of intrinsic information processes in nature (natural computing/morphological computing), and old computationalism based on computer metaphor of the Turing machine, performing symbol processing, which has been rightly criticized as an inadequate model of human cognition.[86,109] As already pointed out, in humans there are two basic cognitive systems, System 1 (reflexive, non-conscious, automatic, intuitive information processing, which is fast) and System 2 (reflective, conscious, reasoning, and decision-making, which is slow).[36,37] Recognizing only symbolic information processing leaves the symbol grounding problem unsolved. Sub-symbolic Symbol 1 data/signal processing provides mechanisms of symbol grounding in deep learning.

Hybrid symbolic-dynamical models[110,111] have been proposed as well, capable of modeling a combination of the two as a reactive-deliberative behavior. According to Ehre,[72] the fast reflexive System 1 can be understood in terms of Rovelli's physical correlations (Shannon's relative information), and it can accommodate for emotion as argued in Ref. 112, while the slow System 2, because of synonymity in the symbol system, introduces an element of choice and indeterminism with higher computational demands. The latter has been addressed in Ref. 113, also addressing the topic of parallel concurrent computation typical of biological systems, for which the Turing Machine model is not adequate.

3.2. *Computation vs. communication in biological systems*

Computation as well as communication involve the transition, transformation, and preservation of information.[114] The relationship between communication and computation has been described by Ref. 115 who argues that they are not conceptually distinguishable. The only difference is that computation concerns actions within a system, while communication is a process of interaction between a system and its environment. Biological systems are open information processing systems in communication with the environment, where the boundary between the system and the environment is dynamic and blurred.

As all other fundamental concepts which are objects of intense research (including information, computation, and cognition), the concept of communication has no generally accepted definition.[37] We use the concept in the sense of computation between systems,[115] that is as exchange of information between the system and the environment. One often thinks of communication as being defined by language with the addition of other symbol systems such as images and sounds. But if we think of a human being with all her senses, then communication takes place on different levels and through many channels that are interacting in the brain. For example, one does not think and feel clearly when disturbed by a constant noise, feeling strong anxiety, or being upset. The whole person participates in communication. Emotional signs may not always be as obvious as the conventional symbolic message, but can be orders of magnitude faster and more important for the receiver, like bodily language, tone of voice, smile, eye gaze, etc.[43,79,112]

To understand communication in a more multifaceted way than the exchange of symbols or signs, it is instructive to look at communication in other, simpler organisms. All living beings communicate.

A cell that is the basic building block of life is a complex communication system. Without communication, life would not be possible.[116] Naturalized communication is based on information defined by structures and computational processes in nature.[117] In the same way as epistemology (knowledge) can be naturalized[26, 78] using natural information processing (natural computation), communication and cognition can be naturalized.

4. Time Aspect of Cognitive Models, Learning and Memory in Naturalized Cognition

Since Turing's work on morphogenesis, the study of morphological computation has focused on spatial structures, even though temporal aspects play equally fundamental roles. Typically, cognitive models assume the mind/brain to be reactive, with information processing starting with a stimulus and ending with a response.[118] However, cells are inherently active, neurons are sustained oscillators, exhibiting electrochemical oscillations even in the absence of stimuli. Input data/information presents stimuli that modulate existing endogenous oscillations.[118] In the book *Rhythms of the Brain*, Buzsaki[93] describes the important role that spontaneous activity of neurons play. Spontaneous firing of neurons is the very basis of human cognition when it comes to its time aspects. A self-organized timing of oscillations has co-evolved as the main organizational principle of neuronal activity. Global computation (on multiple spatial and temporal scales) is enabled by small-world-connectivity of neurons in the cerebral cortex. In a small-world setting, any two nodes are connected through a short sequence of intermediary nodes. The cortical system is in a metastable state, synchronized through weak links between network oscillations in constant interactions. Oscillator frequency determines periods of receiving and transferring information. Based on studies of oscillations, neural computations, and learning, Penagos, Varela, and Wilson[119] proposed that "precisely coordinated representations across brain regions allow the inference and evaluation of causal relationships to train an internal model of the world." Training starts while awake, and processing continues during sleep when periodic nested oscillations induce hierarchical processing of information. Authors suggest that "general inference, prediction and insight" is enabled through periodic states of sleep. Related is the synaptic plasticity of the brain which changes its connections through the long-term potentiation (Hebbian and non-Hebbian), considered to be a basis for learning and memory. Oscillatory behavior is not only the characteristic of the human brain. Similar oscillatory rhythms have been observed in the brains of mice. Being made of oscillators, biological neural networks are able to filter inputs and to resonate with noise. In contrast to the observed oscillatory time behaviors in biological brains, which appear as a result of their physical embodiment artificial neural networks have no such temporal coupling and synchronizing mechanisms. It is an open question how essential oscillatory behavior and metastability are for "fine tuning to the world" and if their function can be obtained in a different way in artificial neural networks.

On the level of unified theories of cognition, time aspect[120] manifests itself in terms of Newell's bands of cognition[121] — the biological "10 millisecond band," cognitive, rational, and social ("long-term") bands. How important is it to have all of them represented and how detailed? Here we talk about the understanding of the temporal aspects of cognition as organized hierarchically in a metastable state, constantly tuning to the environment. Coordination obtained through communication is central for connecting different levels, from molecules to thoughts, in the same coordination dynamics.[34] Through the interplay with the environment this process results in eigenstates.[122] Technological approaches to cognitive models of brain-like computer, based on frequency-fractal computing, have been proposed.[123,124]

5. Conclusions

This chapter presents advances in the understanding of cognitive systems and processes in nature. Interpreting nature in terms of computation (information processing), we can better understand processes of cognition as they function and evolve in living beings, from single cells to complex organisms, and their networks. They can be used to inform future bio-inspired/biomimetic cognitive models in a range of applications, from nano-technology to medicine and robotics.

Novel developments in the understanding of information physics, bioinformatics, information chemistry, computational neuroscience, complexity theory, self-organization, theory of evolution, information and computation as well as embodiment and evolution, support a constructive interdisciplinary framework for cognition in the context of computing nature, where interactions between constituents at different levels of organization lead to complexification of agency and increased cognitive capacities of organisms. Computation in nature/natural computation/physical computation/morphological computation stands for processes of self-structuring of information in a number of organizational levels: physical, chemical, biological, cognitive, and social with networks of communicating agents on every level of organization.[45]

It is important to note the parallel development of our understanding of cognition as a natural phenomenon and its technological implementations that inform each other in a recursive manner.[125,126] At the same time as the knowledge of cognitive processes in living organisms increases, so do also our information processing/ computational models.

We are pointing to the fact that biologically inspired models of cognition such as new developments of connectionist (hybrid deep learning) and dynamical models, including non-neural cognitive systems, deserve a place among cognitive models.

6. Funding

Research supported by the Swedish Research Council project MORCOM@COG Morphological Computing in Cognitive Systems Morcom (2016–2021).

Acknowledgments

The author acknowledges the valuable comments by Zoran Konkoli and Gerard Jagers op Akkerhuis gratefully on the previous version of this chapter.

References

1. P. Thagard, Cognitive science. In *The Stanford Encyclopedia of Philosophy*, E. N. Zalta, editor. Metaphysics Research Lab, Stanford University, 2020, Winter 2020 edition. https://plato.stanford.edu/archives/win2020/entries/cognitive-science/.
2. P. C. Garzon, Plant neurobiology: Lessons for the unity of science. In *Special Sciences and the Unity of Science*, O. Pombo, J. M. Torres, J. Symons, and S. Rahman, editors. Springer, Dordrecht, 2012, 121–137.
3. P. Calvo and A. Trewavas, Cognition and intelligence of green plants. Information for animal scientists. *Biochem. Biophys. Res. Commun.* 564:78–85, 2021. https://doi.org/10.1016/j.bbrc.2020.07.139.Epub.
4. P. Calvo, M. Gagliano, G. M. Souza, and A. Trewavas, Plants are intelligent, here's how. *Ann. Bot.* 125(1): 11–28, 2020. Doi: 10.1093/aob/mcz155.
5. P. Calvo and K. Friston, Predicting green: Really radical (plant) predictive processing. *J. R. Soc. Interface* 14(131): 20170096, 2017. Doi: 10.1098/rsif.2017.0096.
6. E. Ben-Jacob, Bacterial complexity: More is different on all levels. In *Systems Biology- The Challenge of Complexity*, S. Nakanishi, R. Kageyama, and D. Watanabe, editors. Springer, Tokyo Berlin Heidelberg, New York, 2009a, 25–35.
7. E. Ben-Jacob, Learning from bacteria about natural information processing. *Ann. N. Y. Acad. Sci.* 1178:78–90, 2009b.
8. E. Ben-Jacob, Bacterial wisdom, gödels theorem and creative genomic webs. *Physica A.* 248:57–76, 1998.
9. E. Ben-Jacob, Y. Shapira, and A. Tauber, Seeking the foundations of cognition in bacteria. *Physica A.* 359:495–524, 2006.
10. E. Ben-Jacob, Bacterial self-organization: Co-enhancement of complexification and adaptability in a dynamic environment. *Phil. Trans. R. Soc. Lond.* A:315–322, 2003.
11. E. Ben-Jacob, Social behavior of bacteria: From physics to complex organization. *Eur. Phys. J. B.* 65(3):315–322, 2008.
12. W.-L. Ng and B. Bassler, Bacterial quorum-sensing network architectures. *Ann. Rev. Genet.* 43:197–222, 2009.
13. C. M. Waters and B. Bassler, Quorum sensing: Cell-to-cell communication in bacteria. *Ann. Rev. Cell Dev. Bio.* 21:319–346, 2005.
14. S. Schauder and B. Bassler, The languages of bacteria. *Genes Dev.* 15:1468–1480, 2001.

15. J. Vallverdú, O. Castro, R. Mayne, M. Talanov, M. Levin, F. Baluška, Y. Gunji, A. Dussutour, H. Zenil, and A. Adamatzky, Slime mould: The fundamental mechanisms of biological cognition. *BioSystems*. 165(3): 57–70, 2018.
16. F. Baluška and M. Levin, On having no head: Cognition throughout biological systems. *Front. Psychol.* 7:902, 2016.
17. P. Lyon, The biogenic approach to cognition. *Cog. Process.* 7:11–29, 2005.
18. P. Lyon, The cognitive cell: Bacterial behaviour reconsidered. *Front Microbiol.* 6:264, 2015.
19. A. Almér, G. Dodig-Crnkovic, and R. Haugwitz, Collective cognition and distributed information processing from bacteria to humans. In *Proc. AISB Conference Kent*, AISB, 2015. https://www.cs.kent.ac.uk/events/2015/AISB2015/proceedings/socialComp/papers/SACCS-AISB2015_submission_10.pdf.
20. P. Lyon, F. Keijzer, D. Arendt, and M. Levin, Reframing cognition: Getting down to biological basics. *Phil. Trans. R. Soc. B.* 376:20190750, 2021.
21. S. Manicka and M. Levin, The cognitive lens: A primer on conceptual tools for analysing information processing in developmental and regenerative morphogenesis. *Philos. Trans. R. Soc. B.* 374: 2019.
22. M. Levin, The computational boundary of a "self": Developmental bioelectricity drives multicellularity and scale-free cognition. *Front. Psychol.* 10:2688, 2019.
23. M. Levin, F. Keijzer, P. Lyon, and D. Arendt, Uncovering cognitive similarities and differences, conservation and innovation. *Phil. Trans. R. Soc. B.* 376:20200458, 2021.
24. H. Maturana and F. Varela, *The Tree of Knowledge*. Shambala, 1992.
25. J. Stewart, Cognition = life: Implications for higher-level cognition, *Behav. Process.* 35:311–326, 1996.
26. G. Dodig-Crnkovic, Epistemology naturalized: The info-computationalist approach. *APA Newsletter Philos. Comput.* 6(2):9–13, 2007.
27. P. Lyon and F. Kuchling, Valuing what happens: A biogenic approach to valence and (potentially) affect. *Phil. Trans. R. Soc. B.* 376:201907522 0190752, 2021.
28. G. Piccinini, *Neurocognitive Mechanisms: Explaining Biological Cognition.* Oxford scholarship online, Oxford, 2020.
29. S. I. Walker, P. Davies, and G. Ellis, *From Matter to Life Information and Causality.* Cambridge University Press, Kindle Edition, 2017.
30. G. Dodig-Crnkovic and R. Giovagnoli, Representation and reality in humans, other living organisms and intelligent machines. In *Studies in Applied Philosophy, Epistemology and Rational Ethics*, 28, 2017. Doi: 10.1007/978-3-319-43784-2.
31. G. Dodig-Crnkovic, Natural morphological computation as foundation of learning to learn in humans, other living organisms, and intelligent machines. 2020. https://doi.org/10.3390/philosophies5030017.

32. A. Wells, Cognitive science and the turing machine: An ecological perspective. In *Alan Turing: Life and Legacy of a Great Thinker*, C. Teuscher, editor. Springer, Berlin, Heidelberg, 2004.
33. A. Clark, Supersizing the Mind: Embodiment, Action, and Cognitive Extension. Oxford University Press, Oxford, UK. Print ISBN-13: 9780195333213, 2008.
34. S. J. Kelso, G. Dumas, and E. Tognoli, Outline of a general theory of behavior and brain coordination. *Neural Netw.* 37:120–131, 2013.
35. D. Noble, Charles darwin, jean-baptiste lamarck, and 21st century arguments on the fundamentals of biology. *Prog Biophys Mol Biol.* 153:1–4, 2020. Doi: 10.1016/j.pbiomolbio.2020.02.005.
36. D. Kahneman, *Thinking, Fast and Slow.* Macmillan Inprint: Farrar, Straus and Giroux, New York, New York, 2011.
37. T. A. Tjøstheim, A. Stephens, A. Anikin, and A. Schwaninger, The cognitive philosophy of communication. 2020. URL: https://doi.org/10.3390/philosophies5040039.
38. J. Russin, R. C. O'Reilly, and Y. Bengio, Deep learning needs a prefrontal cortex. In *Workshop on "Bridging AI and Cognitive Science"*. ICLR, 2020, 2020.
39. G. Kampis, *Self-Modifying Systems in Biology and Cognitive Science: A New Framework for Dynamics, Information, and Complexity.* Pergamon Press, Amsterdam, 1991.
40. M. Conrad, Molecular and evolutionary computation: The tug of war between context freedom and context sensitivity. *Bio Syst.* 52(1–3):99–110, 1999.
41. G. Witzany and F. Baluška, Life's code script does not code itself. *EMBO Reports.* 13:1054–1056, 2012. Doi: 10.1038/embor.2012.166.
42. G. Witzany, Turing: A formal clash of codes. *Nature.* 483:541, 2012.
43. A. R. Damasio, *The Feeling of What Happens: Body and Emotion in the Making of Consciousness.* Harcourt Brace Jovanovic, San Diego, California, 1999.
44. S. Watanabe, M. A. Hofman, and S. Toru, editors, *Evolution of the Brain, Cognition, and Emotion in Vertebrates.* Springer, Tokyo, 2017.
45. G. Dodig-Crnkovic, Nature as a network of morphological infocomputational processes for cognitive agents. *Eur. Phys. J.* 226:181–195, 2017. Doi: 10.1140/epjst/e2016-60362-9.
46. S. J. Cowley and F. Vallée-Tourangeau, *Cognition Beyond the Brain. Computation, Interactivity and Human Artifice.* Springer International Publishing AG, Cham, 2017. Doi: 10.1007/978-3-319-49115-8.
47. M. Gabriel, *I Am not a Brain: Philosophy of Mind for the 21st Century.* Polity Press, Cambridge, 2017.
48. R. Pfeifer and J. Bongard, *How the Body Shapes the Way We Think — A New View of Intelligence.* MIT Press, Cambridge, MA, 2006.
49. G. Dodig-Crnkovic and R. Giovagnoli, Computing nature — A network of networks of concurrent information processes. In *Computing Nature*, vol. 7. Springer, Berlin, Heidelberg, 2013. Doi: 10.1007/978-3-642-37225-4_1.

50. G. Dodig-Crnkovic and R. Haugwitz, Reality construction in cognitive agents through processes of info-computation. In *Representation and Reality in Humans, Other Living Organisms and Intelligent Machines*, G. Dodig-Crnkovic and R. Giovagnoli, editors. Springer International Publishing, Basel, 2017, 211–235. Doi: 10.1007/978-3-319-43784-2.
51. V. Müller and M. Hoffmann, What is morphological computation? On how the body contributes to cognition and control. *Artifi. Life.* 23(1): 2017.
52. G. Dodig-Crnkovic, Modeling life as cognitive info-computation. In *Computability in Europe 2014. LNCS*, A. Beckmann, E. Csuhaj-Varjú, and K. Meer, editors. Springer, Berlin, Heidelberg, 2014, 153–162. http://arxiv.org/abs/1401.7191.
53. S. Ginsburg and E. Jablonka, *The Evolution of the Sensitive Soul.* MIT Press, Cambridge, MA, 2019. Doi: 10.7551/mitpress/11006.001.0001.
54. R. Brooks, Intelligence without representation. *Artifi. Intell.* 47(1–3):139–159, 1991.
55. E. B. Joyee, A. Szmelter, D. Eddington, and Y. Pan, 3d printed biomimetic soft robot with multimodal locomotion and multifunctionality. *Soft Robotics* 2020. Doi: 10.1089/soro.2020.0004.
56. M. Levin, Life, death, and self: Fundamental questions of primitive cognition viewed through the lens of body plasticity and synthetic organisms. *Biochemical and Biophysical Research Communications*, 2020. In press.
57. M. Minsky, *The Society of Mind.* Simon and Schuster, New York, 1986.
58. G. Terzis and R. Arp, *Information and Living Systems: Philosophical and Scientific Perspectives.* MIT Press, Bradford Books, Cambridge, MA, 2011.
59. E. R. Dougherty, M. L. Bittner, and M. Akay, Epistemology of the cell: A systems perspective on biological knowledge. 2011. https://doi.org/10.1002/9781118104866.
60. V. Braitenberg, *Information — Der Geist in Der Natur.* Schattauer GmbH, Stuttgart, 2011.
61. P. Davies, The demon in the machine. 2019. https://doi.org/10.7208/chicago/9780226669847.001.0001.
62. G. Dodig-Crnkovic, Info-computationalism and morphological computing of informational structure. In *Integral Biomathics: Tracing the road to reality*, P. L. Simeonov, L. S. Smith, and A. C. Ehresmann, editors. Berlin, New York, Springer, 97–104, 2012.
63. G. Dodig-Crnkovic, Information and energy/matter. *Information (Switzerland.* 3(4), 2012. Doi: 10.3390/info3040751.
64. G. Dodig-Crnkovic, Physical computation as dynamics of form that glues everything together. *Information.* 3(4):204–218, 2012. Doi: 10.3390/info3020204.
65. G. Dodig-Crnkovic, Cognition as embodied morphological computation. In *Studies in Applied Philosophy, Epistemology and Rational Ethics.* Springer, 2018. Doi: 10.1007/978-3-319-96448-5_2.
66. K. N. Laland, T. Uller, M. W. Feldman, K. Sterelny, G. B. Müller, A. Moczek, E. Jablonka, and J. Odling-Smee, The extended evolutionary

synthesis: Its structure, assumptions and predictions. *Proc. R. Soc. B.* 282:20151019, 2015. Doi: 10.1098/rspb.2015.1019.

67. E. Jablonka and M. Lamb, Evolution in four dimensions: Genetic, epigenetic, behavioral, and symbolic variation in the history of life. In *Life and Mind: Philosophical Issues in Biology and Psychology.* A Bradford Book, MIT Press, Cambridge, MA, 2014, revised edn.
68. N. Gontier, Evolutionary epistemology as a scientific method: A new look upon the units and levels of evolution. *Theory Biosci.* 129:167–182, 2010.
69. C. Rovelli, Relative information at the foundation of physics. In *It From Bit or Bit From It? The Frontiers Collection.* Springer, 2015, 79–86. Doi: 10.1007/978-3-319-12946-4_7.
70. D. Dennett, *From Bacteria to Bach and Back: The Evolution of Minds.* Norton & Company, New York, 2017.
71. A. C. Ehresmann, Mens, an info-computational model for (neuro-)cognitive systems capable of creativity. *Entropy.* 14:1703–1716, 2012.
72. A. C. Ehresmann, A mathematical model for info-computationalism. *Constr. Found.* 9(2):235–237, 2014.
73. M. J. Schroeder, Dualism of selective and structural manifestations of information in modelling of information dynamics. *Comput Nat.* 7:125–137, 2013.
74. O. Rössler, Endophysics: The world as an interface. World scientific, Singapore, New Jersey, London, Hong Kong, 1998.
75. L. Floridi, Informational realism. *Selected Papers from Conference on Computers and Philosophy.* 37(CRPIT'03): 2003.
76. L. Floridi, A defense of informational structural realism. *Synthese* 161(2):219–253, 2008.
77. G. Rozenberg, T. Bäck, and J. Kok, editors. *Handbook of Natural Computing.* Springer, Berlin, Heidelberg, 2012.
78. G. Dodig-Crnkovic, Biological information and natural computation. thinking machines and the philosophy of computer science: Concepts and principles. URL: https://doi.org/10.4018/978-1-61692-014-2.ch003, 2010.
79. G. Dodig-Crnkovic, Floridi's informational structural realist basis for info-computational modelling of cognizing agents. *J. Exp. Theor. Artif. Intell.* 27(1), 2015. Doi: 10.1080/0952813X.2014.940140.
80. M. Burgin and G. Dodig-Crnkovic, Information and computation — Omnipresent and pervasive. In *Information and Computation, vii–xxxii.* World Scientific Pub Co Inc, New York/London/Singapore, 2011.
81. L. Bull, J. Holley, B. D. L. Costello, and A. Adamatzky, Toward turing's a-type unorganised machines in an unconventional substrate: A dynamic representation in compartmentalised excitable chemical media. In *Computing Nature.* Springer, Berlin, Heidelberg, 2013, 185–199. Doi: 10.1007/978-3-642-37225-4.
82. R. Landauer, The physical nature of information. *Phys. Lett. A.* 217:188, 1996.

83. J. Ladyman, D. Ross, D. Spurrett, and J. Collier, *Everything Must Go: Metaphysics Naturalised.* Clarendon Press, Oxford, 2007.
84. J. M. Lehn, Perspectives in chemistry — Aspects of adaptive chemistry and materials. *Angew. Chem., Int. Ed.* 54:3276–3289, 2015. Doi: 10.1002/anie.201409399.
85. J. M. Lehn, Supramolecular chemistry: Where from? Where to? *Chem. Soc. Rev.* 2017. Doi: 10.1039/c7cs00115k.
86. M. Miłkowski, From computer metaphor to computational modeling: The evolution of computationalism. *Minds Mach.* 2018. Doi: 10.1007/s11023-018-9468-3.
87. A. U. Igamberdiev, Evolutionary transition from biological to social systems via generation of reflexive models of externality. *Progress in Biophysics and Molecular Biology* 2017. Doi: 10.1016/j.pbiomolbio.2017.06.017.
88. J. Torday and W. Miller, Cellular-molecular mechanisms in epigenetic evolutionary biology. In *Cellular-Molecular Mechanisms in Epigenetic Evolutionary Biology.* Springer, 2020. Doi: 10.1007/978-3-030-38133-2.
89. S. Forrest and M. Mitchell, Adaptive computation: The multidisciplinary legacy of john H. holland. *CACM.* 59(8):58–63, 2016.
90. A. Sloman, Meta-morphogenesis: Evolution and development of information-processing machinery. In *Alan Turing: His Work and Impact*, S. Cooper and J. Leeuwen, editors. Elsevier, Amsterdam, 2013, 849.
91. G. Dodig-Crnkovic, The info-computational nature of morphological computing, *Studies in Applied Philosophy, Epistemology and Rational Ethics.* 5: 2013. Doi: 10.1007/978-3-642-31674-6_5.
92. J. Crutchfield, W. Ditto, and S. Sinha, Introduction to focus issue: Intrinsic and designed computation: Information processing in dynamical systems — Beyond the digital hegemony. *Chaos.* 20(037101): 2010.
93. G. Buzsáki, *Rhythms of the Brain.* Oxford University Press, Oxford, UK, 2009. Doi: 10.1093/acprof:oso/9780195301069.001.0001.
94. H. Haken, *Information and Self-Organization: A Macroscopic Approach to Complex Systems.* Springer-Verlag, Berlin, Heidelberg, 2006.
95. H. Haken and J. Portugali, Information and self-organization. *Entropy.* 19:18, 2017.
96. R. Pfeifer, M. Lungarella, and F. Iida, Self-organization, embodiment, and biologically inspired robotics. *Science* 318:1088–1093, 2007.
97. H. Hauser, R. Füchslin, and R. Pfeifer, *Opinions and Outlooks on Morphological Computation.* E-Book, 2014.
98. M. A. Lones, A. P. Turner, L. A. Fuente, S. Stepney, L. S. Caves, and A. M. Tyrrell, Biochemical connectionism. *Nat. Comput.* 2013. Doi: 10.1007/s11047-013-9400-y.
99. A. M. Tyrrell, M. A. Lones, S. L. Smith, and G. B. Fogel, *Information Processing in Cells and Tissues (10th International Conference, IPCAT 2015, San Diego, CA, USA, September 14–16, 2015, Proceedings).* Springer, 2016.
100. A. P. Turner, M. A. Lones, L. A. Fuente, S. Stepney, L. S. Caves, and A. Tyrrell, The artificial epigenetic network. In *IEEE Int. Conf. Evolvable*

Systems, ICES- 2013 Symposium Series on Computational Intelligence, SSCI 2013. IEEE, IEEE, 2013. Doi: 10.1109/ICES.2013.6613284.

101. R. Daniel, J. R. Rubens, R. Sarpeshkar, and T. K. Lu, Synthetic analog computation in living cells. *Nature.* 497:619–623, 2013.
102. G. Piccinini and O. Shagrir, Foundations of computational neuroscience. *Curr. Opin. Neurobiol.* 25:25–30, 2014.
103. S. B. Laughlin and T. J. Sejnowski, Communication in neuronal networks. 2003. https://doi.org/10.1126/science.1089662.
104. G. Buzsáki and A. Draguhn, Neuronal oscillations in cortical networks. 2004. https://doi.org/10.1126/science.1099745.
105. B. B. Averbeck, P. E. Latham, and A. Pouget, Neural correlations, population coding and computation. *Nat. Rev. Neurosci.* 2006. Doi: 10. 1038/nrn1888.
106. S. Sardi, R. Vardi, A. Sheinin, A. Goldental, and I. Kanter, New types of experiments reveal that a neuron functions as multiple independent threshold units. *Sci. Rep.* 2017. Doi: 10.1038/s41598-017-18363-1.
107. J. Hertz, A. Krogh, and R. G. Palmer, Introduction to the theory of neural computation. In *Introduction to the Theory of Neural Computation.* Santa Fe Institute, 2018. Doi: 10.1201/9780429499661.
108. H. C. Barron, H. M. Reeve, R. S. Koolschijn, P. V. Perestenko, A. Shpektor, H. Nili, and R. Rothaermel, Neuronal computation underlying inferential reasoning in humans and mice. *Cell* 2020. Doi: 10.1016/j.cell.2020.08.035.
109. M. Scheutz, *Computationalism New Directions.* MIT Press, Cambridge, MA, 2002.
110. O. Larue, P. Poirier, and R. Nkambou. Hybrid reactive-deliberative behaviour in a symbolic dynamical cognitive architecture. In *Proceedings of the 2012 International Conference on Artificial Intelligence, ICAI.* CSREA Press, Las Vegas, 2012.
111. M. van Bekkum, M. de Boer, F. van Harmelen, Modular design patterns for hybrid learning and reasoning systems. Appl Intell, 51, 6528–6546, 2021. https://doi.org/10.1007/s10489-021-02394-3.
112. R. v. Haugwitz, G. Dodig-Crnkovic, and A. Almér, Computational account of emotion, an oxymoron? In *IS4IS Summit Vienna 2015.* Vienna University of Technology, 2015. http://sciforum.net/conference/isis-summit-vienna-2015/track-triangular.
113. R. Mikkilineni, Going beyond computation and its limits: Injecting cognition into computing. *Appl. Math.* 3:1826–1835, 2012.
114. G. Dodig-Crnkovic, Semantics of information as interactive computation a brief, sketchy picture of infocomputationalism. In *Philosophy*, M. M. W. Neuser and T. Roth-Berghofer, editors. Springer, DFKI Technical Reports, Berlin, 2008, 1–48.
115. B. Broderick, On communication and computation. *Minds Mach.* 14(1): 1–19, 2004.
116. C. E. Maldonado, Hypercomputación biológica y comunicación entre los seres vivos. *Simbiótica.* 3(1):207–229, 2016.

117. M. Burgin, *Theoretical Information Studies: Information in the World.* World Scientific Publishing Co. Series in Information Studies, Singapore, 2020.
118. W. Bechtel, The endogenously active brain: The need for an alternative cognitive architecture. *Philos. Sci.* 17(2):3–30, 2013.
119. H. Penagos, C. Varela, and M. A. Wilson, Oscillations, neural computations and learning during wake and sleep, *Current Opinion in Neurobiology.* 44:193–201, 2017. Doi: https://doi.org/10.1016/j.conb.2017.05.009.
120. J. R. Anderson, Spanning seven orders of magnitude: A challenge for cognitive modeling, *Cogn. Sci.* 2002. Doi: 10.1207/s15516709cog2601_3.
121. A. Newell, *Unified Theories of Cognition.* Reprint edn. Harvard University Press, Cambridge, MA, 1994.
122. H. v. Foerster, *Understanding Understanding: Essays on Cybernetics and Cognition.* Springer, Berlin, Heidelberg, 2003.
123. S. Ghosh, K. Aswani, S. Singh, S. Sahu, D. Fujita, and A. Bandyopadhyay, Design and construction of a brain-like computer: A new class of frequency-fractal computing using wireless communication in a supramolecular organic, inorganic system, *Information* 5(1):28–100, 2014. Doi: 10.3390/info5010028.
124. P. Singh, K. Saxena, A. Singhania, P. Sahoo, S. Ghosh, R. Chhajed, K. Ray, and D. Fujita, A self-operating time crystal model of the human brain: Can we replace entire brain hardware with a 3d fractal architecture of clocks alone? *Information*, 11(5):238, 2020.
125. G. Rozenberg and L. Kari, The many facets of natural computing. *Comm. ACM.* 51:72–83, 2008.
126. J. Bondgard and M. Levin, Living things are not (20th century) machines: Updating mechanism metaphors in light of the modern science of machine behavior. *Front. Ecol. Evol.* 9:147, 2021.

https://doi.org/10.1142/9789811257155_0004

Chapter 4

Are There Traces of Megacomputing in Our Universe

Olga Kosheleva* and Vladik Kreinovich†
University of Texas at El Paso, El Paso, TX 79968, USA
**olgak@utep.edu*
†*vladik@utep.edu*

The recent successes of quantum computing encouraged many researchers to search for other unconventional physical phenomena that could potentially speed up computations. Several promising schemes have been proposed that will — hopefully — lead to faster computations in the future. Some of these schemes — similarly to quantum computing — involve using events from the micro-world, others involve using large-scale phenomena. If some civilization uses micro-world for computations, this will be difficult for us to notice, but if they use mega-scale effects, maybe we can notice these phenomena? In this chapter, we analyze what possible traces such megacomputing can leave — and come up with rather surprising conclusions.

Modern computers are fast. By performing billions of computational steps, we can reasonably well predict tomorrow's weather — and when the prediction is not perfect, the problem is usually not with the computers, but with the fact that we do not have enough weather-related sensors in many geographic areas. On-board computers allow missiles to fly close to the ground at astronomical speeds, without hitting the ground. The recent quarantine enabled billions of people to be connected by reasonably reliable video-connections, helping many people continue to work, to study, and even to enjoy (remotely) their favorite operas.

But for many practical applications, computers are still too slow. For example, a large part of the US is threatened by destructive tornadoes. We still do not have a reliable means to predict where a tornado will be moving. As a result, in tornado-prone ares, alarms sound so often — and usually, with no actual tornado coming — that when the actual deadly tornado does come, people do not react to the warning, do not evacuate — and the consequences are often disastrous. Here, we know the equations that would describe the tornado's dynamics — they are largely the same equations that allow us to predict tomorrow's weather. Experiments have shown that by spending the same computation time — several hours — on a supercomputer, we can predict where a tornado will go in the next 15 min — but that is too late. This is just one example, there are many other problems like that. Many of such problems are related to organic chemistry and biochemistry. So it is not surprising that, e.g., in our university, the main users of high-performance computers are not — as one may think — computer scientists, but folks from the Department of Chemistry and Biochemistry.

How can we make computers faster? Journalists writing about science often express an optimistic belief that human ingenuity will solve all the problems. We are optimistic too, but with computers, we cannot be too optimistic: we are currently reaching the bound set by fundamental physics. This bound is very simple to explain. According to modern physics, nothing can travel faster that the speed of light — i.e., faster than 300,000 km/s, or 300,000,000 m/s. How does this affect computations? A usual laptop on which we are typing this chapter is about 30 cm in diameter, i.e., 0.3 m. This means that for a signal to go from one side of the computer to another, we need 0.3/300,000,000 s, i.e., 0.000,000,001 s. This may sound like a very small time, but even on the cheapest 4 GigaHerz computer, the processor can perform four operations while a signal is still traveling.

To make computers faster, we need to shrink their processing elements even more — this is why we enter the realm of quantum computing; see, e.g.[?] But there is a limit to this shrinkage. Already, a processing element may consist of a few thousand atoms. We can theoretically shrink more, to the level of a single atom — but then what? Then we are stuck.

So what can we do? In this long-term perspective, quantum physics does not seem to help, so let us look at other possible physical phenomena. It would help if we could find a way to speed up all the processes — that will speed up computations as well. Unfortunately, in modern physics, there is no known way to speed up all the processes — there are only ways to slow them down.

According to Einstein's Special Relativity Theory (see, e.g., Refs. **?** and **?**), processes slow down when we travel with a speed close to the speed of light. Actually, they also slow down when we fly on a plane, but that slow-down is so miniscule that only super-precise clocks can detect it, while for the particles in a particle accelerator (that move practically at the speed of light), the time slows down so much that their average decay time increases by orders of magnitude.

According to General Relativity Theory,[?,?] processes also slow down when we are in a strong gravitational field — e.g., near a massive black hole. Yes, they also slow done when a usual gravitational field becomes a little stronger, but this change is also miniscule.

Is the situation hopeless? Good news is that, as the very name of the relativity theory suggests, many things are relative. There is no absolute time with respect to which we want computers to be faster, all we want is that the computers be faster with respect to our time. In other words, what we want is to make sure that computers are in one environment, and we are in another environment, and all the processes in the computer-containing environment should be much faster than all the processes in our environment.

We cannot achieve this by staying on Earth and placing computers somewhere else — the only thing we will then achieve is that, in comparison to our time, computers will be even slower than they are now. But what we can do is leave computers where they are — and place ourselves in situations where time will go slower. If we manage to slow down our own time by a factor of 10, then the same problem that requires 5 h of computation in computer-time will feel like half an hour for us — and this is exactly what we want.

In other words, instead of speeding up computers, we can slow down ourselves, our whole civilization. How can we do it? As we mentioned, there are two ways to do it: we can start traveling with a speed close to the speed of light, and/or we can place ourselves in a strong gravitational field. Let us consider there two options one by one, starting with fast travel. Readers interested in technical details can look into our paper Ref. ?.

We cannot immediately go from 0 to 300,000 km/s: we can only survive the acceleration similar to the Earth's gravitational acceleration of $9.81\,\text{m/s}^2$. So, we need to start going slowly. We cannot travel on the same orbit around the Sun — if we start traveling with too high a speed, centrifugal forces will squeeze us, so we need to go further and further away from the Sun. We cannot simply travel away on a straight line — this way, we will reach a high speed, but by that time, we will be so far away from the left-behind computers that communication time will eat up all the advantages. So, the only way to reach the desired effect is make circles which become wider and wider — in other words, to follow a spiral trajectory.

This acceleration requires a lot of energy. Where can we get this energy? We have to get it as we travel, from the interplanetary and then interstellar particles and gases — and other objects. As we follow this spiral trajectory, we will burn whatever we can along this trajectory, leaving practically nothing. So what will remain? What will remain is empty spaces forming a spiral. Sounds familiar? It should: this is exactly how our own Galaxy and many other galaxies look like.

So maybe the spiral shape of our Galaxy is indeed the trace of an ancient megacomputing civilization? But wait, there is more.

As we have mentioned, another way to speed up is to place oneself in a strong gravitational field — e.g., near a massive black hole. At first, we can use existing black holes, but what if we want to perform even faster computations? The only way to do that is to make the black hole bigger and bigger — thus increasing its gravitational field and slowing us down even more. So how can it be known that a supercivization used this idea to perform megacomputations? By observing a humongous black hole that our previous astrophysical theories did not explain. Sounds familiar? It should. According to modern astrophysics, there is indeed a very massive black hole in the center of our Galaxy — and in the centers of many other galaxies. So maybe these black holes are also traces of megacomputing civilizations?

Who knows?

Acknowledgments

This work was supported in part by the US National Science Foundation grants 1623190 (A Model of Change for Preparing a New Generation for Professional Practice in Computer Science) and HRD-1242122 (Cyber-ShARE Center of Excellence).

References

1. M. A. Nielsen and I. L. Chuang, *Quantum Computation and Quantum Information.* Cambridge University Press, Cambridge, UK, 2000.
2. R. Feynman, R. Leighton, and M. Sands, *The Feynman Lectures on Physics.* Addison Wesley, Boston, MA, 2005.
3. K. S. Thorne and R. D. Blandford, *Modern Classical Physics: Optics, Fluids, Plasmas, Elasticity, Relativity, and Statistical Physics.* Princeton University Press, Princeton, New Jersey, NJ, 2017.
4. O. Kosheleva and V. Kreinovich, Relativistic effects can be used to achieve a universal square-root (or even faster) computation speedup. In *Fields of Logic and Computation III*, A. Blass, P. Cegielsky, N. Dershowitz, M. Droste, and B. Finkbeiner, editors. Springer, 179–189, 2020.

https://doi.org/10.1142/9789811257155_0005

Chapter 5

Communication, Information, Energy, and Music

Dawid Przyczyna[*,§], Marcin Strzelecki[†,¶], Gisya Abdi[*,‖], Lulu Alluhaibi[*,**], Kapela Pilaka[‡,††] and Konrad Szaciłowski[*,‡‡]

[*] *Academic Centre for Materials and Nanotechnology, AGH University of Science and Technology, al. Adama Mickiewicza 30, 30-059 Krakow, Poland,*

[†] *Faculty of Composition, Interpretation and Musical Education, Krzysztof Penderecki Academy of Music in Kraków, ul. Św. Tomasza 43, 31-027 Kraków, Poland*

[‡] *Beýik Türkmenbaşy Laboratory of Dutar Music, Faculty of Folk Instruments, Maya Kuliyeva Turkmen National Conservatory, Pushkina 22, 744000 Ashgabat, Turkmenistan*

[§] *dawidp@agh.edu.pl*

[¶] *marcin.strzelecki@amuz.krakow.pl*

[‖] *agisya@agh.edu.pl*

[**] *lalluhaibi@agh.edu.pl*

[††] *pilaka@saz.akademiyasy.tm*

[‡‡] *konrad.szacilowski@agh.edu.pl*

In this brief chapter, we would like to take the reader on a meandering journey through the fields of information, energy, communication, and music, as well as the places where these domains intersect and percolate, even in such distant fields as religion and politics. As it turns out, contemporary and seemingly abstract concepts related to information processing can be conveyed in an accessible form through music. To spice things up, the potpourri of ideas presented in this chapter is seasoned with pictures and descriptions of postage stamps and various unexpected and serendipitous associations, involving information processing, history, music, and philately.

All science is either physics or stamp collecting.
Ernest Rutherford

Ohne Musik wäre das Leben ein Irrtum.
Friedrich Nietzsche

No se preocupe, tome café y fume un buen cigarro.
Javier Pereira

井の中の蛙大海を知らず
Japanese proverb

Communication between organisms is a ubiquitous phenomenon, both at intraspecies and interspecies levels in all kingdoms: *Archaea*,[1] *Bacteria*,[2] *Protista*,[3] *Fungi*,[4] *Plantae*,[5] and *Animalia*.[6] Surprisingly, primitive communication was detected even between individual virions.[7,8] All these organisms possess both intracellular, intraorganismic, and transorganismic communication protocols, however, the most complex and interesting ones, from the point of view of information theory, are those between individual organisms. In most cases, the intracellular and intraorganismic, communication is based on signaling molecules, the same concerns most of the interorganismic and interspecies communication protocols. Communication in general can be described as a sign-mediated interaction between at least two living entities, which share the common repertoire of signs representing a form of natural language. These signs may be combined according to syntactic rules in various contexts (according to pragmatic rules) and used to convey biologically relevant information. Almost all kingdoms of life use molecules as the only available communication tool, whereas animals add vocal and visual communication tools to their repertoire of available signs.

In humans these evolutionary novelties dominate, almost completely, over the molecular language, however, "molecular senses" of olfaction and gustation are still significantly important, including sex pheromones as a still disputed example of signaling molecules.[9,10] Most of the animals use senses of vision and hearing for most of their communication purposes. Whereas our own (human) senses seem to be impaired (as compared with some predatory birds), their ability to process signals is still amazing. We have also developed unique ways of communication: music and language, manifested sonically as speech, and graphically as writing. These tools provide an unprecedented opportunity to communicate language and emotions using graphical symbols and aesthetic, religious, and cultural feelings via organized sounds of different parameters like pitch, duration, and timbral qualities, arranged in melodic, rhythmic, and harmonic (tonal) patterns.

Music is the only form of natural communication that is created and perceived only by humans (however, studies on animals indicate some aspects of sensitivity to music). Music belongs to human universals, i.e., elements, patterns, features, or notions that are common to all human cultures worldwide, however, according to some opinions, it does not convey any biologically-relevant information. According to Guerino Mazzola: "music embodies meaningful communication and mediates physically between its emotional and symbolic layers".[11] The importance of music is exemplified by the discovery of Palaeolithic musical instruments (Figure 1). Although probably music at early times had no direct effect on the economy or a reproductive success, it may have provided a medium of social integration. However, nowadays the influence of *muzak* on our decisions in supermarkets and retail centers is undeniable and proves its productive impact on the economy. As of today, music is one of the most ubiquitous human activities independent of any social and cultural attributes or intellectual abilities.

The importance of music in modern society is unquestionable — composers, performers, and musical instruments are leading motifs of numerous postage stamps (Figure 2), along with monarchs, dictators, religion, nature, and sport. Interestingly, in the past, postage stamps (first introduced in 1840 by United Kingdom) were considered as a very effective medium of communication. Therefore, each political or territorial change as well as other (often tragic) events were (and still are) immediately reflected on postage stamps (Figures 3 and 4). For instance, year 1985 was announced as the European Year of Music

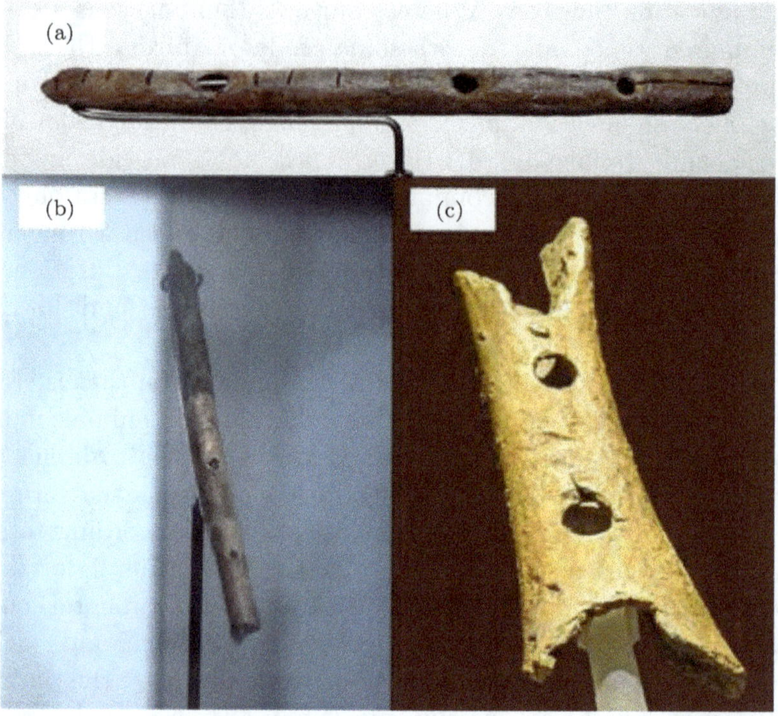

Fig. 1. Palaeolithic musical instruments: upper Palaeolithic from Geissenklösterle (a), middle Palaeolithic flute, ca. 35,000–40,000 years, (b), and Neanderthal flute from Divje Babe Cave (Slovenia, 55,000 years ago, c). Photos courtesy José-Manuel Benito (a), Marco Ciamella (b), and Jean-Pierre Dalbéra (c).

(this fact was commemorated by numerous series of stamps issued by European countries), and 2019 was announced as the Smithsonian Year of Music.

Music and language are created, developed, and processed in distinctly different neural structures but have some common features: they are specific forms of communication and they have specific syntax and vocabulary — i.e., they have a set of elements (words or notes) and a set of rules (grammar or harmony and counterpoint) that govern the appropriate combination of these elements. Some kinds of music, like European tonal music, have strict syntax, some others (like dodecaphonic music) may be strictly organized while at

Fig. 2. Prominent musical personalities: Narayan Gopal Gurucharya, Frederic Chopin, Sissel Kyrkjebø, and Sayed Darwish portrayed on postage stamps of Nepal, Soviet Union, Norway, United Arab Republic.

the same time lacking the audible regularities. Finally, there exist also genres like avant-garde, experimental music, and anti-music movements, which aims at breaking traditional regularities. Such exceptions and declared negation of musical syntax also confirms the existence of one. Music is a domain of human artistic and entertaining activity, but also a field of vigorous study. Similarly to information, music is a notion very difficult to define in precise terms, dislike speech, it is not meant for *direct* communication purposes, especially of biological importance.[12] Conversely, it is meant to trigger various emotional responses in recipients due to aesthetic feelings. *Indirectly*, music can convey information related to feelings of love or hate (Figure 5), and this already has biological significance — especially when reinforced feelings of love are directed toward a partner (potential for a new life, John Paul Young — *Love Is In The Air*). Likewise, a different spectrum of emotions (which we only mention) can affect a person's well-being, and the author's goal could be to convey this message to the audience (e.g., Bob Marley — *Three Little Birds*). The songs can also help listeners through a sense of belonging (Pink Floyd — *Another Brick in the Wall*) or by feeling similar

Fig. 3. Important political changes depicted on postage stamps: German occupations of Poland, Romania, and Belgium during the First World War, German occupation of Luxembourg during the Second World War, The Russian Revolution of 1917, Egyptian Revolution (1952), formation of a puppet state of Manchukuo (1932), socialistic revolution in Bavaria (1918), independence of Uzbekistan and Ukraine (1991), Jordanian annexation of the West Bank (1948), the fall of the Nazi Germany (1945), French occupation of Saarland and Palatinate (1945), formation of the Soviet Occupation Zone (1945), and Upper Silesia plebiscite (1921).

tragedies (Led Zeppelin — *Baby I'm Gonna Leave You*). Emotional signals transmitted by music may dissuade some listeners from committing suicide, which is of fundamental biological significance (Bo Burnham — *Kill Yourself*). Therefore, music should be considered as a powerful communication channel for emotions.[13]

Music's enormous potential to induce a wide variety of emotional states is the subject of recent researches aimed at modeling these complex processes computationally.[14] However, such models still deal with relatively basic emotions compared to human capability to recognize subtle emotional nuances.[15] Automatic, large-scale music emotion recognition finds its applications in classification of big music

Fig. 4. Polish postage stamp commemorating victims of Polish Air Force Flight 101 crash on April 10, 2010.

databases, music on the internet, and music present in radio and television broadcasts.[16]

Music's emotional potential supports empathy and charity (Figure 6). As an extraordinary example, let us mention The Great Orchestra of Christmas Charity Foundation (https://en.wosp.org.pl/, Figure 7), which, as it has been doing for the last 30 years, unites millions of people in Poland. The main vehicle of this action is actually the one-day music festival, when millions of dollars are collected every year.

Fig. 5. Feelings of friendship, love, and hate presented on postage stamps of Norway, France, Israel, and North Korea.

On the other hand, music can be regarded as a series of acoustic waves, the frequencies of which conform with the rules mentioned above. The acoustic wave is a potentially high capacity medium for transferring information. Sound waves may be modulated in order to convey digital information (like the old-times dial up internet), singular symbols (DTFM), or images (SSTV technique). Encoding information in sound is a subject of a general audio-based steganography. Amateur radio transmissions with simple CW modulation make use of frequency sub-bands 200–500 Hz wide, which allows many simultaneous channels within the acoustic range. Digital communication protocols, like FT4 or FT8, are capable of packing even more information within the acoustic bandwidth.

Sound becomes a powerful communication medium when it comes to music. As a complex audio signal, well suited to the specific human perceptual and cognitive limitations, music may communicate extremely subtle messages such as feelings of love or hate, and their countless variations, moods, and even philosophical questions.[17] On the social level, music is a group identity device of importance hard to overestimate. From suburban rap to national anthems, from the sentimentally patriotic Chopin's *Mazurkas*[18] to Serbian nationalist songs,[19] music indicates social and political affiliation (Figure 8).

Fig. 6. Postage stamps promoting charity: Bahraini famine and war stamps, German Red Cross stamp issued by Protectorate of Bohemia and Moravia, Iraqi stamp for victims in Palestine, Chilean stamp promoting Lions International, Iraqi stamp with surcharge supporting flood victims, Peruvian stamp promoting fight against tuberculosis, Iranian stamp with surcharge for earthquake victims, and a German stamp with a charity surcharge, which was issued with the tag saying: "*Gutes tun, mit Briefmarken helfen*" (Do good, help with postage stamps).

Interestingly, postage stamps are (and were in the past) also used to enhance political or national affiliation. National emblems, flags, or images of country leaders are frequently used for this purpose (Figure 9). One of the very first things the new Kosovo government did in 2008 was to announce a competition for the national anthem, won by Mendi Mengjiqi, who graduated in music composition from the Kraków Academy of Music. It should be noted, however, that the first postage stamp of Kosovo had been issued a couple of years earlier (in March 2000, Figure 10). Along with national and political

Fig. 7. Postage stamps promoting charity: Polish stamps promoting The Great Orchestra of Christmas Charity (GOCC, in Polish *Wielka Orkiestra Świątecznej Pomocy, WOŚP*), the biggest, non-governmental, non-profit, charity organization in Poland.

Fig. 8. Examples of music and philately as ways of enhancing national and political identity: national anthems (Israel, Bolivia, Haiti, and Andorra) and flags (Peoples Republic of China, Unites States of America, Republic of China (Taiwan), Syria, and Finland).

Fig. 9. Postate stamps as a medium of creating national identity. Emblems: Russian Empire, Soviet Union, Poland, Free City of Danzig, Finland, and country leaders: Queen Victoria (UK), Sun Jat-sen (Republic of China), Chairman Mao Tse Tung (Peoples Republic of China), King Mohammed Zahir Shah (Afghanistan), Mustafa Kemal Atatürk (Turkey), and Shah Reza Pahlawi (Iran).

Fig. 10. The second issue of Kosovo (United Nations Interim Administration Mission in Kosovo) postage stamps from 2001.

affiliation, postage stamps are used in an opposite way — to unify nations and social groups (Figure 11).

So music carries on subtle emotions, complex concepts, like national identity, and philosophical ideas, but also is often considered as having particular influence on both physical and mental health. From magical shamanic dances, David's harp calming King Saul's madness down, to the wide and intensively developing field of

Fig. 11. Postage stamps as a tool of social unification.

contemporary music therapy. The role of music in medical therapies was for the first time described by Abū Bakr Muammad ibn Zakariyyā al-Rāzī,[20] Persian polymath, physician, and alchemist, most known for pioneering the distillation of alcohol (Figure 12). Such a "magical" function of music brings it close to a religious one. From ancient tribal dances to contemporary religious rituals — everywhere a great devotion to proper music and its proper performance may be observed. Particular belief and institutional churches elaborate own music, which identifies and distinguishes believers like biblical "shibboleth". Music connects with "supranatural" forces (spectacular "dancing Darwish" rituals) and saves those, who sing proper psalms (like in biblical story on salvation of Shadrach, Meshach, and Abednego, thrown into a fiery furnace by a really cruel King Nebuchadnezzar II) (Figure 13).[21] On the other hand, music was banned by the Taliban regime of Afghanistan on the basis of falsified (and also misunderstood) *hadith* (*khabar*) saying "*Whoever listens to a songstress, molten lead will be poured into*

Fig. 12. Abū Bakr Muammad ibn Zakariyyā al-Rāzī (865–925).

Fig. 13. Postage stamps, like music, are an efficient vehicle for religious feelings (here represented by Polish stamps commemorating catholic radio station "Radio Maryja", Israeli stamp depicting menorah and Iranian stamps commemorating The Islamic Revolution and depicting Seyed Hassan Shirazi, a famous Islamic scholar. Despite quite opposite attitudes toward music, Polish and Iranian religion fundamentalists share similar views on human rights, freedom of press, and independence of science and education.[24]

his ears on the Day of Judgement".[22] Interestingly, some Western musical instruments originate from arabic ancestors (e.g., *rebab* → violin and *oud* → lute). Interestingly, the Taliban's actions against music were also reflected in an almost complete abandonment of the issuing and use of postage stamps.[23]

At the other extreme of the spectrum of music-religion relationship, Kurdish Yarsanism is located. Yarsani people regard music as a way to express their religious feeling and some musical instruments (e.g., *tambur*) are regarded as sacred objects. The Yarsani believe the Sun and fire are holy entities and follow the principles of equalization, purity, righteousness, and oneness, which leads some researchers to find Mithraic roots in this religion. They also have many things in common with Zoroastrians.

Back to par-Islam era, the female singer known as the qaynāt (singing girls) performed music at scattered taverns as well as in noble households in the Arabian Peninsula region.[25] Also, many competitions of poetry and musical performances were organized in Arab marketplaces, such as Souk Okaz in Taif city, which was a major gathering place for Arabs to define tribe rules, settle disputes, pass judgments, reach agreements, sign treaties and truces, host sporting competitions and races, poetry competitions, and religious gatherings. It has been compared to the ancient Greek institution of the panegyric in this regard.[26]

During the Islamic era, between 960 and 970 A.D., Abu al-Faraj al-Isfahani wrote *Kitāb al-Aghānī* (The Book of Songs), an encyclopedic collection of poems and songs that spanned over 10,000 pages and contained over 16,000 verses of Arabic poetry. In current editions it is contained in almost 20 volumes.[27] These examples prove the importance of musical expression in Islamic culture and tradition. Therefore, the Taliban attitude to all kinds of music is difficult to understand.[28] It has been perfectly stated by Joseph Alagha: "While Hizbullah embraces 'resistance art' and encourages purposeful music and artistic expressions as pious entertainment, the Taliban censor music and restrict artistic activities, considering them innovations (bid'ah) that distract from the practice of 'authentic Islam' and 'true worship' ".[29]

The Middle East has been an important source of musical instruments for other parts of the world. Bagpipes, guitar, lute, oboe, tambourine, viols, and most zithers have a Middle Eastern origin. Throughout the region, religion has been a common factor in uniting peoples of different languages, cultures, and nations. Despite three major languages and associated cultural differences, the Middle Eastern music (music of the Arabic-, Turkish-, and Persian-speaking worlds), can be seen as a single great tradition because of the unifying element of Islam. The fact that Islam has historically found music problematic has resulted in relatively little ceremonial music, but it has not held back secular and folkloric music. With the advent of Islam in the 7th century A.D., as well as other Persian cultural elements, Persian music became a formative element in what has since become “Islamic civilization”. Persian musicologists and musicians overwhelmingly ruled the musical life of the Eastern Islamic Empire. Baghdad, the current capital of Iraq, became the center of Persian music, and many musicians who were once considered to be Arabs are now known to have been Iranians.

The first written evidence of Persian music goes back to the Sasanian Period (226–643 CE). Iranian folk music refers to the music transmitted through generations among the people of Iran, often consisting of tunes that exist in numerous variants. The variance of the folk music of Iran has often been stressed, in accordance to the cultural diversity of the country’s ethnic and regional groups (Figure 14). It also influenced areas and regions that are considered part of the *Greater Iran.* Due to the exchange of musical science throughout history, many of Iran’s classical melodies and improvization modes are related to those of its neighboring cultures. Indigenous Iranian musical instruments used in traditional music include string instruments such as the chang (harp), qanun, santur, rud (oud, barbat), tar, dotar, setar, tanbur, and kamanche, wind instruments such as the sorna (zurna, karna), ney, neyanban, and percussion instruments such as the tompak, kus, daf (dayere), naqare, and dohol, shared by many cultures and countries like Iran, Azerbaijan, Uzbekistan, Armenia, Georgia, Tajikistan (Iranian Plateau), and other countries in the Caucasus region.

Fig. 14. Rich Persian musical traditions depicted on Yemeni postage stamps from 1971.

Traditional Kurdish music is very distinctive from Arabic and Turkish music, has not been affected by Islamic beliefs, and has largely retained its Iranian origins, with musical instruments played by Kurds (men and women; young and old) to express the reality of the lives of the zealous Kurdish tribes and their beliefs. For instance, tanbur (with a history of over 5,000 years) is currently the musical instrument used in Yarsan rituals, and practitioners venerate tanburs as sacred objects. In this particular case, music and religious identity are convoluted into an unseparable sociocultural fabric.

Why information contained in music is often difficult to extract? Most probably because it is not entirely there. Marvin Minsky points out: "*Our culture immerses us in it [music] for hours each day.*"[30] For years, we learn how to link certain sounds and certain meanings together. Finally, it is our mind, where the information is stored. Music just pushes the buttons. Particular cultures create physical and psychoacoustical mechanisms which in turn form a layer of possibilities. Depending on the perceiving mind, the same physical stimuli — let's say a dissonance — may trigger different values and meanings rooted in a specific environment. Today we may artificially recognize a dissonance *per se*. To model the mechanisms of its aesthetic function is another, fascinating and urgent research challenge.

Music is a very well-organized structure. Even the denial of the existence of such a structure, conceptually declared by the author, proves the existence of a specific "musical language" with appropriate grammar, syntax, and vocabulary — the harmony, rhythmical patterns, timbres, and their mutual relations. Therefore, not every combination of sounds should be considered as music. This indicates that music (like hardcore pornography) may be considered as a kind of an emotional communication of the "*I know it when I see it*" type.[31] Despite that, specific fractal signatures derived from compositions can be assigned to specific genres. The simplest musical message, melody, can be defined as an appropriate time sequence of quantized frequencies, usually noted as a musical score (Figure 15).

These frequencies, called steps, are strictly defined by tuning systems. Most of the musical systems are founded on a concept of the octave: an interval between the frequencies of f and $2f$. In other words, an octave is an interval between the first and second harmonics of the harmonic series. Therefore, it is considered as a natural phenomenon that has been referred to as the *basic miracle of music*, the use of which is *common in most musical systems*.[32] There exist many different tuning systems[33] as well as octave divisions (like Balinese and Javanese gamelan systems, Figure 16). Other musical systems, both traditional (e.g., the Middle East, India, and Far East) as well as modern experimental musical genres, use different intervals, including division of the octave into 4, 5, 7, 34 (to name only a few possibilities) or even 96 equal steps, leading to the whole musical tuning continuum.[34]

The very strict structural rules and mathematical relations are naturally embedded in musical structures.[35] Current progress in computer sciences, machine learning, and artificial neural networks significantly influences musical creativity.[36] Unconventional computing is a natural consequence of research toward new computational paradigms and application of new materials and systems as

Fig. 15. Examples of musical scores: *Die Meistersinger von Nürnberg* (R. Wagner), *2nd Brandenburg Concerto* (J.S. Bach), and *Kimjongilia* (a song praising "an exceptionally great man whom all the world's people revere, His Excellency Kim Jong Il who is contributing to friendship between the Korean and Japanese peoples and to the cause of world peace", by U Chang-Heui), as depicted on German and North Korean postage stamps.

Fig. 16. Tjlempung, sasando, kultjapi, hape, gangsa, rebab, totobuang, and kalintang: traditional musical instruments used in gamelan music.

computational platforms.[37] Therefore, some time ago Eduardo Miranda initiated the multidisciplinary research and creative activity in the field at the border of music and unconventional computing.[38] Among the newest computational paradigms, reservoir computing is one of the latest discoveries.

Reservoir computing is a computational paradigm that explores the internal dynamics of physical systems for information processing. In principle, any physical system with internal dynamics can serve as a foundation for reservoir computing.[39–41] Dynamics at the edge of chaos renders a perfect medium for computation as it is the most sensitive for any perturbation of external signals (input data). Graphical representation of such dynamic behavior usually resembles tuğra, a calligraphic signature of a sultan, frequently found on older postage stamps of the Ottoman Empire, Saudi Arabia, and Egypt (Figure 17) or other complex graphical patterns presented on postage stamps as a decorative motif (Figure 18). A fading memory feature is another requirement for reservoir computing. It provides the lack of

Fig. 17. Sultans' and Royal tuğras on old Ottoman (1863-1910), Saudi (1934), Egyptian (1938), and Turkish (1963) postage stamps.

Fig. 18. Visualization of a complex attractor on United Nations block of stamps (1993).

dependence of the reservoir's dynamics on past history, as beautifully described by Mao Tse-tung: On a blank sheet of paper free from any mark, the freshest and most beautiful characters can be written, and most beautiful pictures can be painted.[42]

Reservoirs must be also equipped with input and output ports. Any physical stimulus altering the internal dynamics can be considered as a carrier of information.[40,41] The output in turn monitors the internal state of a part of the reservoir and is the only trainable (in the sense of machine learning) element of the whole device. Whereas the construction of a reservoir computer following the description given above may pose significant difficulties, a simplified scheme based on a single nonlinear node equipped with a delayed feedback loop may be equally efficient from the computational point of view, however much easier to build and operate. In such systems, the input signal (stream of data) circulates in a feedback loop and undergoes gradual changes (Figure 19). The evolution of this signal serves as the reservoir state and is used for the generation of the output. This computational scheme has inspired us to compose and perform a piece of music inspired by the concept of reservoir computer.

The *Reservoir Study No. 1* (Marcin Strzelecki, 2019) scored for two electric guitars, cello, piano four hands, and electronics is a composition in repetitive minimalism style. It is based on 555 ms delay feedback loop, which repeats and transforms music being played by a consort of musicians. These repetitions and transformations result in harmonic and timbral fluctuations of particular aesthetic quality. Numerous repetitions (both in the score and also added by the feedback) reflect dynamic changes inside the reservoir computer.

The *Reservoir Study* opens with a short improvization on the keyboard (Figure 20(a)), that can be understood as an input for computation. These improvisations are followed by a slowly evolving tune, each bar is repeated four times and subsequently bars introduce subtle harmonic and timbral changes, as a very suggestive illustration of a revolution of signal in a feedback loop (Figure 20(b)). In the middle part this, regular pace is suddenly broken and followed by another improvization loaded with glissandi and irregular rhythmic patterns. Then the reservoir has reached the chaotic state

Fig. 19. Circulation of goods and information as a decorative motif of postage stamps of Germany, France, Switzerland, and Bulgaria.

(Figure 20(c))! The musical chaos slowly calms down and the regular pattern is reborn — the reservoir has reached the final state, which is the end of computation (Figure 20(d)). Final chords represent the output layer generating the result of computation. The world

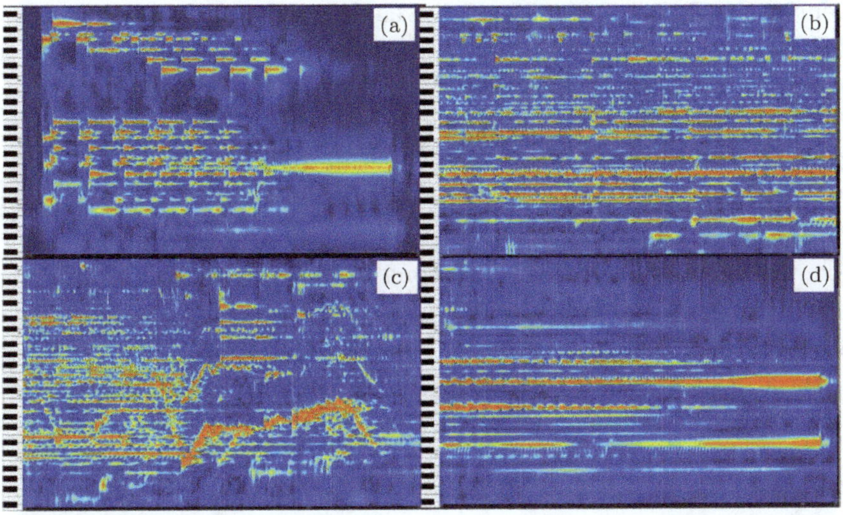

Fig. 20. Spectrograms calculated from the audio path of the *Reservoir Study No. 1* world premiere recording.

premiere performance of this piece was given by The Nano Consort (Dominika Peszko, Piotr Zieliński — piano, Dawid Przyczyna, Kacper Pilarczyk — guitars, Marcin Strzelecki — keyboard, Konrad Szaciłowski — cello) in Krakow Opera House, September 16th, 2019. Original recording is available as a supplementary video to our recent paper, Ref. 43.

The use of machine learning and artificial neural networks in study of music is a broad research topic involving manipulation, classification, or generation of its elements. In general, machine learning (ML) and artificial neural networks (ANN) approaches are based on feature selection from the sample data-set (also called training set) to construct mathematical models which then can be used to classify, predict, or generate data of similar character. These are some of the methods used in the multidisciplinary field of Music Information Retrieval (MIR), which is closely related to business.[44] One of the most obvious examples is music recommendation based on the songs you listened to (used by YouTube, Spotify, iTunes, etc.). In this case, specific features of the songs are important, such as the melodies, harmonies, or rhythms used, as well as to which genres these songs belong and even what instruments were used. The separation of specific tracks and their analysis also enables the automatic generation of karaoke recordings by removing the vocal track. It doesn't sound like rocket science, but it's worth noting that the global karaoke market is worth around 10 billion USD.

Another option is automatic music transcription, which can be useful, for example, for the music teaching industry, where we would like to adjust the scale of a song to the vocalists' abilities, or allow instrumentalists to practice playing scales and chords of given songs in different keys. There is also functional music called muzak (also known as "elevator music"), a background music created not for aesthetic experiences, but for a profit: it's played in shops to increase sales through inducing relaxed state in customers or to regulate emotions in a boring factory environment at assembly lines. Now, thanks to projects like MuseNet and Jukebox from OpenAI, we can have access to an endless amount of this type of music.[45] Reading from their website: MuseNet is "a deep neural network that can generate 4-minute musical compositions with 10 different instruments, and can combine styles from country to Mozart to the Beatles". Only a few notes are needed to train the system as state by the authors on their website. On the other hand, thc Jukcbox project aims to generate music from scratch, trained on the basis of pieces of music from a specific style.

But what if the superb capabilities of the artificial intelligence could be used to deepen our understanding of the experience of music itself? Just like in material sciences, maybe this approach could point us into new and exciting discoveries. Studies on the emotional response to various musical pieces could be a starting direction.[46] Analysis of melodic, harmonic, and rhythmic components of such samples could form a basis for something like a blueprint of an "ideal" music piece. One can argue that such tasks were undertaken by the historical geniuses of music and we can say that for their time they may have succeeded. But as everything "flows", music evolves. The world is much more complex than when Beethoven (Figure 21) was composing his 9th symphony. Demand for music that people from all generations could identify with is never ending. Now we could apply

Fig. 21. Ludvig van Beethoven (1770–1827).

new tools for such purposes. It is possible that purely fabricated music may lose its organic character, so perhaps the hybrid approach is the right way for artists to take advantage of the new possibilities offered by artificial intelligence.

In the near future, technology based on the analysis of emotions of listeners to maximize the aesthetic musical experience could take an unprecedented turn, the basis of which can be Brain–Computer Interface (BCI) devices. Currently, many different companies are undertaking the construction of BCI, but the most popular one is Neuralink — a company founded by Elon Musk, which presented an already working device connected to the brain of a mouse.[47] What if the brainwave analysis responsible for the aesthetic experience of music could be done in real time? What if a musician could record the perfect melody he has in mind without having to use any instruments? Time will show whether this type of technology will only be used in accordance with its original purpose (assisting victims of paralysis, sensory or neural dysfunction, etc.) or perhaps to enhance our everyday human experience of reality. It's truly exciting that we are living in times where technology improves not only our physical well-being but could also improve our creative capabilities, products of which could be responsible for our stronger emotional connections. This example illustrates the close relationship of contemporary music with unconventional computing, especially with the novel computational paradigms. It also shows how areas seemingly unrelated to art can become an inspiration for it and can be better understood thanks to it.

One should be, however, very careful with exploring the culture–technology and religion–technology interface.[48] Some mutual inspiration can be very beneficial for both disciplines, but too deep an infiltration or fusion of ideas from one field to the other one may lead to utmost absurdity, with Martian Liturgical Calendar,[49] quantum psychology,[50] or extreme pancomputationism[51,52] as the most striking examples. Some studies at the interface of religion and technology, however, initiate a valuable discourse on morality and eschatology,[53] whereas the others bring only hassle and misconceptions. Such accounts are harmful from both religious and scientific points of view: e.g., application of Einstein's special theory of relativity[54] to explain Eucharistic miracles[55] (sic!). Alternatively, such publications and activities can be regarded as a part of scientific folklore.[56–59] When politics is involved in the scientific process, the outcome may have a bitter aftertaste,[60,61] and involvement of politics in music leads to a dramatic decrease of artistic level of broadcasts, as it can be observed in National Polish Television, which extensively promotes *disco polo*, a musical genre originating from "sidewalk music" or "backyard music", with simple melodies and ribald lyrics. After a gradual decline in popularity, it resurged in 2007 due to the specific socio-political climate in Poland.[62–67]

We are in the middle of another technological revolution and reservoir computing is only one example of various revolutionary computational paradigms that have recently emerged. Development of new algorithmic and hardware solutions, quantum computing, and molecular information processing has opened the whole universe of novel computational technologies.[37,68–73] The current revolution in computing technology can be compared to other technological

Fig. 22. Industrial Revolution: The Great Laxey Wheel — an impressive technical structure from 19th century erected on the Isle of Man.

revolutions of the past: Industrial Revolution of the 18th century (Figure 22) and postal revolution, initiated in UK, in 1840. At the end of the 20th century, the rapid development of information and communications technology (ICT) as well as telecommunications techniques began. Information has become a very sought-after commodity. Currently, most people cannot imagine life without a mobile phone, which is also a miniature computer with a lot of computing power, equipped with a high-resolution camera and fast internet connection. The beginnings of the revolutions leading to the emergence of the information society should be sought already in the times of the industrial revolution — then the first invention revolutionizing the transmission of information — a postage stamp — was created.

UK postal tariffs were extremely complex and the postage was usually paid by the recipient. On May 6, 1840, an unusual innovation was introduced — the shipment was paid in advance by the sender, the price was predetermined regardless of the route, and a new type of security — a postage stamp — was used to pay

Fig. 23. Inventor of the postage stamp concept, Sir Rowland Hill (1795–1879).

Fig. 24. The first postage stamp of the world: Penny Black (1840) and her younger sisters: Penny Red (1841) and Two Pence Blue (1841).

for shipments. The originator of this form of shipping charges was Sir Rowland Hill, a teacher and postal clerk (Figure 23). The fact that the idea has worked may be proved by the circulation of the first stamp — the Penny Black, amounting to 68,808,000 pieces. Shortly thereafter, red and blue (two pence) versions were circulated, and more countries soon adopted the same convention (Figure 24). Paradoxically, the history of modern computers also begins in the

Fig. 25. Michael Faraday (1791–1867) and Charles Babbage (1791–1871).

times of the industrial revolution. The foundation of electronics had been formulated by Michael Faraday, whereas the English engineer Charles Babbage (Figure 25) designed and tried to build mechanical computing devices: a differential machine (finally built in 1989–1991) and an analytical machine that was never put into practice. The conceptual assumptions of these machines served as inspiration for John von Neumann in the work on computational architectures, which are still the basis for the operation of modern computers. Further development of electrotechnics and electronics has facilitated the development of information technologies (Figure 26).

The modern world is very different from Victorian England. In the world based on information, its efficient acquisition, processing, and storage is of the utmost importance (Figure 27). Today's microprocessors are devices of unprecedented complexity, and metal–oxide semiconductor field effect transistors (MOSFETs) are considered the most-manufactured objects in human history — the total number of components produced is approaching 13 sextillions (1.3×10^{22}). A modern microprocessor contains up to four billion transistors.

However, the quick processing and storage of information is extremely energy-intensive. It is estimated that in 2016 the data

Fig. 26. From Marconi's spark coils to complex semiconductor-based circuitry: progress of electronics depicted on US postage stamps.

Fig. 27. An analogy of classical mail and electronic communication as presented on Gibraltar postage stamps of 1984.

storage and processing centers consumed more than 416 TWh of electricity, which is more than the total energy consumption in the UK in the same period (300 TWh). It is also estimated that energy consumption and costs related to energy flow account for 3.2% of the anthropogenic emission of carbon dioxide. The amount of energy used by data centers continues to double every 4 years, which means they have one of the fastest growing carbon footprints of all information technology data. Therefore, an intense search for energy-saving techniques for storing and processing information, including unconventional computing, as well as research on renewable energy sources are so desirable (cf. Figure 32).

Nature has very effectively solved the problem of energy efficiency of information processing. The most powerful information processing system is the human brain. The average brain is made up of 10–20 billion neurons, and the cerebellum adds another 55–70 billion neurons (Figure 28). Each of the neurons can make up to 10,000 synaptic connections with its neighbours. While the brain typically accounts for only 2% of body weight, it consumes about 20% of resting energy requirements. Despite the enormous energy efficiency of neural systems, which are in stark contrast to computer systems, the brain is still the most energy-demanding organ in the human body. These facts stimulate intensive research into neuromimetic devices and alternative ways to obtain energy and eliminate the carbon footprint.

Existing computer technologies are largely based on semiconductor materials and their development is an example of an exceptionally successful history. However, despite their unprecedented achievements, the current classical computing paradigms cover only a small subset of all computational possibilities. There is a very wide class of computational substrates and paradigms that use dynamic physical systems and can be successfully used as computer media (Figure 29).[74,75]

Unconventional computing, along with the promise of solving difficult, or otherwise unsolvable computational problems, may offer

Fig. 28. António Egas Moniz (1874–1955) was one of the pioneers researching on brain physiology, and the development of the technique of lobotomy won him an award in 1949, the Nobel Prize in Medicine.

Fig. 29. Fungi can serve as an efficient computational medium.

Fig. 30. Virtuous circle of computing: Transition from natural information processing (biomolecules) to electronics and back to molecular substrates illustrated by a collage of Israeli, German, and USA postage stamps.

a way to energy-effective computation. Especially memristors, a working horse and at the same time the Holy Grail of modern electronics and information technology, may offer much more energy-effective computation (to the best of our knowledge memristor, in contrast to resistor, cf. Figure 30, has not been portrayed on any postage stamp yet). Assuming that the switching takes place at the potential of 1 V (very reasonable assumption, in the case of Schottky barrier memristors switching takes place above 0.5 V) and each unit cell at the interface can trap 1 electron, it gives the energetic cost of switching of 1 eV/unit cell at the interface. An average unit cell wall area for materials studied in this context is ca $1\,\text{nm}^2$. Therefore, energetic cost of switching of 1eV/nm^2 can be regarded as an upper limit for one-electron switching process in molecular Schottky junctions. For a device with the footprint of $100\,\text{nm} \times 100\,\text{nm}$ (reasonable size for electron beam lithography fabrication) it yields a figure of 1.6 fJ, which is much less than an energetic cost of a single operation of biological synapse (10 fJ).

It is still much higher that the energetic cost of transistor switching (0.1 fJ),[76] but the synaptic process brings about much more advanced information processing. Moreover, memristors are especially well suited for signal processing in dynamic systems utilizing a concept of reservoir computing. Furthermore, they can serve as neuromimetic elements providing functionalities like long term potentiation (LTP) and depression (LDP), spike-timing-dependent potentiation (STDP), spike-rate-dependent plasticity (SRDP), and pulse pair facilitation (PPF).[77] Observed progress in unconventional computing (transition from electronic logic gates to molecular logic gates and finally to the plethora of post-Boolean computational paradigms implemented in alternative computing media) seems to be a return to the roots. Natural processes of computation (or more generally information processing and storage) involve DNA and RNA molecules. Technological revolution has transferred computation to inorganic substrates (vacuum tubes and later silicon). Unconventional computing brings in back to its origins — molecular materials and even living organisms (Figure 30).

The various concepts on unconventional computing even included *Codex Seraphinianus*, an abstract encyclopedia of an imaginary world, written in imaginary language and using imaginary script (Figure 31).[78] Disturbingly abstract but appealingly beautiful illustrations show imaginary plants, animals, and machines — the latter in some cases resemble logic structures of neural networks and even reservoir computers.[79,80] Interestingly, this visionary book was written in the late seventies of the 20th century, when the concepts of reservoir computing had not yet been formulated.

Coming back to the roots may have marked benefits: reduction of environmental pollution, increased energetic efficiency, and direct utilization of renewable energy sources (Figure 32). This in turn may contribute to the improvement of living standards, comfort,[81–83] and longer life expectancy (Figure 33).[84] Therefore, the research on unconventional, sometimes very exotic computational paradigms is not in vain. It may even be recognized as stupid, but it is a "productive stupidity" — it is focused on learning and discoveries

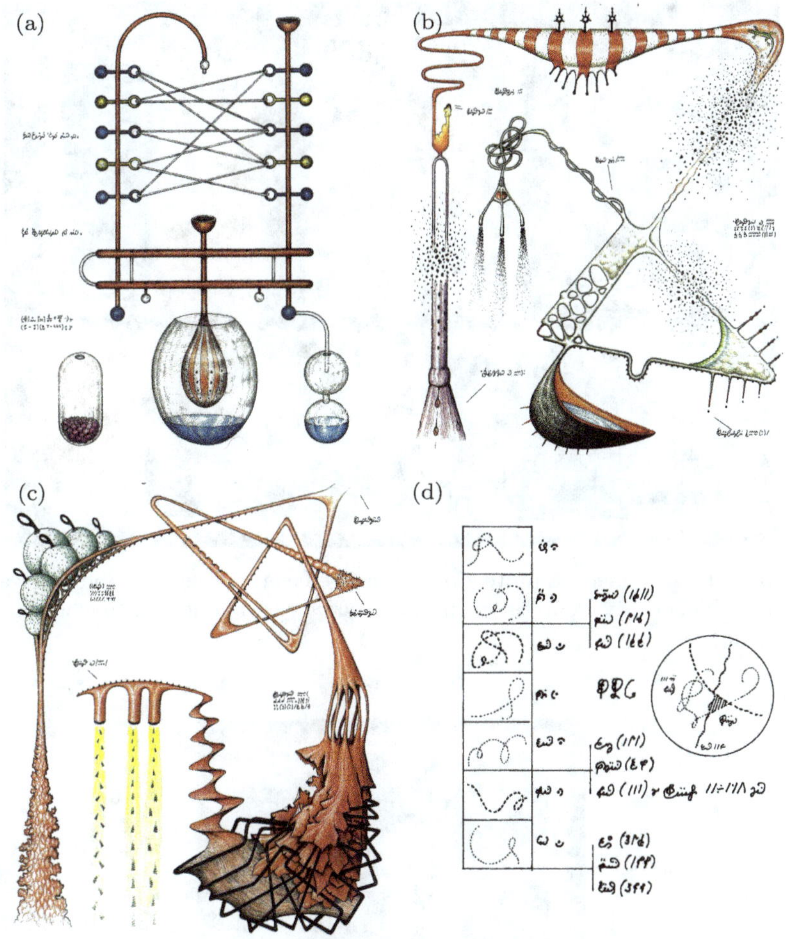

Fig. 31. At the science/arts interface: Neural network (a), reservoir computers (b, c) as well as trajectories and phase space separation (d) in imaginary worlds of Luigi Serafini. Adapted from *Codex Seraphinianus.*[78]

of unexpected relations.[85] It stimulates development of all possible fields of science from psychiatry to philately and cosmology, brings about progress in classical computing technologies, establishes new pathways in arts and develops unexpected links between various fields of human activity.

Fig. 32. Environmental protection, energy-efficient economy, and utilization of renewable energy sources: main prospective advantages of unconventional computing.

Fig. 33. Javier Pereira: The man who allegedly lived for 167 years.

Acknowledgments

The authors thank Dr. Magdalena Wytrwał-Sarna, Dr. Michał Szuwarzyński, and Dr. Tomasz Mazur for their support during bibliography search and Professors Marek Przybylski, Piotr Bała, Andrzej Bernasik, and Zbigniew Pasek for numerous stimulating discussions. This work was in part supported by National Science Centre (Poland) within MAESTRO (UMO-2015/18/A/ST4/00058) and PRELUDIUM projects (UMO-2018/31/N/ST5/03112) as well as by the program "Excellence initiative — research university" for the AGH University of Science and Technology. DP has been partly supported by the EU Project POWR.03.02.00-00-I004/16.

References

1. G. Witzany, *Biocommunication of Archaea.* Springer International Publishing, 2017.
2. M. Miller and B. Bassler, Quorum sensing in bacteria. *Annu. Rev. Microbiol.* 55:165–199, 2001.
3. G. Witzany and M. Nowacki, *Biocommunication of Ciliates.* Springer International Publishing, Cham, 2016.
4. G. Witzany, *Biocommunication of Fungi.* Springer Science+Business Media, Dordrecht, 2012.
5. G. Witzany and F. Baluška, *Biocommunication of Plants.* Springer Verlag, Berlin, 2012.
6. G. Witzany, *Biocommunication of Animals.* Springer Science+Business Media, Dordrecht, 2014.
7. Z. Erez, I. Steinberger-Levy, M. Shamir, S. Doron, A. Stokar-Avihail, Y. Peleg, S. Melamed, A. Leavitt, A. Savidor, S. Albeck, G. Amitai, and R. Sorek, Communication between viruses guides lysis–lysogeny decisions. *Nature* 541:488–493, 2017.
8. S. Abedon, Look who's talking: T-even phage lysis inhibition, the granddaddy of virus-virus intercellular communication research. *Viruses* 11:951, 2019.
9. C. J. Wysocki and G. Preti, Facts, fallacies, fears, and frustrations with human pheromones. *Anatom. Record A.* 281:1201–1211, 2004.
10. K. Grammer, B. Fink, and N. Neave, Human pheromones and sexual attraction. *Eur. J. Obstetrics Gynecol. Reprod. Biol.* 118:135–142, 2005.
11. G. Mazzola, M. Mannone, Y. Pang, M. O'Brien, and N. Torunsky, *All About Music: The Complete Ontology: Realities, Semiotics, Communication, and Embodiment.* Cham, Springer, 2016.
12. J. G. Roederer, *The Physics and Psychophysics of Music: An Introduction.* Springer Science & Business Media, New York, 2008.

13. P. Juslin, *Musical Emotions Explained. Unlocking the Secrets of Musical Affect.* Oxford University Press, Oxford, 2019.
14. M. Barthet, G. Fazekas, and M. Sandler, *Music emotion recognition: From content- to context-based models.* In *From Sounds to Music and Emotions. 9th International Symposium, CMMR 2012 London, UK, June 19–22, 2012, Revised Selected Papers*, M. Aramaki, M. Barthet, R. Kronland-Martinet, and S. Ystad, editors. Springer-Verlag, Berlin, Heidelberg, 2013.
15. E. Schubert, S. Ferguson, N. Farrar, D. Taylor, and G. E. McPherson, The six emotion-face clock as a tool for continuously rating discrete emotional responses to music. In *From Sounds to Music and Emotions. 9th International Symposium, CMMR 2012 London, UK, June 19–22, 2012, Revised Selected Papers*, M. Aramaki, M. Barthet, R. Kronland-Martinet, and S. Ystad, editors. Springer-Verlag, Berlin, Heidelberg, 2013.
16. J. Grekow, *From Content-based Music Emotion Recognition to Emotion Maps of Musical Pieces.* Springer International Publishing AG, Cham, 2018.
17. T. Adorno, *Philosophy of New Music.* University of Minnesota Press, Minneapolis, 2019.
18. Z. Mach, National anthems: The case of Chopin as a National Composer. In *Ethnicity, Identity, and Music: the Musical Construction of Place*, M. Stokes, editors. Berg, Oxford, 1994.
19. I. Biddle and V. Knights, *Music, National Identity and the Politics of Location: Between the Global and the Local.* Ashgate, London, 2007.
20. R. R. Shane Tubbs, M. Shoja, M. Loukas, and J. Oakes, Abubakr Muhammad Ibn Zakaria Razi, Rhazes (865–925 AD). *Childs. Nerv. Syst.* 23:1225–1226, 2007.
21. *Daniel 3, 1-30*, In *Holy Bible. King James Version.* Intellectual Reserve, Inc., Salt Lake City, 2000.
22. *Fatwa No. 285310.* www.islamweb.net, 2015.
23. F.-J. Pütz, Das Postwesen in Afghanistan: Exotisch und spannend. *Philatelie Postgeschichte.* 337:46–51, 2012.
24. A. Nowak and S. Obirek, *Gomorrah. Power, Fear and Money in the Polish Church (in Polish).* Agora, Warszawa, 2021.
25. G. Fisher, *Arabs and Empires Before Islam.* Oxford University Press, Oxford, 2015.
26. A. Alghamdi, Segmentation by visitor motivation in cultural festivals: A case study of the souk okaz festival. *J. Bus. Econ.* 11:2180–2202, 2014.
27. G. Sawa, *An Arabic Musical and Socio-cultural Glossary of Kitab Al-aghani.* Brill, Leiden, 2015.
28. V. C. Müller, *Music censorship in Afghanistan before and after the Taliban*, In *Shoot the Singer! Music Censorship Today*, M. Korpe, editor. Zed Books, London, 2004.
29. J. Alagha, Jihad through 'music': The taliban and hizbullah. *Perf. Islam.* 1:263–289, 2012.
30. M. Minsky, Music, mind, and meaning. *Comput. Music J.* 5(3):28–44, 1981.
31. P. Gewirtz, On "I know it when I see it". *Yale Law J.* 105(4):1023–1047, 1996.

32. P. Cooper, *Perspectives in Music Theory: An Historical-Analytical Approach.* Dodd, Mead, 1973.
33. R. Chuckrow, *Historical Tuning of Keyboard Instruments: Theory and Practice: Pythagorean Through Equal Temperament.* Rising Mist Publications, Briarcliff Manor, NY, 2006.
34. E. Blackwood, *The Structure of Recognizable Diatonic Tunings.* Princeton University Press, Princeton, 2014.
35. G. Loy, *Musimathics: The Mathematical Foundations of Music*, vol. 1. MIT Press, London, 2011.
36. G. Nierhaus, *Algorithmic Composition: Paradigms of Automated Music Generation.* Springer Science & Business Media, Vienna, 2009.
37. A. Adamatzky, *Advances in Unconventional Computing.* Springer, Cham, 2016.
38. E. R. Miranda, *Guide to Unconventional Computing for Music.* Springer, Cham, 2017.
39. V. Athanasiou and Z. Konkoli, On mathematics of universal computation with generic dynamical systems. In *From Parallel to Emergent Computing*, A. Adamatzky, S. Akl, and G. Sirakoulis, editors. CRC Press, Boca Raton, 2019.
40. N. Fresco and M. Miłkowski, Mechanistic computational individuation without biting the bullet. *Brit. J. Phil. Sci.* 72:431–438, 2021.
41. A. Rueger and W. D. Sharp, Simple theories of a messy world: Truth and explanatory power in nonlinear dynamics. *Brit. J. Phil. Sci.* 72:93–112, 1996.
42. M. Tse-tung, *Quotations from Chairman Mao Tse-tung.* Xinhua Bookstore, Hohhot, 1967.
43. D. Przyczyna, P. Zawal, T. Mazur, M. Strzelecki, P. L. Gentili, and K. Szaciłowski, In-materio neuromimetic devices: Dynamics, information processing and pattern recognition. *Jpn. J. Appl. Phys.* 59(5):050504, 2020.
44. B. Sturm, The state of the art ten years after a state of the art: Future research in music information retrieval. *J. New Music Res.* 43:147–172, 2014.
45. P. Dhariwal, H. Jun, C. Payne, J. W. Kim, A. Radford, and I. Sutskever, Jukebox: A generative model for music, *arXiv.* 2005.00341, 2020.
46. A. M. Bhatti, M. Majid, S. M. Anwar, and B. Khan, Human emotion recognition and analysis in response to audio music using brain signals. *Comp. Human. Behav.* 65:267–275, 2016.
47. E. Musk, An integrated brain-machine interface platform with thousands of channels, *J. Med. Internet Res.* 21:e16194, 2019.
48. C. L. Carvalho, Drunkenness, tattoos, and dirty underwear: Jeremiah as a modern masculine metaphor. *Cathol. Bibl. Quarterly.* 80:597–618, 2019.
49. P. Karocki, Martian liturgical calendar (in Polish). *Ruch Bibl. Liturg.* 68(4):333–351, 2015.
50. T. Rodowski, *Resocialisation through Sacred Art in the Context of Quantum Psychology (in Polish).* Difin, Warszawa, 2017.
51. P. Polak and R. Krzanowski, Deanthropomorphized pancomputationalism and the concept of computing. *Found. Comput. Decision Sci.* 44(1):45–54, 2019.

52. V. C. Müller, Pancomputationalism: Theory or metaphor? In *Philosophy, Computing and Information Science*, R. Hagengruber and U. Riss, editors. Pickering & Chattoo, London, 2013.
53. J. Traphagan, SETI, evolutionary eschatology, and the Star trek imaginary. *Theology Sci.* 19(2):120–131, 2021.
54. A. Einstein, *Über die spezielle und die allgemeine Relativitätstheorie.* Friedrich Vieweg und Sohn, Braunschweig, 1919.
55. Z. Jacyna-Onyszkiewicz, E. Sobaniec-Łotowska, S. Sulkowski, K. A., and M. Rucki, Eucharistic miracle from the scientific perspective (in Polish). *Teologia i Człowiek.* 43(3):81–98, 2018.
56. A. Sokal, Transgressing the boundaries: Towards a transformative hermeneutics of quantum gravity. *Social Text.* 46–47(1):217–252, 1996.
57. R. Robert W Yeh, L. Valsdottir, M. Yeh, C. Shen, D. Kramer, J. Strom, E. Secemsky, J. Healy, R. Domeier, D. Kazi, and B. Nallamothu, Parachute use to prevent death and major trauma when jumping from aircraft: Randomized controlled trial. *British Med. J.* 363:k5094, 2018.
58. L. Wang, Z. Sofer, and M. Pumera, Will any crap we put into graphene increase its electrocatalytic effect? *ACS Nano.* 14:21–25, 2020.
59. L. Wang, Z. Sofer, and M. Pumera, Beyond guano-doped graphene. *ACS Nano.* 14:2555–2556, 2020.
60. E. Ceauşescu, *Polimerizarea stereospecifică a izoprenului.* Editura Academiei Republicii Socialiste România, Bucharest, 1979.
61. Repeat after me. *Nature* 488:253, 2012.
62. M. Matyjewicz, Cultural competences of young people in the disco-polo era. Towards the pedagogy of shamelessness (in Polish). *Ars Inter Culturas.* 7:263–272, 2018.
63. D. J. Sendler, Similar mechanisms of traumatic rectal injuries in patients who had anal sex with animals to those who were butt-fisted by human sexual partner. *J. Forensic Legal Med.* 51:69–73, 2017.
64. P. Czarnek, Natural law in europe — The foundation or forgotten reality (in Polish). *Biul. SAWP KUL.* 13:27–45, 2018.
65. J. Zamana, The style of leadership of Jaroslaw Kaczynski. Retrospective analysis (in Polish). *Crit. Stud.* 2:201–213, 2016.
66. E. Kot, *Cizia, dupek* i *poczwara*: How young people are describing people (based on chosen dictionaries) (in Polish). *Conv. Linguist.* 7:54–61, 2013.
67. K. Boleba and T. P. Bocheński, *Z ciebie lania jak z kozich cycków kastaniety.* Animalisms in youth's sociolect based on *Miejski slownik slangu i mowy potocznej* (in Polish). *Maska.* 22:95–110, 2014.
68. K. Szaciłowski, *Infochemistry. Information Processing at the Nanoscale.* John Wiley & Sons, Chichester, 2012.
69. E. Katz, *Biomolecular Information Processing.* Wiley-VCH, Weinheim, 2012.
70. E. Katz, *Signal-Switchable Electrochemical Systems.* Wiley-VCH, Weinheim, 2018.
71. E. Katz, *Molecular and supramolecular information processing.* Wiley-VCH, Weinheim, 2018.

72. A. Adamatzky, *Advances in Physarum Machines. Sensing and Computing with Slime Mold.* Springer International Publishing, Cham, 2016.
73. S. Stepney, S. Rasmussen, and M. Amos, *Computational Matter.* Springer, Cham, 2018.
74. A. Adamatzky, Towards fungal computer. *Interface Focus.* 8:20180029, 2018.
75. M. M. Dehshibi and A. Adamatzky, Electrical activity of fungi: Spikes detection and complexity analysis. *Biosystems.* 8:104373, 2021.
76. V. Zhirnov and L. Gammaitoni, *Minimum energy of computing, fundamental considerations.* In *ICT — Energy — Concepts Towards Zero — Power Information and Communication Technology,* G. Fagas, L. Gammaitoni, D. Paul, and G. A. Berini, editors. Intech Open, London, 2014.
77. S. Klejna, T. Mazur, E. Wlaźlak, P. Zawal, H. S. Soo, and K. Szaciłowski, Halogen-containing semiconductors: From artificial photosynthesis to unconventional computing. *Coord. Chem. Revs.* 415:213316, 2020.
78. L. Serafini, *Codex Seraphinianus.* Rizzoli, New York, 2013.
79. J. Choueiri and F. Türetken, Interiors of memories: A study of personal memories based on the works of Luigi Serafini and Georges Perec. *Interstices J. Arch. Relat. Arts.* 18:64–72, 2017.
80. M. Mangialardi, Luigi Serafini e il *Codex Seraphinianus.* Con intervista all'artista. *Bollettino Telematico dell'Arte.* 802:1–11, 2016.
81. M. K. Bakkevig and R. Nielsen, Impact of wet underwear on thermoregulatory responses and thermal comfort in the cold. *Ergonomics* 37:1375–1389, 1994.
82. Y. Wang, X. Qian, W. Zhang, and H. Liu, Heat and moisture comfort of knitted fabrics for underwear. *Adv. Mater. Res.* 332–334:890–893, 2011.
83. J. Phillips and P. Phillips, History from below: Women's underwear and the rise of women's sport. *J. Pop. Cult.* 27:129–148, 1993.
84. M. Terpstra, The correlation between sustainable development and home hygiene. *Am. J. Infection Control.* 29:211–217, 2001.
85. M. A. Schwartz, The importance of stupidity in scientific research. *J. Cell. Sci.* 121(1):1771 (2008).

https://doi.org/10.1142/9789811257155_0006

Chapter 6

Swarms of Microscopic Agents Self-assemble into Complex Bodies

Bruce MacLennan

Department of Electrical Engineering and Computer Science, University of Tennessee, Knoxville 37996, USA

maclennan@utk.edu

Nature produces living bodies of magnificent beauty and complexity, which are not the work of an external artist, but of a self-organized dance of a vast number of microscopic cells communicating and coordinating with each other. We can learn the dance scores from nature and use them to control massive swarms of microscopic artificial agents to assemble complex structures organized at many hierarchical levels, from the nanoscale up to the macroscale. Simulations demonstrate the application of these ideas in artificial intelligence and artificial life.

1. The Problem

As Richard Feynman said, "there is plenty of room at the bottom,"[1] and we are discovering ways to organize matter and processes at the microscopic scale — and even at the nanoscale — that have macroscopic effects that are relevant to technology, medicine, and art. But to date, nanotechnology has been limited to the creation of simple structures with a relatively small number of components, or to structures with a large number of components with either a highly regular or a random organization. For many purposes, however, we need to be able to create complex structures hierarchically organized at many spatial scales from the microscopic to the macroscopic. Consider these examples.

The contemporary, most successful approaches to artificial intelligence are inspired by the brain; they strive to model massive neural networks at a level of abstraction that preserves their information processing capabilities while omitting irrelevant biological detail. Research in cognitive neuroscience and artificial neural networks has revealed that much of the flexible, context-sensitive information processing of our brains depends on large numbers of massively interconnected neurons (8.6×10^{10} neurons in a human brain, with perhaps 10^{15} connections). Moreover, these neurons are not arranged and interconnected in either a regular or a random pattern; there are complex organizations both within individual brain regions and among the regions. All this within a volume of 1,100 to 1,300 cm^3, and consuming about 20 W of power. If we want to achieve human-scale artificial intelligence, it is reasonable to suppose that we may need to construct artificial brains similarly organized from the microscale to the macroscale (Figure 1).

Continuing the example of artificial intelligence, we now understand much better the importance of embodiment as a foundation for intelligence.[3–5] Brains have evolved to control physical bodies in their physical environments, although, of course, we also use them for abstract thought and contemplation. Nevertheless, control of our mechanically complex bodies by means of rich sensorimotor feedback provides the foundation from which language and abstract thought can emerge. Large numbers of sensory neurons embedded in our

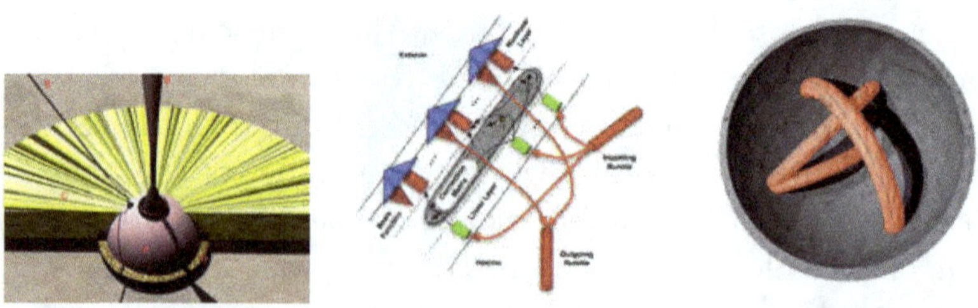

(a) (b) (c)

Fig. 1. Depictions of a large-scale neurocomputer.[2] (a) External connections to artificial cortex. (b) Computational microstructure of artificial cortex. (c) Example of three massive fiber bundles connecting functional regions of cortex.

muscles, joints, and skin provide rich, high-dimensional input that allows our (quite slow) neurons to coordinate complex behavior in real time. Future robots with physical competence comparable to that of animals will be facilitated by similarly complex artificial sense organs, skin, and muscles (actuators). Although macroscopic in size, they will be complex structures of many microscopic parts. For example, the human eye has approximately 100 million retinal cells, which preprocess the visual image in real time, reducing its dimension so that it can be transmitted on about one million optic nerve fibers. It is not unreasonable to suppose that an artificial eye, with similar visual acuity and information processing capacity to ours, will have similar complexity.

Such robotic devices will have important applications in medicine, such as prosthetic limbs and sense organs. Surely, we would like prosthetic eyes with the sensitivity and acuity of natural eyes, and prosthetic hands with the dexterity and sense of touch of natural hands, but these require vast numbers of components to be assembled in precise ways.

2. Artificial Morphogenesis

Assembling many millions of components into complex systems, structured from the nanoscale up through the macroscale, might seem impossible, but developmental processes in embryos prove that it is possible. An embryo develops from a single cell, which divides repeatedly producing an exponentially growing cell mass. These cells differentiate and orchestrate a complex dance that assembles the tissues, organs, and limbs of a complete organism (3.7×10^{13} cells in a human adult). A fetus may comprise a trillion cells, which have self organized into innumerable structures at many spatial scales spanning five orders of magnitude. In this robust process, cells migrate, following chemical way stations to distant destinations, signaling each other to coordinate their movement and trigger context-sensitive differentiation into hundreds of cell types. Masses of cells flow viscously, forming intricately structured tissues that stretch, fold, and grow. This is our inspiration.

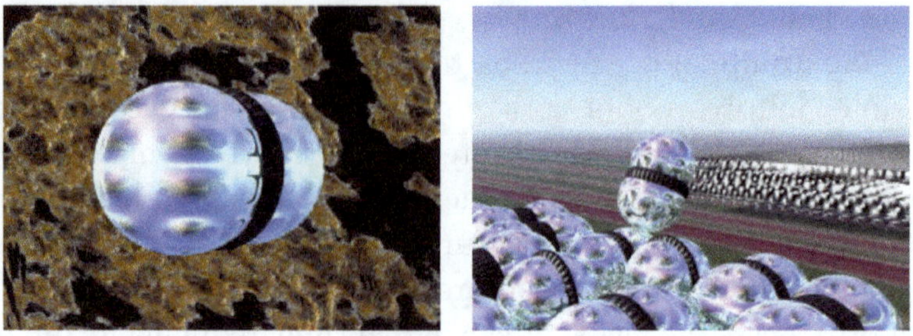

(a) (b)

Fig. 2. Artistic depiction of microrobots. (a) A single microrobot attached to a surface. (b) A swarm of microrobots assembling into a layer of an artificial tissue.

Artificial morphogenesis applies the principles of embryological morphogenesis to coordinating very large numbers of simple agents to assemble complex hierarchical structures.[6,7] It may be considered one approach to *morphogenetic engineering*, which applies principles of morphogenesis to the creation of form.[8,9] In artificial morphogenesis, very large numbers of microscopic agents cooperate to assemble themselves and inanimate microscopic components into a desired structure (Figure 2). Under some conditions, they can disassemble such a structure in order to reassemble it into a new form.

The agents, which may be wholly artificial or produced by synthetic biology, have relatively simple capabilities for motion, signaling, and information processing and control. Agents need to be able to signal their immediate neighbors, but also more distant agents in order to coordinate their behavior. Cells in a developing embryo accomplish this by emitting chemical morphogens, which diffuse in the intercellular environment and can be detected by other cells. Likewise, in artificial morphogenesis, agents communicate by chemical and other signals. Agents also need to be able to move, either through a fluid medium or as a viscoelastic mass. This movement is accomplished by a variety of simple mechanical mechanisms. Finally, the behavior of agents is governed by relatively simple, primarily analog control mechanisms.

Progress in microrobotics is ongoing, and we anticipate that there will soon be sub-millimeter-scale autonomous robots with the capabilities required for artificial morphogenesis. Another attractive option is to use the techniques of synthetic biology to modify microorganisms to implement the required agents. Genetic regulatory circuits are essentially analog control mechanisms, and they can be modified to implement the behavior required of a morphogenetic agent. Another advantage of biological agents is that they can reproduce, thus eliminating the need to manufacture artificial agents in large numbers.

Agents move in coordinated masses and differentiate under the influence of signaling substances that diffuse in the environment. Since our goal is to have very large numbers of very small agents moving *en masse*, we describe morphogenetic processes by partial differential equations (PDEs), which treats the agent mass as a continuous fluid or tissue. Embryologists often use PDEs for the same reason, and we can use their equations in artificial morphogenesis. Mathematically, we treat our agents as infinitesimal particles moving under the influence of forces and morphogen concentrations. This approach helps to ensure that our algorithms scale up to very large numbers of very small agents, but also keeps them largely scale-invariant, that is, independent of the exact size of the agents relative to the macroscopic object being constructed.

3. Examples

Since the required microscopic agents are not yet available, we test our morphogenetic algorithms through simulation.[10,11]

Our first example (Figure 3) shows how an indefinite but very large number of agents can be coordinated to lay down neural fiber bundles between selected regions of an artificial brain.[12] In this morphogenetic algorithm, the bundles are grown one at a time. A massive swarm of agents (equal in number to the number of nerve fibers in the bundle) is injected at the origin region, and they follow the gradient of a morphogen diffusing from the destination region, depositing fiber material as they go. Once a bundle has been created, its material absorbs the attractant, and this causes agents to steer around already created bundles, avoiding collisions. Moreover, we

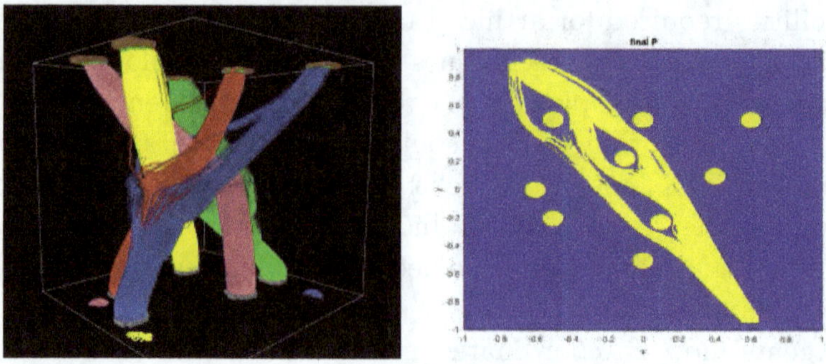

(a) (b)

Fig. 3. (a) Simulation of swarms of 5000 agents depositing neural fiber bundles between randomly chosen origins and destinations. (b) Simulation of massive swarm of agents creating paths around obstacles from lower right to upper left.

can allow fiber bundles to split to go around obstacles when that is required.

Our second example (Figure 4) shows how natural morphogenetic processes — in this case spinal segmentation — can be used for both similar and different applications in artificial morphogenesis, in this case, assembling an insect-like robot body frame.[13] The example exploits the idea that in morphogenesis, patterns in time can create patterns in space. In this case, we use the clock-and-wavefront model of spinal segmentation, which was first proposed in 1976 but finally confirmed in 2008.[14,15] We use this process to assemble the spine of a robot, which is similar to its function in vertebrate development, but we also use it to assemble segmented legs, which develop differently in nature. Thus, we are using a natural process both for a purpose that it served in nature (spinal segmentation) and also for a purpose it does not serve in nature (leg segmentation). The number and lengths of the segments are parameters that we can control in each case.

In brief, morphogenesis proceeds as follows. Agents are recruited to assemble between the spine and the tail bud, which is moved rightward. Both the tail bud and completed spinal segments produce morphogens which diffuse into the undifferentiated spinal region. Periodically (and this is the temporal patterning), a pacemaker in the tail bud produces a pulse of a third morphogen, which is propagated

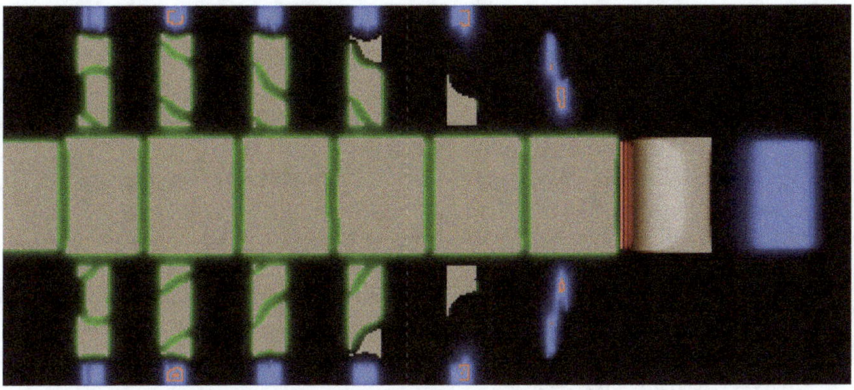

Fig. 4. Simulation of assembly of insect-like robot body frame using clock-and-wavefront process. Head end to left, tail end to right. Red color denotes wave of segmentation morphogen propagating to left, which has just passed through and differentiated the right-most tan segment. Similar processes are simultaneously assembling and differentiating leg segments.

through the tissue toward the head. As it passes through a region of relatively low concentrations of the first two morphogens, it leads to differentiation of a new spinal segment. The length of the segments is determined by the ratio of the growth rate and the pacemaker frequency; the number of segments is controlled by the product of pacemaker frequency and the growth time. This same process is used to grow segmented legs on the spinal segments.

4. Conclusions

Morphogenetic processes in nature can be imitated and adapted to control massive swarms of microscopic agents to assemble complex, hierarchically structured systems. Describing these processes by partial differential equations describing masses of infinitesimal particles helps to ensure that they scale up to very large numbers of microscopic agents, which is what will be required for the self-assembly of very complex structures, organized from the microscale up to the macroscale. In this way, we may hope to produce artifacts of a similar sophistication to those in nature.

References

1. R. P. Feynman, There's plenty of room at the bottom. *Eng. Sci.* 23:22–36, 1960. https://resolver.caltech.edu/CaltechES:23.5.1960Bottom.
2. B. J. MacLennan, The U-machine: A model of generalized computation. *Int. J. Unconv. Comput.* 6(3–4):265–283, 2010.
3. A. Clark, *Being There: Putting Brain, Body, and World Together Again.* MIT Press, Cambridge, MA, 1997.
4. G. Lakoff and M. Johnson, *Philosophy in the Flesh: The Embodied Mind and its Challenge to Western Thought.* Basic Books, New York, 1999.
5. M. Johnson and T. Rohrer. We are live creatures: Embodiment, *American pragmatism, and the cognitive organism.* In *Body, Language, and Mind,* J. Zlatev, T. Ziemke, R. Frank, and R. Dirven, editors. vol. 1, Mouton de Gruyter, Berlin, 2007, 17–54.
6. B. J. MacLennan, Morphogenesis as a model for nano communication. *Nano Commun. Netw.* 1(3):199–208, 2010. Doi: 10.1016/j.nancom.2010.09.007.
7. B. J. MacLennan, The morphogenetic path to programmable matter. *Proc. IEEE.* 103(7):1226–1232, 2015. Doi: 10.1109/JPROC.2015.2425394.
8. R. Doursat, H. Sayama, and O. Michel, A review of morphogenetic engineering. *Nat. Comput.* 12(4):517–535, 2013. Doi: 10.1007/s11047-013-9398-1.
9. H. Oh, A. Ramezan Shirazi, C. Sun, and Y. Jin, Bio-inspired self-organising multi-robot pattern formation: A review. *Rob. Auton. Syst.* 91:83–100, 2017. Doi: 10.1016/j.robot.2016.12.006.
10. B. J. MacLennan and A. C. McBride, Swarm intelligence for morphogenetic engineering. Technical Report arXiv:1908.08787, Dept. of Electrical Engineering & Computer Science, University of Tennessee, Knoxville, August, 2019. https://arxiv.org/abs/1908.08787.
11. B. J. MacLennan and A. C. McBride. Swarm intelligence for morphogenetic engineering. In *Swarm Intelligence: From Social Bacteria to Humans,* A. Schumann, editor, Chapter 2, Taylor & Francis/CRC Press, Boca Raton, 2021, 9–53.
12. B. J. MacLennan, A morphogenetic program for path formation by continuous flocking. *Int. J. Unconv. Comput.* 14:91–119, 2019.
13. B. J. MacLennan, Coordinating swarms of microscopic agents to assemble complex structures. In *Swarm Intelligence, Vol. 1: Principles, Current Algorithms and Methods*, Y. Tan, editor, PBCE 119, Chapter 20, Institution of Engineering and Technology, Stevenage, UK, 2018, 583–612.
14. J. Cooke and E. C. Zeeman, A clock and wavefront model for control of the number of repeated structures during animal morphogenesis. *J. Theor. Biol.*, 58:455–476, 1976.
15. M.-L. Dequéant and O. Pourquié, Segmental patterning of the vertebrate embryonic axis. *Nat. Rev. Genet.*, 9:370–382, 2008.

https://doi.org/10.1142/9789811257155_0007

Chapter 7

Kakiwari: The Device Summoning Creativity in Art and Cognition

Yukio Pegio Gunji[*,‡] and Kyoko Nakamura[†,§]

[*]*School of Fundamental Science and Engineering, Waseda University, 3-4-1 Ohkubo, Shinjuku-ku, Tokyo 169-8555, Japan*

[†]*School of Design, Kyushu University, 4-9-1 Shinobaru, Minami-ku, Fukuoka 815-8540, Japan*

[‡]*yukio@waseda.jp*

[§]*kyoko608@gmail.com*

Creativity is not a code, but the process of summoning something from outside our cognitive world, and the device to realize creation is called *kakiwari*. Here, we show that qualia are creations in cognition, and that *kakiwari* is expressed as quantum-like logic in terms of lattice theory. We also demonstrate the role of *kakiwari* in art, Japanese painting, and one of the authors, Nakamura, shares her own artwork, whose theme is *kakiwari* itself.

1. Introduction

The notion of creativity is intrinsically personal and neither public nor objective. Creating something belongs to "myself" who creates, but not to people in general. Although "myself" is not privately closed[1] since I live within environments, the pressure, suffering, and pleasure of creating is what I feel, but not what society feels. Notwithstanding that, asking what creativity is makes us consider creativity in public and as an objective form in nature. Since creativity is regarded as unnamed, someone in "creativity for someone" can be replaced by anything, which leads to the question, "Does artificial intelligence (AI) realize creativity?".[2–4]

Here, we argue for "creativity in me" in a universal way. Talking about "creativity in me" and "for me" is not personal, and can be universally argued. Although the content of my tasting is destined to be personal, how to deal with the outside or how to compare the inside with the outside is not personal and can be universally argued. Since creativity is nothing but exploring the outside, even creativity "in me" entails the universal.

We call the domain that is outside what one perceives "the outside". Any definable, describable things are destined to fall under the "inside" of one's world. In contrast, it is impossible to perceive phenomena and matters outside the perceived world because they cannot be perceived.[5,6] By using the notion of the outside, one can describe emergent properties and/or creation. If there is no notion like the outside, then one is destined to assume that something arises from nothing. In introducing the outside, one can consider the outside, which cannot be perceived, but exists and can yield the core of creation.[7]

In this chapter, we first describe creation with respect to accessing the outside or summoning something from the outside, compared to the notion of the creative code. In particular, we examine the pioneering work on creativity by Boden[2,3] and a recent study by Sautoy[4] referring to artificial intelligence. Second, we show that creation in cognition can be described as a quantum-like structure.[8] This also indicates that creation in cognition and/or perception implies the generation of qualia.[9,10] Finally, we identify creation as the process by which something is summoned from the outside in a cultural anthropological event,[11] and exhibit the artwork made by one of the authors, inspired by that event.

2. Creativity not a Code, but Refers to Summoning the Outside

Boden classified creativity into three forms: combination, exploration, and qualitative transformation.[2,3] Combinational creativity is described as making unfamiliar combinations of familiar ideas. Since the combination of what is newly obtained already exists, but was not experienced by an observer, Boden first considers that combinational creativity is not genuine creativity. The creativity of exploring is described as a surprising idea resulting from exploring conceptual space, which is a set of structured styles of thought. The idea is thought being newly created if it is surprising as an example of an unexpected general type. In chess, there are possible moves in advance, and a player manages to examine the potential space of the moves and to find an optimal move for a given aspect. This can be a typical example of exploring creativity. In addition to combinational creativity, Boden first considers that the creativity of exploring is not genuine creativity.

While Boden claims that combinational creativity and the creativity of exploring can be implemented by AI, she investigates whether the creativity of transforming the space — which could be regarded as genuine creativity — could be implemented by a computer. Since AI programs can transform their own conceptual space by changing their own rules, some interesting ideas that are novel (even for humans) might be obtained. The evaluation of selecting certain ideas is a key notion for novel and interesting ideas, and is implemented through evolutionary computing. Thus, one might conclude that only AI can create new ideas without humans. Boden, however, claims that creativity depends on the underlying values influenced by contemporary culture and science, and that "values" are ambiguous and vague, and cannot be pinned down precisely.

Insofar as one pays attention to values in thinking about creativity, one notices that values always contribute to creation, not only to the creativity of transformation, but also to the creativity of exploring and combinational creativity. In combinational creativity, a new combination of words that used to be uttered in everyday life could entail novel meaning accompanied by new value, which is exactly what poets do. Although a word like "artificial" classically implies "unnatural", it has recently come to imply the machine-like. In the creativity of exploring, the same move of the pawn for the same aspect might be altered from a mere ordinary move to an artistic move. In that sense, the value of realizing creativity cannot be controlled, and it might come outside what one can precisely describe or control. Creation arises from the outside.

While the mathematician Sautoy adopts Boden's attitude toward creativity,[4] he denies that creation arises from the outside. According to Sautoy, artists frequently say, "I was inspired by the outside," but this is just a feigned attitude. It is just a way of showing as if creating is something mysterious. Sautoy, however, sways between AI threat theory and AI harmless theory in viewing the success of alpha-go search engines such as Google, the algorithmic evolution of recommendation systems, success in resurrecting Rembrandt, and the evolution of proof algorithms. Finally, while Sautoy recognizes that the creativity of AI is located along the same lines as human creativity, he concludes, "At this moment, human creativity is not invaded by AI." Even at this time, AI technologies have taken away many jobs from various workers. Likewise, at the time of the industrial revolution, many workers lost their jobs to steam engines and/or automatic looms. Similarly, "creativity for someone" is invaded by AI.

Sautoy's problematic statement relies on careless thinking and not paying attention to whose creativity is involved. The creativity Sautoy considers is the creativity hidden in nature that can be assumed in science. This is called creativity in nature. Sautoy asserts that creativity in nature exists in principle, and that mathematics can approach creativity in nature as the code of creativity. This is why he considers that AI — whose algorithm is expressed as mathematics — will approach creativity in nature. Human creativity seems to be implemented in organisms based on the DNA program. Such an implementation implies that creativity evolves through the secretion of dopamine in the brain. One cannot control the boundary condition of the secretion of dopamine, because it is strongly related to value in the sense of Boden, and because it is opened to the outside of the boundary condition, which is precisely described. This leads to the definition of creativity as being dependent on the person. Ignoring the indeterminacy of boundary conditions and values leads to the loss of personality in creativity. Although we cannot describe creativity without indeterminacy, we will always confront the aspect in which we would create something. We are destined to focus on creativity "for me" because creativity is not something to be described but to be experienced.

The next problem arises with "what I am". What is the definition of "myself", which has both body and consciousness? Instead of "myself", Sautoy considers what a human is. A human has consciousness and is destined to die. According to Sautoy, our will is inside of our consciousness, and we have a tendency to express our will toward the outside. This is the first reason for creativity. Additionally, we are destined to die. Thus, we want to leave our expressions after we die. That is the second reason for creativity. Sautoy tends to relate creativity with consciousness. Most artists never think that artworks result from their will inside of their consciousness. What is created is summoned from the outside of the artist; this is called "inspiration", which Sautoy criticized.

Not only art, but also everyday life, does not result from free will as the ultimate cause. Assume that you feel, by yourself, that you want to look at a book in front of you. You can feel your hand approaching the book, triggered by your own will. However, preference and awareness of the book unconsciously appear from external cues, and the readiness potential in your brain is fired. Finally, you think you want to read the book in the form of post-diction.[12,13] Thus, your will is just retrofitted to the neural reaction of the readiness potential. One cannot say that your decision results from your free will. Similarly, creativity does not result from free will.

The core of creativity is dependency on value that cannot be described, and is openness in the boundary condition to create. Therefore, creativity is not described as the code of creativity. Creativity is a process not to be described in public, but to be experienced personally. Both Boden and Sautoy believe that AI is strongly relevant for creativity and can be used as a tool for humans to create something. These considerations do not respond to how one can create something or how one can experience creativity.

Here, we shift from creativity to be described to creativity to be experienced. The diagram, shown in the inner box in Figure 1, reveals the code of creativity relating the inside ("my desire" to

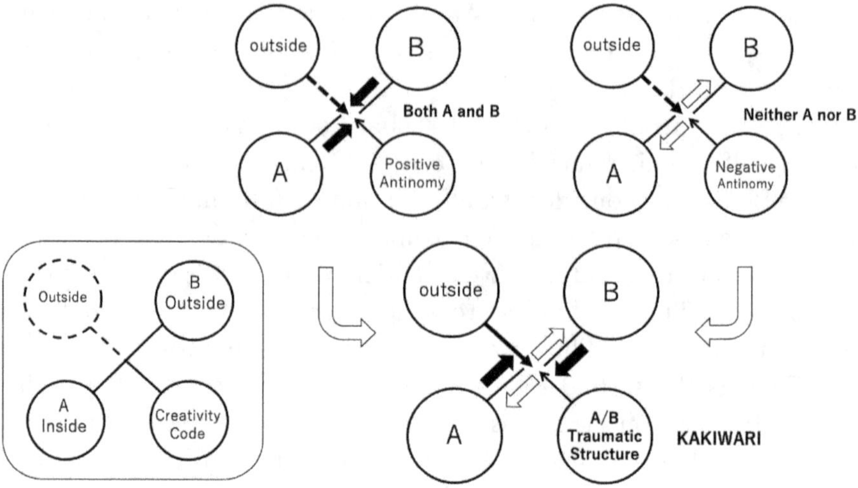

Fig. 1. *Kakiwari* to summon the outside compared to the code of creativity.

create something) to the outside (the production made of materials, words, or symbols). Since the code of creativity represents the internal insight of artists, players, and scientists, the outside — located outside the framework, in which insight and representation are indicated — cannot be summoned.[14–16]

Now, we generalize a pair of the inside and the outside to a pair of A and B. It is easy to see that the inside and the outside might be in binary opposition and may be related to each other. Creativity as experience is defined as the process of summoning something from the outside via a special device, *kakiwari*, which is defined by the A/B traumatic structure[14,15] (Figure 1). Originally, we call the board-like mountains of Japanese painting *kakiwari* ("stage setting" in Japanese). The reason we call the A/B traumatic structure *kakiwari* is discussed in a later section.

The A/B traumatic structure is described as the coexistence of positive A/B antinomy and negative A/B antinomy. Since the positive A/B antinomy allows for both A and B, A can be close to B and vice versa in the framework in which A and B are indicated. Since the negative A/B antinomy allows for neither A nor B, A is separated from B and vice versa in the framework. Thus, coexistence of the positive and negative antinomy entails disequilibration between A and B, which brings forth the crack in the framework. Something is summoned from the outside toward the crack, which implies creativity (Figure 1). Therefore, *kakiwari* plays a role as a device by which something is summoned.

The reason we call coexistence of the positive and negative A/B antinomy, the traumatic structure, is based on the entanglement of A and B. The patient suffering trauma frequently feels not only like a victim, but also like a perpetrator, even if he or she is just the victim and cannot be the perpetrator.[17] If the reason for the damage is too irrational and too unreasonable, the patient is destined to fall into the meta-cognition of the victim. Therefore, meta-cognition could lead to jumping up to the framework of the binary opposition of the victim and perpetrator, and brings forth the entanglement of the victim and perpetrator. It is very difficult to find such a complex situation consisting of two components, such as a pair of victims (A) and perpetrators (B), except for trauma. Thus, we call it the A/B traumatic structure.

Kakiwari can be illustrated by contact improvization.[18] A pair of A and B is replaced by a pair of passive and active attitudes via bodily contact. This is a kind of practice for the body's expression. First, a dancer is taught to dance by keeping in contact with another dancer via a part of the body such as the back. Since a dancer autonomously dances by touching another person via the back, he or she dances not only passively but also actively. In that stage, the situation allows for positive passive/active antinomy because both passive and active attitudes are permitted. After mastering both passive and active attitudes, the dancer is separated from another dancer and is taught to dance freely. The dancer, in this stage, can feel that he or she dances in a thoroughly passive attitude as if he or she were moved by the air surrounding him or her. Thoroughly passive attitudes[19] are different from passive attitudes in a pair of passive and active attitudes because they are independent of contact. In this vein, the thoroughly passive attitude is summoned from the outside of the framework of a pair of passive and active attitudes defined via contact. This shows the creation of the dance with a thoroughly passive attitude.

3. Creativity via *Kakiwari* in Cognition: Qualia

3.1. *Kakiwari as a binary relation*

If a pair of A and B in Figure 1 is replaced by an object and representation, creativity in cognition is argued through *kakiwari*, which is object/representation. During an infant's development, he or she experiences the first event in which an object, such as a cat, is perceived as the representation of "cat". While there is simply correspondence between an object and its representation, an infant must perceive his or her own reality or qualia of a cat when he or she perceives a cat. How are the qualia generated? Qualia are not attributes of an object and do not belong to representation.[9] Since relating an object with its representation can be regarded as computation, it can be embedded in psychological phenomena. In contrast, qualia are interpreted as something outside psychological phenomena. Phenomenal consciousness, including qualia, is thought to supervene with psychological consciousness in which cause–effect relations are described.[9] In that sense, whenever qualia are perceived, they are summoned from the outside of where objects and representations are indicated.

Here, we propose that qualia are generated through *kakiwari*, the object/representation traumatic structure. Since the positive object/representation antinomy allows for both an object and representation, it implies approaching the object, moving toward representation, and vice versa. In contrast, the negative object/representation antinomy allows for neither an object nor representation; it implies separating an object form representation and vice versa. Thus, coexistence of positive and negative antinomy generates a crack between an object and representation in the framework of psychological consciousness, which summons qualia from the outside of psychological consciousness. This leads to the supervenience of psychological and phenomenal consciousness (Figure 2).

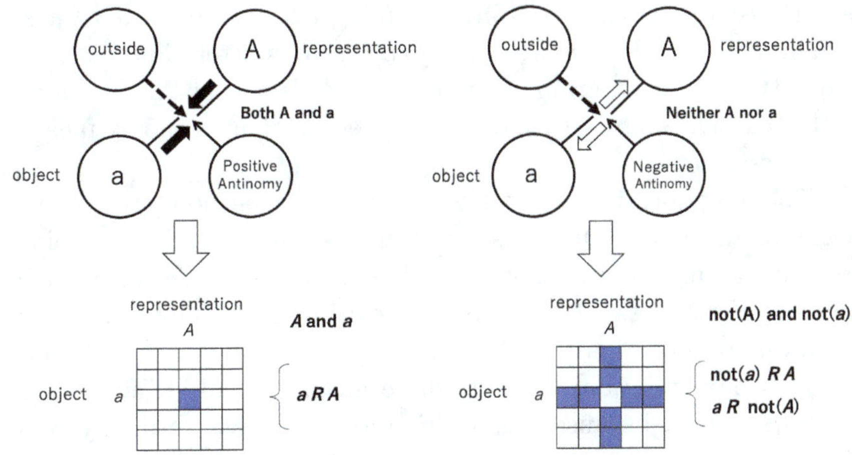

Fig. 2. Positive and negative object/representation antinomy and their relations.

Consider an extreme case of positive and negative antinomy in the form of a binary relation (Figure 2). Let a set of objects be S and a set of representations be M. It is assumed that an object has a unique representation and that the cardinalities of S and M are the same. A binary relation, $R \subseteq S \times M$ is defined to satisfy both positive and negative object/representation antinomy. Let an object be a and its representation be A. Any object is expressed by a lowercase letter, and the corresponding representation is expressed by an uppercase letter. Since positive antinomy implies that a is related to A, and vice versa, positive antinomy is expressed as aRA for any object $a \in S$. It is clear that the correspondence between an object and its representation leads to aRA being expressed as a diagonal element in R, as shown at the bottom left of Figure 2. In the binary relation in Figure 2, xRY is expressed as a blue cell at (Y, x), and no relation is expressed as a blank cell.

In contrast, since the negative object/representation antinomy implies that a is separated from A and vice versa, it can be expressed as a related to any representations, instead of A and A related to any object rather than a. If rows of the binary relation in Figure 2 are represented by X_1, X_2, A, X_3, X_4 and the columns are denoted by x_1, x_2, a, x_3, x_4, the relation allowing the negative antinomy is expressed as a set consisting of aRX_1, aRX_2, aRX_3, aRX_4, and x_1RA, x_2RA, x_3RA, x_4RA. It is expressed as a relation, as seen in the bottom right diagram of Figure 2. It is clear that the coexistence of the positive and negative object/representation antinomy falls into a contradiction.

The next question arises: How can positive antinomy coexist with negative antinomy? One possible answer is differentiation in a binary relation (Figure 3). It is easy to see that the positive a/A antinomy for any a in S is expressed as a diagonal relation, as displayed in the left-hand diagram of Figure 3. If the negative antinomy is superimposed in the binary relation of the positive antinomy, all elements of the relation vanish, because the positive and negative antinomy cancel each other out. The way to avoid canceling is to divide the domain of negative antinomy from the domain of positive antinomy. The marginal area surrounding the relation of positive antinomy in the central bottom part of Figure 3 is a binary relation of negative antinomy. Note that there is no corresponding representation for each object in the domain of the negative antinomy. Since the negative

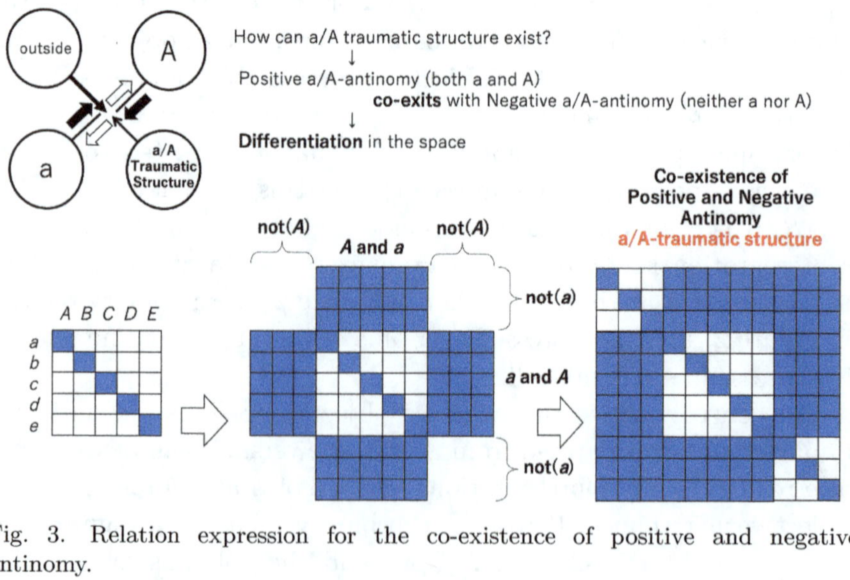

Fig. 3. Relation expression for the co-existence of positive and negative antinomy.

antinomy implies that a in S has a relation to any representations except for A, called not(A), an object a has a relation to any representations in the domain of the negative antinomy.

Hence, the marginal area surrounding the domain of the positive antinomy consists of blue cells at (Y, x) representing xRY, as seen in the diagram of the central bottom part of Figure 3. In addition to the domain of not(A), the domain of not(a) can be defined. The relation of the negative antinomy is called the background, since that domain lies at the fringe of the diagonal relation representing the positive antinomy. Due to differentiation, *kakiwari* — which coexists with positive and negative object/representation antinomy — can be expressed as a cross-like pattern, as seen in Figure 3. If any pair of an object and its representation is expressed as *kakiwari*, the object/representation antinomy, the relation between objects and representations is expressed as a binary relation repeating the cross-like structure. Finally, one can say that *kakiwari*, the traumatic object/representation structure, consists of diagonal relations and a background, as portrayed in the right bottom diagram in Figure 3.

It is remarkable that the binary relation of *kakiwari* does not imply the structure of qualia. Although positive and negative antinomy are visualized as diagonal relations and backgrounds, they belong to psychological consciousness. In contrast, qualia — belonging to phenomenal consciousness — is not visualized in the form of binary relations. While a binary relation implies a device to summon qualia from the outside of psychological consciousness, qualia are generated in a non-visualized gap between the diagonal relation and the background.

3.2. *A lattice structure derived by rough set*

Here, we evaluate a structure of logic hidden in the traumatic structure in terms of lattice theory. For that purpose, we first describe the approximation by a rough set.

Figure 4 shows the definition of a rough set and the approximation with respect to the rough set. The diagram in the inner

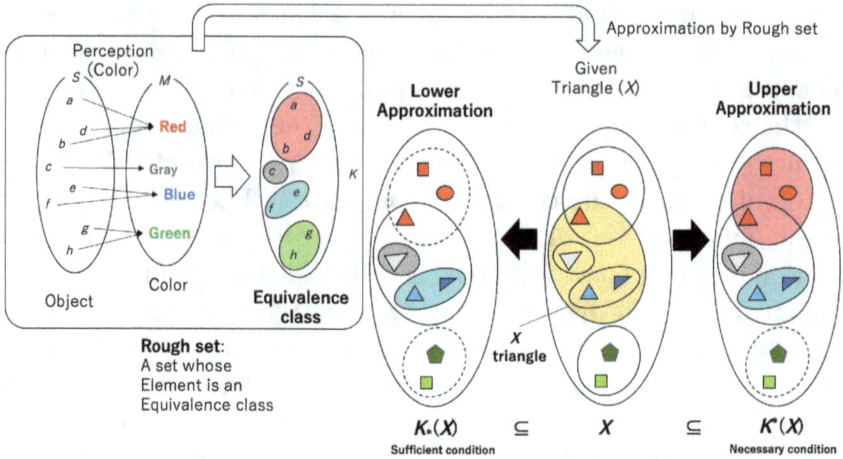

Fig. 4. A rough set as a set of equivalence classes and approximation by rough set.

box of Figure 4 displays a map, $f : S \to M$, representing a perception with respect to colors, where S is a set of objects and M is a set of representations (colors). The equivalence relation K is defined as follows. For any object x, y, xKy if — and only if — $f(x) = f(y)$. It can be straightforwardly verified that K is an equivalence relation. For any object x, an equivalence class of x is defined by $[x]_K = \{y \in S | xKy\}$. A set of objects is divided into equivalence classes that are disjoint from each other. Thus, a set of all equivalence classes is called a partition. A rough set is a set whose elements are equivalence classes with respect to K. In the inner box of Figure 4, $[a]_K = \{a, b, d\} = [b]_K = [d]_K$, $[c]_K = \{c\}$, $[e]_K = \{e, f\} = [f]_K$, and $[g]_K = \{g, h\} = [h]_K$. In a rough set, elements of an equivalent class are indiscernible from each other. It is assumed that an agent equipped with K cannot distinguish a from d, and that the agent recognizes just the red.

By using a rough set, one can approximate any subset of a set of objects. Assume that each object of a set of objects has two kinds of attributes — shape and colors — as depicted in Figure 4. For instance, element a is assumed to be a red square, and element

e is a blue triangle. It is easy to see that due to the equivalence relation, K, a partition with respect to colors, is obtained. Let X be a subset of S; that is, a set of triangles. Rough set approximation with respect to K implies the concept of a triangle in terms of colors. There are two kinds of rough set approximations: the lower approximation, represented by $K_*(X)$ and the upper approximation, represented by $K^*(X)$. The lower approximation of X is defined by a set of elements of equivalence classes included by X, such that $K_*(X) = \{x \in S | [x]_K \subseteq X\}$. The upper approximation of X is defined by a set of elements of equivalence classes that have common elements with X, such that $K^*(X) = \{x \in S | [x]_K \cap X \neq \emptyset\}$. In the case of Figure 4, for $X = \{b, c, e, f, K_*(X) = [c]_K \cup [e]_K = \{c, e, f,$ and $K^*(X) = [b]_K \cup [c]_K \cup [e]_K = \{a, b, c, d, e, f\}$. Since it is straightforwardly verified that $K_*(X) \subseteq X \subseteq K^*(X)$, one can see that $K_*(X)$ is a sufficient condition for X, and that $K^*(X)$ is a necessary condition for X.

By using the upper and lower approximations, one can collect a subset satisfying a necessary and sufficient condition, such that $L_{\mathrm{B}} = \{X \subseteq S | K_*(K^*(X)) = X\}$. Note that suffix B represents Boolean algebra, to be discussed later. It is easy to see that any equivalence class $[x]_K$ satisfies $K_*(K^*([x]_K)) = [x]_K$. Thus, any equivalence class is an element of L_{B}. Indeed, any union of equivalence classes, X — including S itself and the empty set — satisfies $K_*(K^*(X)) = X$. This implies that L_{B} is a power set of a set of equivalence classes, and that L_{B} is a set lattice. A lattice is a kind of algebra derived from a partially ordered set. An ordered set is equipped with the order relation between elements. The order relation, $\leq$, is defined by reflective law, $x \leq x$; antisymmetric law, if $x \leq y$ and $y \leq x$, then $x = y$; and transitive law, if $x \leq y$ and $y \leq z$, then $x \leq z$. A lattice, L, is defined by an ordered set such that for any elements $x, y \in L$, meeting x and y, joining x and y are also elements of L. The meat of x and y, represented by $x \wedge y$, is defined such that if $z \leq x$, $z \leq y$, then $z \leq x \wedge y$. The joining of x and y, represented by $x \vee y$, is defined such that if $x \leq z$, $y \leq z$, then $x \vee y \leq z$. Since elements of $L_{\mathrm{B}} = \{X \subseteq S | K_*(K^*(X)) = X\}$ are subsets of

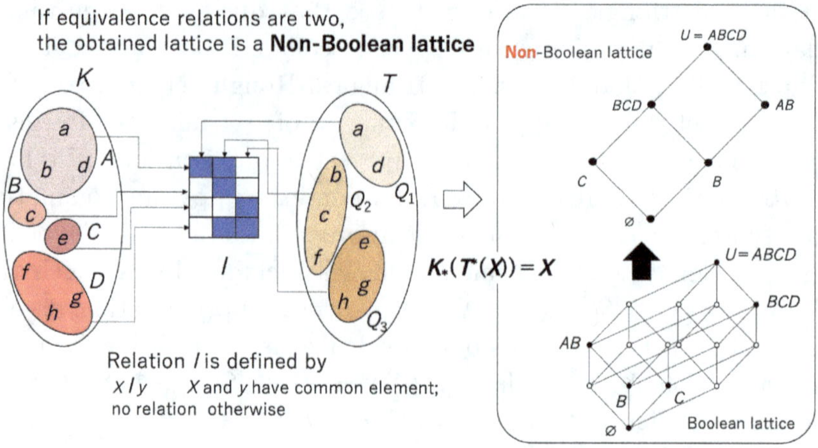

Fig. 5. Non-Boolean lattice obtained by $L = \{X \subseteq S | K_*(T^*(X)) = X\}$.

S, the order relation is defined here by inclusion. In particular, in a set lattice, joining coincides with union and meeting coincides with inclusion. It is known that a set lattice is Boolean algebra.

If an equivalence relation is uniquely determined, both the upper and lower approximations are obtained by using the same equivalence relation. There are, however, ambiguous situations such that multiple equivalence relations coexist. Figure 5 reveals a situation in which there are two equivalence relations: K and T. Equivalence classes derived by K are named A, B, C, D, while those derived by T are named Q_1, Q_2, Q_3. Under this situation, one can construct the necessary and sufficient condition by using two different equivalence relations, K and T, such that $L = \{X \subseteq S | K_*(T^*(X)) = X\}$. L is a lattice that is generally a non-Boolean lattice, and lattices of a collection of subsets satisfying a fixed point, $K_*(T^*(X)) = X$, $T_*(K^*(X)) = X$, $K^*(T_*(X)) = X$, $T^*(K_*(X)) = X$, are isomorphic to each other. This implies that the structure of a lattice is the same whether the upper or lower approximations are exchanged and K and T are exchanged.[20,23]

In Figure 5, a lattice is obtained as $L = \{X \subseteq S | K_*(T^*(X)) = X\}$. It is easy to collect subsets satisfying a fixed point by using a relation, I defined by the following: If X is an equivalence class derived by K, and y is an equivalence class derived by T, and they share common elements of a universal set, XIy; otherwise, there is no relation between two equivalence classes. Since $A = \{a, b, d\}$ and $Q_1 = \{a, d\}$, they have the common element a, which leads to AIQ_1. For $Q_3 = \{e, g, h\}$, there is no common element between A and Q_3. Thus, there is no relation between A and Q_3. A lattice, $L = \{X \subseteq S | K_*(T^*(X)) = X\}$, and a power set (Boolean lattice) obtained from a set of equivalence classes, $A = \{a, b, d\}$, $B = \{c\}$, $C = \{e\}$, $D = \{f, g, h\}$, are presented in the form of a Hasse diagram. The loss of some elements from a power set realizes a non-Boolean lattice of L. In addition to Figure 5, elements of a lattice are abbreviated by removing the symbol of the union.

In the next section, we recall a binary relation revealing the traumatic structure. In the above section, we describe that the inside (artist motivation) is related to the outside (production made of materials) by the code of creativity. We describe perception from a set of objects to a set of representations in the same framework, in which objects are on the outside and representations are generated on the inside of an observer. However, such a dichotomy of the inside and outside is superficial, since the outside is described by relating with the inside. This leads to the insight that a dichotomy of the inside and outside constitutes a framework to be deconstructed. This implies that both an object and a representation are equivalence classes, while a universal set is unknown. Under such a situation, the binary relation between the object and representation is not founded by sharing elements of a universal set, but is determined ad hoc. This suggests that the relation between objects and representations is determined by itself, although it cannot be determined in principle. This situation is expressed by the traumatic structure mentioned in Section 3.1.

3.3. *Kakiwari as a lattice, almost disjoint union of Boolean algebras*

Recall that in *kakiwari* the binary relation shows the traumatic object/representation structure. It consists of multiple diagonal relations and a background. Figure 6 indicates that any diagonal relation reveals Boolean algebra, even if the diagonal relation is surrounded by a background. A binary relation is assumed to be a relation between equivalence classes derived by equivalence relation K, and equivalence classes derived by equivalence relation T. The former corresponds to an object, while the latter corresponds to representation. Under that assumption, a collection of unions of equivalence classes satisfying $K_*(T^*(X)) = X$ leads to a lattice.

In Figure 6, objects are named, $a_1, a_2, \ldots, a_{12}$, and representations are named, $A_1, A_2, \ldots, A_{12}$. The relation consists of 3×3, 5×5, and 4×4 diagonal relations and a background. It is easy to see that the $n \times n$ diagonal relation is transformed to 2^n-Boolean algebras, even if the diagonal relation is surrounded by a background. Assume that the 3×3 diagonal relation is isolated from other relations (i.e., without a background). It is verified that each equivalence class satisfies $K_*(T^*(X)) = X$ such that $K_*(T^*(a_1)) = K_*(A_1) = a_1$, and that any union of equivalence classes also satisfies the fixed point, such that $K_*(T^*(a_1 \cup a_2)) = K_*(A_1 \cup A_2) = a_1 \cup a_2$ and $K_*(T^*(a_1 \cup a_2 \cup a_3)) = K_*(A_1 \cup A_2 \cup A_3) = a_1 \cup a_2 \cup a_3 = u_1$.

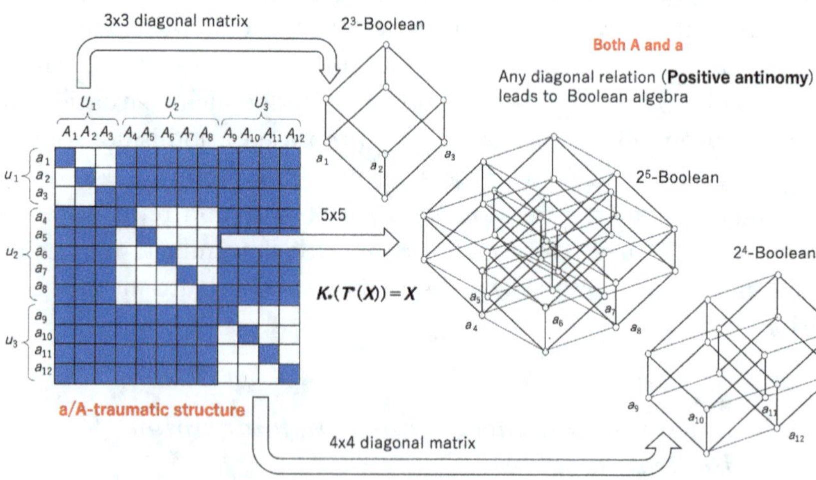

Fig. 6. Constructing lattices corresponding to positive antinomy.

The empty set also satisfies the fixed point such that $K_*(T^*(\oslash)) = \oslash$. For the sake of convenience, we abbreviate some unions of equivalence classes such as $a_1 \cup a_2 \cup a_3 = u_1$, $a_4 \cup \cdots \cup a_8 = u_2$, $a_9 \cup \cdots \cup a_{12} = u_3$, and $A_1 \cup A_2 \cup A_3 = U_1$, $A_4 \cup \cdots \cup A_8 = U_2$, $A_9 \cup \cdots \cup A_{12} = U_3$. Next, recall the diagonal relation surrounded by a background, as seen in Figure 6. It is easy to note that units of equivalence classes satisfy the fixed point, such that $K_*(T^*(a_1)) = K_*(A_1 \cup U_2 \cup U_3) = a_1$. Note that the calculation of the upper approximation leads to $U_2 \cup U_3$, which is a background. Even if $T^*(a_1)$ is accompanied by the background, $A_1 \cup U_2 \cup U_3$ is distinguished from any other row. Therefore, $K_*(A_1 \cup U_2 \cup U_3)$ is uniquely determined. However, $T^*(u_1) = U_1 \cup U_2 \cup U_3 = U$, and then $K_*(U) = U$. This implies that the largest elements of 2^3-Boolean algebras are shared by all 2^3-, 2^5-, and 2^4-Boolean algebras. The least element is also shared. It is easy to observe that the union of equivalence classes in different diagonal relations does not satisfy the fixed point, such that $K_*(T^*(a_1 \cup a_5)) = K_*(U_1 \cup U_2 \cup U_3) = u$. Finally, all elements of 2^3-, 2^5-, and 2^4-Boolean algebras appear in the underlying lattice, except for the least and the greatest elements, and no other unions of equivalence classes are obtained.

Figure 7 reveals the final construction of a lattice derived from the binary relation, demonstrating the traumatic structure. Since the least and greatest elements are common for all Boolean algebras, a lattice, finally obtained, is not an almost disjoint union of Boolean algebras, and is called a disjoint union of Boolean algebras (ADBA).[22] ADBA is a necessary condition for an orthomodular lattice. If the Hilbert space is equipped with an orthomodular lattice, it implies quantum logic. In that sense, ADBA is regarded as quantum-like logic. Although one can realize classical logical thinking in each Boolean algebra in ADBA, it fails if terms in different Boolean algebras are mixed. ADBA has the property that a complement of a set is not uniquely determined.

In a lattice implementation, the positive antinomy is visualized as local logic corresponding to Boolean algebras, and the negative antinomy is visualized as glue (the least and greatest element)

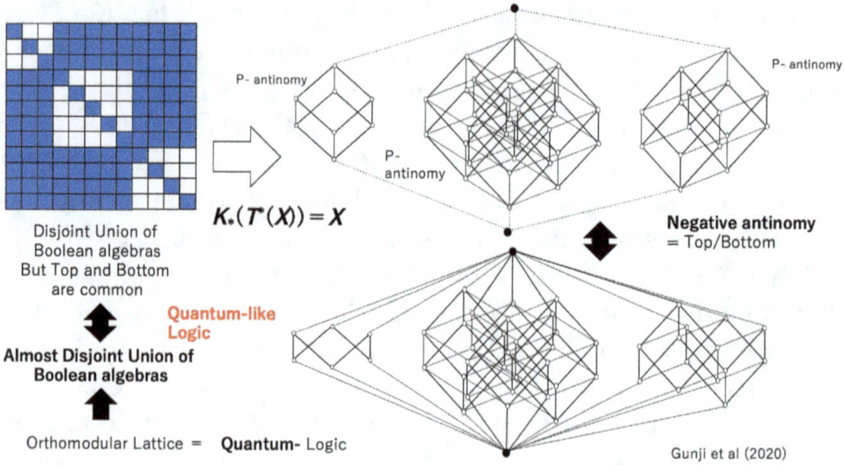

Fig. 7. An almost disjoint union obtained from the traumatic structure.

pasting together Boolean algebras. Such pasting brings forth intrinsic indeterminacy in cognition.

4. Creativity in Art, with and Through *Kakiwari*

One of the authors, Nakamura, found a situation in which both positive and negative antinomies could coexist in Chilean *animita*, a kind of grave marker. An *animita* is a small house the size of a doghouse that is widely worshiped in Chile. If you drive along the road in Santiago, Chile, you will see many *animitas*. Some are elaborate miniature houses with triangular roofs and glass windows, others are dome-shaped shrines, some are overgrown with plants, and some are extravagant in design. Many offerings accompany them such as stuffed animals, plastic bottles, artificial flowers, fresh flowers, rosaries, candlesticks, portraits, framed pictures of supposed saints, and small plaster figures. *Animita* is a word used in Chile and comes from the word *ánima*, meaning soul. The most commonly held theory is that the *animita* is a house built for the souls of those who have died tragically, such as unexpected deaths.[11] It is believed that souls who die a tragic death cannot find their way to heaven and wander in torment, so the *animita* is a stopover for wandering souls in the middle of the human and soul worlds. The *animita* is built at the site of death, separate from the grave. Since unexpected deaths are often caused by traffic accidents, *animitas* are often found along roads.

The *animita* can be classified into two types.[21] Upon death, it is often the deceased's next of kin who first installs the *animita*, and it is the next of kin who prays regularly. Since the deceased is of no relevance to the general public, it is not something visited by an indefinite number of people. Such an *animita* is considered an "impersonal *animita*". It is a grave marker for an individual who worships the fullness of a person's life, and the truth of death does not cut off the relationship with the deceased. On the other hand, there is the "personal *animita*", which is visited and prayed at by an unspecified number of people who are not related to the deceased. The personal *animita* itself becomes "personalized" independent of the deceased. For example, the most famous personal *animita* in Santiago is that of Romualdito, which attracts worshipers from Chile and other countries. One feature of the personal *animita* is a plaque's dedication by the worshipers, upon which words of gratitude are inscribed. On these plaques, we see the transition from impersonal to personal. Most of the personal *animitas* are of deceased who had tragic deaths. In the case of Romualdito, it is said that he was assaulted and killed on San Borja Street, Santiago, in 1930. The extraordinary power of the tragic soul (i.e., the tragic nature of a death) gave rise to the folk belief that it can call forth miracles. The *animita*, initially impersonal, becomes an object of worship once it is discovered that it can evoke miracles that bring benefits.[11] As a result, many worshipers come and light candles to receive benefits, such as healing from illness. When people sanctify the *animita* beyond the tragic soul as part of their faith in God, the *animita* begins to take on a personality of its own from its role as a grave marker for the dead individual.

When the *animita*'s transition from impersonal to personal occurs, it is believed that the soul of the dead begins to respond to the prayers of worshipers, and then mediates between worshipers and saints.[11] However, we found a mechanism that is different from the theory that the *animita* becomes personalized (created) by transforming into an interface for mediation of the outside, such as saints, as a factor that manifests the sacred. This is the "*kakiwari*-zation" of animita. *Kakiwari* is an expression of mountain scenery that can be found in Japanese art from ancient times, especially in the *Rimpa* school, and its realism contains a different view from the reality of perspective, which has been commonly used since the Renaissance until the present.[15,16] The semicircular shape of the mountain range is a flat representation, such as a papier-mâché scrawl of a stage background or a paper puppet (Figure 8 on page 157–158).

Unlike the realism of perspective that shows the overside of the mountain, *kakiwari*'s mountain blocks the overside of the mountain. From the viewpoint of *kakiwari*, there is no perceivable world on the other side that can be seen from this side. *Kakiwari's* mountains are not concerned with a simple perspective because *kakiwari* does not have a back side. *Kakiwari* strips away the concept of distant/ nearness from the landscape, allowing for the complete outside behind it; rather, *kakiwari* reveals the outside. *Kakiwari* is a device that concretely realizes the coexistence of positive and negative antinomies. A series of mountains, like a sheet of paper — which is abstract but forms a real landscape — acts as a positive antinomy to demonstrate the connection between the relationship between this side and that side, while at the same time connecting to the outside, which means disconnection in the sense of being inaccessible. In this vein, it realizes the coexistence of positive and negative antinomies, connecting while disconnecting.

We found the same method as *kakiwari* in the way the *animita* is personalized. When the death of a deceased person is so tragic, only the tragedy of the death is remembered by people, and the life of the deceased — even the details of the death — are forgotten. In this way, only the death is focused on. In fact, even in the *animita* of Romualdito, the descriptions of the dead that are loved are believed to be different from people. Romualdito was widely believed to be a boy when he may have been 41 years old. It is believed that he may have been mistaken for a boy because of his mental disorder.[11]

He has been worshiped without knowing for sure how he died or what he looked like. Even the names inscribed on the plaques are slightly different for each worshiper. At least 20 names are given including Romualdito, Raymundito, Reinaldo, Remialdito, Reynaldo, Rinaldo, Romualdo, and so on. The worshipers pray to the same *animita* under different names, regardless of the deceased being worshiped. The number of worshipers has increased, and many *animitas* are erected for only a single person such as Romualdito.

Impersonal *animitas*, originally built to think of the dead, are considered to have power in the grandeur and misery of death, and are seen as sacred. Rumors may spread that worshipers have benefited from the power of death. At this time, by focusing only on death, the meanings of before and after life, related to the deceased of the *animita*, are stripped away. This implies that the *animita* was originally built as a grave marker by which the history of the deceased (before death) was recalled and the soul was mourned (after death), and that the *animita* is accepted both before death and after death. This is positive antinomy. Personalization is realized by forgetting who the deceased is, or by evaporating before death and after the death of the deceased, which implies negative antinomy. Thus, the *animita* is personalized via *kakiwari*, a traumatic structure (Figure 9). In other words, like the aforementioned mountain of a Japanese painting, the positive antinomy of the *animita* — which is strongly connected with the pre- and post-mortem meanings of the deceased — is decolorized because of its intensity, resulting in a negative antinomy. This is the "*kakiwari*-zation" of the animita. As an interface that simply acts as a mediator, the coexistence of positive and negative antinomies cannot occur. Since the *animita* becomes the body of *kakiwari* that realizes coexistence, each worshiper can independently encounter the sacred individual. It is not a public Romualdito with a predetermined meaning, but a myriad of personal "Romualdito things" for "me" that mean nothing to others, except for the individual worshiper. Because the *animita* is a "*kakiwari*-ized" personal *animita*, a faith-gathering *animita* is created, and the worshipers benefit from its creativity.

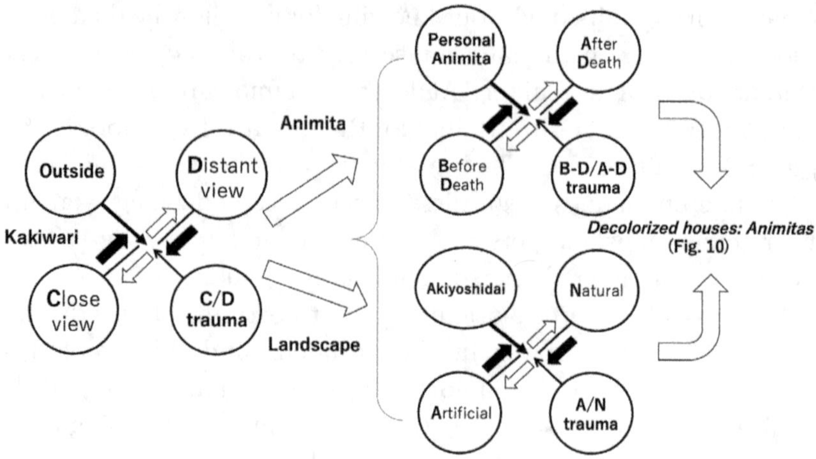

Fig. 9. Structure of *kakiwari* regarding personal *animita*.

Kakiwari's point of view — which decolorizes the concept of perspective and summons the outside — shows the mechanism of the *animita*'s creativity as personalization by *kakiwari*. Nakamura depicts this structure, in which the personal *animita* is summoned by decolorizing the meaning of the dead before and after death as a landscape of "*kakiwari*-ized" *animitas* (Figure 8). The work shown in Figure 8, *Decolorized houses: Animitas*, is set in Akiyoshidai, Yamaguchi Prefecture, Japan. Inspired by the landscape of limestones lining the karst plateau of Akiyoshidai, Nakamura conceived of *animitas* of *kakiwari* as landscape paintings. When the *animita* loses its relationship with the dead and the life of the dead is decolorized, the dead evaporate and are stripped of their meanings and concepts of life before and after death. In this way, the *animita* becomes empty, or in other words, it becomes the body of *kakiwari*, demonstrating its potentiality as the outside entity and becoming personal. This is where the real pleasure of refining trauma lies. Nakamura is currently researching the *kakiwari* of the human spirit, which she depicted as *kakiwari* girls inspired by the rituals and spirits of lost people in the southernmost part of Chile (Figure 10).

Fig. 8. NAKAMURA Kyoko, *Wait for Flower-Holding Birds*, color on silk, hanging scroll, 153 × 50 cm, 2018. Painting mounted by Sōgadō, Tokyo, Japan.

Fig. 8. (*Continued*)

Fig. 10. NAKAMURA Kyoko, *Decolorized houses — Animitas*, color on silk, hanging scroll, 88.5 × 85 cm, 2020. Painting mounted by Sōgadō, Tokyo, Japan.

Fig. 10. (*Continued*)

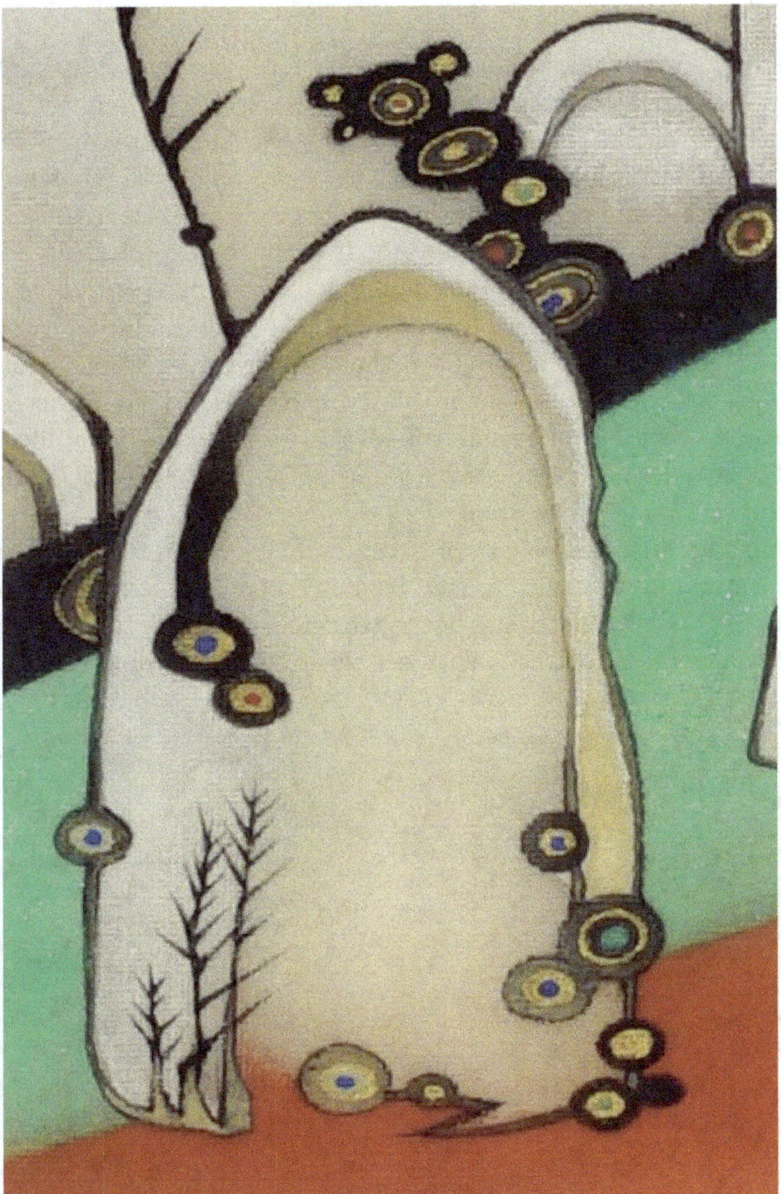

Fig. 10. (*Continued*)

5. Conclusion

We start from the statement of which form of creativity is intrinsically personal and neither public nor objective. If creativity is considered objective, the problem of whether AI has creativity makes sense. This idea leads to creativity that can be described as a code. Our stance — which opposes the idea of a code of creativity — claims that creativity is not to be described, but to be experienced.

Creativity to be experienced does not lead to the problem of what creativity is, but to the problem of how creativity can be experienced. Since creation is destined to be an emergent event, creation is nothing but accessing the outside of our perceived world. While creation looks like an unusual event, it is always found in everyday life. One of the most common creations is feeling reality or feeling qualia in perception and/or cognition.

If matching the object of a cat with its representation implies understanding the cat, it can be implemented on a conventional computer. Only matching never contains reality or qualia. While one manages to match an object with representation, the object

Fig. 11. NAKAMURA Kyoko, *Kakiwari girls — Tanu*, color on silk, hanging scroll, each 84.5 × 30 cm, 2020. Painting mounted by Sōgadō, Tokyo, Japan.

Fig. 11. (*Continued*)

Fig. 11. (*Continued*)

Fig. 11. (*Continued*)

is perpetually separated from representation. The former aspect is called positive object/representation antinomy, and the latter is called negative antinomy. The device satisfying both positive and negative antinomy is called *kakiwari*, which carries the traumatic structure. We think that *kakiwari* destroys the framework consisting of objects and representations, and that something is summoned from outside of the framework (i.e., qualia).

Here, we define *kakiwari* as a binary relation; that is, nothing but the aspect in which qualia are generated. By using rough set lattice techniques, the binary relation is transformed to a specific lattice called a nearly disjoint union of Boolean algebras; that is, quantum-like logic.

Creation through *kakiwari* is illustrated through Japanese paintings of the *Rinpa* school, which is the etymology of *kakiwari*. Creation through *kakiwari* is also illustrated by an anthropomorphic event: *animita* in Chile. In particular, personalization of the *animita* yields an excellent illustration of creation via *kakiwari*. On the one hand, *animita*, a grave marker, satisfies the time before and after the death of the deceased; on the other hand, some *animitas* acquire evaporation of the time before death and after death, which satisfies neither before death nor after death. Only *animitas* satisfying both positive and negative antimony can become personalized and become objects of faith.

One of the authors, Nakamura, took *kakiwari* and *animitas* as the theme of painting and created her own art pieces, which contain the significance of creativity in a double-layered fashion because they are *kakiwari* in their own right and the theme of *kakiwari* is also *kakiwari*. Viewers are trapped in double-layered *kakiwari* and can create their own emotions when confronting artwork.

Acknowledgment

This work was supported by JSPS JPJS00120351748 and KAKENHI 18K18478.

References

1. S. Kripke, *Wittgenstein on Rules and Private Language: An elementary Exposition.* Harvard University Press, Cambridge, USA, 1982.
2. M. A. Boden, Creativity. In *Artificial Intelligence, Handbook of Perception and Cognition*, M. A. Boden, editor. Academic Press, New York, USA, 1996, 267–291.
3. M. A. Boden, Creativity in a nutshell. In *The Creative Mind; Myths and Mechanisms.* Routledge, Oxford, UK, 2004, 1–10.
4. M. D. Sautoy, *The Creativity Code: Art and Innovation in the Age of AI.* Belknap Press, Harvard, USA, 2019.
5. Q. Meillassoux, *Après la finitude. Essai sur la nécessité de la contingence,* Éditions du Seuil, 2006 (Japanese translation by Chiba, M. *et al.*, 2016, Jinbun sho-in), Tokyo, Japan.
6. M. Gabriel, *Warum es die Weld nicht gibt.* Ullstein Buchverlage, 2013 (Japanese translation by K. Shimizu, 2018, Kodansha Metie), Tokyo, Japan.
7. Y. P. Gunji, K. Nakamura, Psychological origin of quantum logic: An orthomodular lattice derived from natural-born intelligence without hilbert space. *BioSystems.* 215–216, 104649, 2022.
8. Y. P. Gunji, K. Nakamura, M. Minoura and A. Adamatzky, Three types of logical structure resulting from the trilemma of free will, determinism and locality. *BioSystems.* 2020. DOI: 10.1016/j.biosystems.2020.104151.
9. D. J. Chalmers, *Conscious Mind: In Search of a Fundamental Theory.* Oxford University Press, New York, NY, 1996.
10. D. J. Chalmers, Panpsychism and panprotopsychism. In *Consciousness in the Physical World: Perspectives on Russelian Monism*, T. Alter and Y. Nagasawa, editors. Oxford University Press, New York, USA, 2015, 246–276.
11. O. Plath, *L Animita: Hagiografía Folclórica.* Fondo de Cultura Económica, Santiago, Chile, 2012.
12. B. Libet, *Mind Time.* Harvard University Press, Cambridge, MA, 2004.
13. S. Shimojo, Postdiction: Its implications on visual awareness, hindsight, and sense of agency. *Front. Psychol.* 5:00196, 2014.
14. Y. P. Gunji, *Natural Born Intelligence (Tennen Chinou).* Kodansha Metie, (in Japanese), Tokyo, Japan, 2019.
15. K. Nakamura and Y. P. Gunji, *TANKURI: Shooting Creativity.* Suisei-sha, 2018 (in Japanese), Tokyo, Japan.
16. K. Nakamura, De-creation in Japanese painting: Materialization of thoroughly passive attitude. *Philosophies.* 6(2):35, 2021.
17. B. van der Kolk, *The Body Keeps the Score: Brain, Mind, and Body in the Healing of Trauma.* Viking Press, 2014 (Japanese translation by Shibata, H; Sugiyama, T., Kinokuniya shoten, Tokyo, Japan, 2016).
18. C. Pallant, *Contact Improvisation, An Introduction to a Vitalizing Dance Form.* McFarland & Comp. Inc. Pub., 2006.

19. G. Agamben, *Bartleby, la formula della creazione, Quodlibet*, 1993. (Japanese translation by Takahashi, K. Getsuyo-sha, Tokyo, Japan, 2015).
20. Y. P. Gunji and T. Haruna, A non-boolean lattice derived by double indiscernibility. *Transactions on Rough Sets XII*, 211–225, 2010.
21. Y. Mihara, Animita Worship in Chile and Argentina. *Rev. Inquiry Res.* 43:119–144, 1986 (in Japanese).
22. K. Svozil, *Randomness and Undecidability in Physics*. World Scientific, Singapore, 1993.
23. Y. Y. Yao, A comparative study of formal concept analysis and rough set theory in data analysis. *Lect. Notes Comput. Sci.* 3066:59–68, 2004.

https://doi.org/10.1142/9789811257155_0008

Chapter 8

What Universal Turing Machines Might Tell Us

Maurice Margenstern

Université de Lorraine,
LGIPM, 3 rue Augustin Fresnel, BP 45112,
57073 METZ, Cedex 03, France
margenstern@gmail.com
maurice.margenstern@univ-lorraine.fr

The present chapter tries to establish a comparison between some kind of universal Turing machines with viruses, like influenza virus or SARS-Cov2 virus.

1. Introduction

It is certainly a challenge to compare Turing machines, possibly universal ones, with such biological phenomena as viruses. In one of my papers, sorry, I do not remember in which one, I wrote that if we encode one of the smallest universal Turing machines in a DNA way, we get something whose size is much smaller than the smallest virus. As far as a universal Turing machine has an unpredictable behavior, the parallel we indicated suggested that the behavior of a virus is more unpredictable. At that time, I did not argue more than those few words which, clearly, did not raise any noise. As far as at the present moment people are more concerned with what we know about viruses, it can be interesting to go back to that comparison and to discuss whether it is relevant or not.

In Section 2, I sketchily recall what a Turing machine is with as less as possible technicalities. Readers knowing those definitions may skip that section. In Section 3, I also sketchily remind the reader about universal Turing machines and about the consequence of those in terms of algorithmic unsolvability, which is performed in Section 3.1. In Section 3.2, I mention what is usually meant by small universal Turing machines. Eventually, in Section 3.3, I define what I call a **pedagogical** universal Turing machine. In Section 4, I discuss whether the comparison indicated in the previous paragraph is relevant or not. In Section 5, we put a temporary end to the dispute.

2. Turing Machines

For a reader familiar with Turing machines, I just mention that, in the paper, I only consider deterministic Turing machines with a single head and a single bi-infinite tape. I also assume the initial configuration to be finite, which means that outside some segment of the tape, all its squares contain the **blank symbol**, indicating that the corresponding square does not contain any information.

A Turing machine is an ideal object you have to imagine with possibly the help of paper and of a pencil. The basic component of a Turing machine is its tape which is divided into **squares**. There are as many squares as it is needed at each step of the computation which follows a discrete clock whose count starts with the first step of the computation. The computation is materialized by the behavior of a **head** over the tape. At each tip of the clock, the machine head is in some **state**, think to a state of mind as Turing indicated in his fundamental paper, see Ref. 1, and the head is looking at the content of a square which is called the **scanned** square. According to what the head can see in the scanned square and according to its state, the head performs three actions: firstly, it replaces the content of the scanned square by another content, possibly the same one; secondly it moves to a neighboring square, either to right or to left or it remains scanning the same square; thirdly, the head changes its head to a new one. The content of a square is called a **letter** which belongs to some finite **alphabet**, we denote A_M, including the blank symbol. The state of the head belongs to a certain finite set we call the **set of states**, denoted by Q_M. From those assumptions we can see that the three actions above indicated can be gathered in what is called an **instruction**. As far as an instruction depends on two pieces of information, namely a letter and a state, it is possible to arrange into a table the set of instructions which may be applied by the head to the tape. Such a table is called the **program** of the machine. Among the instructions, it is expected that at least one instruction contains the **halting** state, which orders the machine to stop the computation. If it is not the case, once started, the computation goes on for ever.

We have no room in that paper to display tables of Turing machines. We refer the reader to Ref. 2. In place of tables, we shall display configurations of the tape when we explain the working of a particular Turing machine.

Let us provide here a simple example.

The Turing machine A we here consider starts from the following initial configuration:

$$\_ * |^n * |^m * \_ \tag{1}$$

In that configuration, $\_$ represents the blank symbol. We can see two blocks of $|$: one with n symbols, the others with m ones. The blocks are on both sides delimited by an $*$. The goal of A is to compute $n+m$. A simple solution consists in replacing the $*$ between both blocks of $|$ by a $|$ and then erase the rightmost $*$ and replace the rightmost $|$ by an $*$. However, A performs that computation without erasing its initial data, as far as the final configuration is

$$\_ * |^n * |^m * |^{n+m} * \_ \tag{2}$$

I say that A is a **courteous** Turing machine: it does not destroy the data and it places the result beside the data, to its right-hand side, separated from it by a suitable symbol. Moreover, it is required that there is a square σ such that the head never goes to the left-hand side of σ.

The working of the machine is rather simple: it copies each $|$ of the blocks to the right-hand side of the rightmost $*$. The copy consists in marking the copied symbol and removing the mark when the head proceeds to the next symbol to be copied. When it is completed, the machine writes the rightmost $*$ of (2) and it stops its computation on the leftmost $*$ of (2).

3. Universal Turing Machines

Let us now turn to universal Turing machines. What does *universal* mean? A Turing machine is called **universal** if it is able to simulate any other Turing machine. Let M be a Turing machine and C be an initial configuration for M. The **computation** of M starting from C is a sequence $\{C_k\}_{k\geq 0}$ of configurations for M such that $C_0 = C$ and C_{k+1} is obtained from C_k by the application of the single instruction of M which can be applied to C_k. The computation is a finite sequence if and only if the last configuration in the sequence contains the halting state. Such a sequence is called a **final configuration**. A machine U is said to be universal if, for any Turing machine M and for any initial configuration C for M, there is a configuration K of U such that the computation $\{K_n\}_{n\geq 0}$ of U starting from K contains a sub-sequence $\{K_{n_k}\}_{k\geq 0}$ such that $K_{n_k} = \tau(C_k)$ where τ is a encoding of configurations for M and of its program which does not depend either on C or on M.

We stress that in its definition, a universal Turing machine U must be able to simulate *any* Turing machine M, even if the number of states of M is bigger than that of U, even if the size of A_M is bigger than the size of A_U. Turing proved that it is possible and the key point for that is the construction of τ and of K.

For that purpose, K is split into two parts: to the left, the encoding of the program of M, to the right, the encoding of the configuration of M. The program of M is displayed on the tape state by state, the area devoted to a state being marked by a left-hand side `X`. In between two consecutive `X`s, the instructions associated to the state marked by the left-hand side `X` are displayed according to the order of A_M, each instruction being delimited by a left-hand side `Y`. The configuration of M is displayed to the right-hand side of the area devoted to the program of M. It displays the squares of the configuration of M, in the same order of that configuration, each square being delimited by a left-hand side `U`. The size of a square of M is defined by the distance between two consecutive $\mathtt{U}$, except for the last square where the second $\mathtt{U}$ has to be replaced by the blank of U. Note that $\mathtt{X}$, $\mathtt{Y}$, and $\mathtt{U}$ belong to A_U. The letters of A_M and the states of Q_M are encoded by the binary representation of their rank in A_M and Q_M, respectively. If P is the number of letters in A_M, the size of a square of M on the tape of U is the size of the binary representation of P plus one. In an instruction, the encoding of the new state is the binary representation of its rank in Q_M.

3.1. *Consequences of universality*

For the reader who is not familiar with Turing machines, I recall the main theorem which was proved by Turing in his fundamental paper.[1]

Theorem 1 (Turing, see Ref. 1). *There is no machine Turing H which applied to any machine M and to any initial configuration C of M in its alphabet would say whether the computation of M starting from C eventually halts or not.*

The result is often called the undecidability or algorithmic unsolvability of the halting problem. Although elementary, I skip the proof of that important theorem. Intuitively, the theorem entails that most problems about the behavior of a Turing machine cannot be solved by a Turing machine as far as for most of such problems, if we could solve them with the help of a Turing machine, we could solve the halting problem. Informally, I shall say that the behavior of a Turing machine on its tape is **unpredictable**. We can explain the undecidability of the halting problem as follows. With the help of a universal Turing machine, we can construct such a Turing machine W, starting from a configuration containing a code of M, of C, an initial configuration i for M, and an integer k, the Turing machine writes 1 if the machine M stopped its computation after k steps and the W writes 0 if it is not the case. Clearly, M does not stop if and only W always writes 0. We can check that M stopped if it is the case after waiting long enough. But the opposite situation can be observed if and only if we have observed the computation for all steps k, which is impossible.

From the proof of Theorem 1, we get the following result:

Theorem 2. *If U is a universal Turing machine, the computation of U over $\tau(U)$ does not stop, where τ is the encoding used by U for the simulation of Turing machines computing on their initial configurations.*

Note that from the existence of a universal Turing machine, we can construct infinitely many universal Turing machines. Indeed: assume that we have the program of a universal Turing machine U. We can construct a Turing machine U_1 which computes A for a given initial configuration and then computes U starting from another part of the final configuration reached by A. Clearly, U_1 is also a universal Turing machine. We can thus construct a sequence $\{U_n\}_{n\in\mathbb{N}}$ of Turing machines such that $U_0 = U$ and U_{n+1} is composed with U_n and A exactly as was done with U and A. Clearly, the program of U_{n+1} is bigger than that of U_n so that the programs of the U_ns are bigger and bigger. Accordingly, it is easy to construct bigger and bigger programs.

In Section 3.3, we shall see that it is possible to construct U with 87 states and 16 letters. In Section 3.2, we shall see that we can get a similar result with much less states and much less letters.

3.2. *Small universal Turing machines*

At the end of Section 3.1, we have seen that it is easy to make a universal Turing machine with a program as big as we wish. Can we make smaller and smaller universal Turing machines? The answer cannot be an infinite sequence of universal Turing machines with smaller and smaller programs. Certainly because the program has at least one instruction and it must have a bit more according to the following result:

Theorem 3 (see Ref. 3). *There is a Turing machine which says for any Turing machine M with two letters and two states whether*

M stops or not its computation starting from any given starting configuration.

Accordingly, there are finitely many machines between the universal machine with 87 states and 16 letters and the non-universal ones with two states and two letters. It is true, but for each such machine there are infinitely many possible initial configurations so that it is impossible to check that possibility. Moreover, from Theorem 2, we know that for any universal Turing machine U there is always a configuration on which its computation does not stop. For a machine about which we do not know whether it is universal or not, checking the above is impossible.

However, there are universal Turing machines with much less than 87 states and much less than 16 letters. Reaching the smallest universal Turing machines was a source of important works from 1950. The best result up to now is probably given in Ref. 4. The interested reader can find an account of that race in Ref. 5 and a more recent one in Ref. 6. We have to note that the machines of Ref. 4 do not directly simulate a Turing machine. It simulates another system of computation where that latter system computes a Turing machine. It is what can be called a *delayed* computation of a Turing machine. Explaining how such an intermediate system is built implies too many technicalities so I skip them referring the reader to Ref. 4. I also keep the consideration of configurations for such machines as they are encodings of the intermediate system used to simulate a Turing machine.

An interesting machine for the discussion of Section 4 is a machine of Ref. 4 which has 4 letters and 6 states. It means 24 instructions, in fact one of them is empty. According to the encoding we suggested in Section 3, its program is encoded by a small number of letters, at most 200.

3.3. *Pedagogical universal Turing machine*

As indicated in the name **pedagogical universal Turing machine**, the machines we consider under that terminology define something which should be easily understood. It is impossible to provide a formal definition for that notion: what is indeed easily understandable? Such a notion is clearly subjective. However, I think that if somebody actually knows and understands what a universal Turing machine is, that person can understand the working of the pedagogical machine U we give in the present section.

We mainly describe its working. We do not give the table which can be found in Ref. 2.

We again remind the reader that we consider a deterministic Turing machine with a single tape and a single head. At the beginning of the section, we stressed the feature that a universal Turing machine U must be able to simulate any Turing machine M, whatever the size of the alphabet A of M and whatever the number N of states of M. We assume the reader is familiar with the description and the notations introduced in the beginning of Section 3.

In that description, we defined a uniform size for the representation of the letters of A_M which coincide with the size of the representation of a square of the tape of M. That uniformity allows us to simplify a copying process we consider in what follows. That size is connected with the binary representation of a number. However, the unary representation of the number of a letter or of a state has its advantage: it allows us to simplify the programming of the locating problem of an instruction or of a state. It is a reason why the program of U contains a sub-program K to convert the binary representation of a number k into its unary representation.

When U starts its working, it first prepares the work of K. Let P be the maximum number between the size of A_M and that of Q_M. P is computed in unary by counting the Ys in between two consecutive Xs and then, on the same area where that count was performed, it computes the number of Xs. We get the situation depicted in (a) of (3) where the unary representation of P is made with 1s. In that example, we assume that M has 8 letters and 15 states.

Then, to the left-hand side of that area, U computes the unary representations of the powers of 2 which are not greater than P, see (b) of (3). Those representations are superposed on the highest one, marking each power by the auxiliary symbol d, see (b) of (3). We call that area the **power scale**. When looking at a number, the rank of a letter in the scanned cell or the number of the new state in an instruction, K replicates its binary encoding onto the power scale, there marking the powers to be copied.

```
        F 1 1 1 1 1 1 1 1 1 1 1 1 1 1 1 1 S     (a)
d 0 0 0 d 0 d d F 1 1 1 1 1 1 1 1 1 1 1 1 1 1 1 S  (b),   (3)
```

where `1`, `F`, and `S` are appropriate markers belonging to the alphabet A_U of U.

Imagine that we have to consider state 13. Its binary representation is `1101` so that K transforms (*b*) of (3) into (*a*) of (4). The `d` in (*b*) of (4) means that the corresponding power of 2 will not be copied. Starting from the highest power of 2 marked by `e`, also belonging to A_U, K successively replaces `e` by `d`, each time copying the corresponding power of 2 onto the `1`s, replacing `1` by `h` as long as needed. The corresponding three steps are illustrated by (*b*), (*c*), and (*d*) of (4). Note that the power marked by `d` in (*a*) of (4) is not copied, according to the binary representation of 13. We can see in (*d*) of (4) that we have 13 `h`s to the right-hand side of `F`.

```
e 0 0 0 e 0 d e F 1 1 1 1 1 1 1 1 1 1 1 1 1 1 1 S  (a)
d 0 0 0 e 0 d e F h h h h h h h 1 1 1 1 1 1 1 1 S  (b)
d 0 0 0 d 0 d e F h h h h h h h h h h h 1 1 1 1 S  (c)   (4)
d 0 0 0 d 0 d d F h h h h h h h h h h h h h 1 1 S  (d).
```

The rest of the program of U performs the execution of one step of the computation of M: from the current configuration to the next one. Such a sequence of applied instructions is a **cycle** in the execution of U. Note that the numbers of instructions executed between two different cycles may be different. At the beginning of a cycle, there are two `W`s on the tape of U. One of them stands in the area of U devoted to the program of M. It replaces the `X` delimiting the instructions associated to the current state of M. The other `W` stands in the area devoted to the current configuration of M. It replaces the `U` delimiting the scanned square.

A cycle is split into the following seven stages:

1. Get with K the unary representation of the scanned letter.
2. Mark with `W` the instruction to be executed.
3. Overwrite the scanned letter by the letter of the instruction.
4. Perform the move of the head.
5. Get with K the unary representation of the new state.
6. Replace the current `W` by `X`.
7. Locate the `X` of the new current state of M.

When stage 7 is completed, a new cycle starts. Note that copying and locating some place are basically based on the same algorithm. There is a source and a target. The source contains the information needed to reach the target: each symbol of the source is successively marked and unmarked when the next symbol is examined. In the place of the target, a similar marking is performed for locating, there is no unmarking in the target in case of copying.

In Ref. 2, the table for the program of U is given. It was checked by a computer program which counted 10,351 letters to encode its program according to the above indications.

4. Comparison: Relevant or Not?

Let us now turn to what is raised in the introduction. What can be the basis of a comparison between a universal Turing machine and a biological virus? Note that a computer virus is very different: a computer virus has a malevolent intention, a natural virus has no will. What we call will when speaking of a natural virus is the result of natural selection on its evolution. As an example, if we say that a natural virus tries to adapt to its hosts, that "trying" is indeed the result of selection: the variant of the virus which is the least malevolent to its host has the best chance to survive.

The basis of our comparison is, I think, more profound. It is grounded on the complexity of the things we consider and on their behavior, mainly the working of a universal Turing machine and the behavior of a virus at a molecular level.

The unpredictability of the behavior of a universal Turing machine is a consequence of Theorem 1. A corollary of that theorem is the impossibility to construct a Turing machine that would say whether a given Turing machine launched on a given configuration reaches another fixed in an advanced configuration.

In Ref. 2, we proceed to the same discussion but we introduce a new ingredient. The pedagogical universal Turing machines presented in Section 3.3 and explicitly given in the just quoted papers is translated in those papers into what is there called an RNA-universal Turing machine.

The universal machine U considered in Section 3.3 is built on an alphabet A_U constituted of 16 letters. The genome of a DNA-virus is a very long chain of nucleotides of four kinds: adenine, cytosine, guanine, and thymine denoted by `A`, `C`, `G`, and `T`, respectively, given in alphabetic order. In the case of an RNA-virus, the composition of the genome is similar: we have also four kinds of nucleotides, the same ones with the exception of thymine which is replaced by uracil, denoted by `U`. We later refer to the alphabet {`A`,`C`,`G`,`U`} as the **RNA-alphabet**. With those four letters, we may encode the alphabet of U. For example, we can define the following correspondence, encoding a letter of U by two letters of the RNA-alphabet:

S	X	Y	U	T	F	W	_	L	R	0	1	h	d	e	Z
UU	AA	CC	GG	UA	UC	UG	CG	AC	AG	CA	CU	GA	GC	GU	AU.

(5)

Let us see how such a universal RNA Turing machine works, say *RNA* universal Turing machine, denote it by *rnaU*. It is assumed that the data used by *rnaU* are a translation of the encoding of M on the tape of U translated according to (5).

The principle of transforming U into *rnaU* is simple: each state of U contains 16 instructions. With a four-letter alphabet, five states at most are needed to sort the instructions corresponding to those of U. The execution of an instruction by *rnaU* requires to write down two letters in place of the two scanned by that machine. That may require one, two, exceptionally four more states as a backward step may be required to overwrite a two-symbol pattern. Most often a state is called by just a single other one from a single instruction. The calling instruction I has a move, say $\mu \in \{\mathtt{L}, \mathtt{R}\}$. Most often too, the majority of the instructions under the state s called by I

have the move μ. Most often too, in the calling state of I, the non-glide also have the move μ. A detailed analysis of that translation of states from U into states of $rnaU$ is performed in Ref. 2 to which the reader is referred.

The number of letters of the code of U is 10,351 so that its translation in the RNA-alphabet yields 20,702 letters. Note that the genome of the SARS-Cov2 virus contains around 31,000 letters, each one being materialized by nucleotides. The genome of that virus has 15 genes, we shall go back to that point in Section 5.

The RNA-universal machine requires 2,303,033 steps corresponding to the 106 steps of the courteous addition simulated by U. It was checked by a computer program I devised for that purpose. Compared to the 1,143,717 steps of U, the RNA-universal machine requires a little more than twice that number of steps.

Clearly, the complexity of the RNA-code obtained from the code of U is significantly less than the size of the SARS-Cov2. It is comparable to the size of an influenza virus. At first glance, as the behavior of U is unpredictable we could conclude that, *a fortiori*, the behavior of the SARS-Cov2 is also unpredictable. Although the history of the pandemia confirms that point up to now, is that comparison of code sizes a relevant argument?

One might certainly retort: what does a Turing machine have to do with a virus? Those objects are incomparable.

I add an objection: the tape of a Turing machine is a linear thread like in the case of any code written within a segment of the tape even if the segment does not contain any blank square. A virus has a complex $3D$-structure which is not at all represented by the Turing tape.

I can retort to the objection that a Turing tape is an abstract object. What is important, in the tape, is its linear order. Whether the tape is folded in whichever way is meaningless for the abstract computation. For RNA or DNA strands, the way they are folded may be important as far as elements which are far away from each other in the linear order may become neighbors thanks to the folding. But, that means more complication on the side of viruses. Consequently, I do not consider that objection as serious. Moreover, the $3D$-folding which makes molecules close neighbors has a kind of counterpart in the Turing technology: in U, most instructions of a state refer to a rather neighboring state according to their numbering, if not the immediate successor or the immediate predecessor. Other references to distant states less frequent may be looked as jumps which make close states distant with respect to their numbering. Accordingly, the jumps perform what the folding does. Also, the possible change of folding has its counterpart with the fact that executed jumps do not necessarily follow their order in the program.

Another objection could be the fact that in the concept of a Turing machine, at least those considered in the present chapter, the tape is infinite in both directions, which has no meaning in a biological context. I answered to that objection in underlying that the configuration of a Turing machine is finite at each step of its computation. Remember that property allows us to construct universal Turing machines. Now, a finite segment which can be continued step by step as long as needed has its relevance in a biological context. Another point on the side of the relevance is the consideration of courteous machines only as simulated machines. We mentioned that theoretically that restriction does not alter the generality of the result as far as a non-courteous machine can be simulated by a courteous one. But such a limitation on the tape means that it has a beginning which is also the case for RNA-strands of a virus.

The complexity is not the single argument in favor of the unpredictability of the behavior or evolution of viruses. The other reason I have to consider that comparison as relevant lies in the behavior of U. What are its actions if we only look at the motion of its head? We can mainly see two basic features. In many cases, the head runs over the configuration looking after some pattern which is expected to be somewhere on the tape. It is most probably something which also happens within a cell during a replication process. That latter feature points at the other main activity of the Turing machine: it copies some portion of the tape over another portion of the tape. What does a virus do? It takes advantage of the copying possibilities contained in a cell to replicate itself as it cannot do that by itself only. Copying and searching something are the basic actions of a Turing machine and those phenomena are also at work at a biomolecular level. But there is a third point. A code by itself can do nothing. It must be treated by a Turing machine in the case of abstract computations, and it must be treated by the cell machinery in case of the replication of a virus. The computation of a Turing machine is performed in the head of the human being who created it. The same person may create a computer program in order to perform that computation, in particular to check the correctness of the machine he/she built. It is the reason why at several places I clearly distinguished between a Turing machine and its encoding.

There are simple models of computations, more simple but more abstract than the Turing machine, which are based on that copying process like Post systems, for example, see Ref. 7 for a short description. I have no room to consider them here, but it is significant that such models are used in order to make the machines of Ref. 4 capable of universal computations. Universality lies in the data, too. Which makes us return to the code.

Another point about Turing machines and viruses is the contrast with errors. In the abstract world in which they are living, Turing machines are error-free objects which work on error-free data without error during the computation. A Turing machine exactly does what is written in its program. Clearly, such a perfect behavior is far away from what happens in a biological context. However, things are not that different if we look at Turing machines as objects created by human brains. How does a human being create a Turing machine in order to solve a given problem? He/she writes down the program and submits it with the data to a simulator of Turing machines. Whatever the simulator is, there are several trials before the human being obtains a machine he/she can trust as error-free. Those trials are interesting as far as there are usually errors during that tuning process. There are errors in the data which are not completely conformal to the format which they are supposed to obey. There are also errors in the Turing machine which does not behave as expected. Sometimes, the Turing machine halts because no instruction corresponds to what it reads under its current state. Sometimes, too, the Turing machine runs for so long without halting that its creator thinks it never halts. Looking carefully, the human being detects an error in his/her program or in the data he/she wrote. Both cases may also happen together. Accordingly, data and program are muted until they arrive to a sufficiently stable form. That real life process of creating a Turing machine looks like some evolution process of real life. Things occur as just described as far as nobody can prove the correctness of a program, as far as the controlling tools are guaranteed to work just by long enough trials. We reach here the same limit to the power of a universal Turing machine: there are (infinitely many) problems which they cannot solve, as known from Theorem 1.

Another interesting point consists in the following statistics: among the 1,652 instructions of the RNA universal Turing machine (4 letters and 413 states), 958 of them are glides: the head does not overwrite the scanned square. Also, among those 1,652 instructions, 542 of them are halting instructions, so that we remain with 152 instructions which overwrite the scanned square with another letter: it means less than 10% of the instructions make changes on the tape. It looks like the situation in the genome where active coding elements are a small minority, which does not mean that the others are useless as is the case in the RNA universal Turing machine: only 45 instructions among the non-halting one does not change the state of the head. Moreover, among those 45 instructions, 24 of them replace the letter of the scanned square by another one. The big number of changes of states comes from the fact that U has 16 letters in its alphabet while $rnaU$ has only 4 of them at its disposal. In Section 3.3, we distinguished nine groups of states: the one which counts P, the one which constructs the power scale, and then a group for each stage of a cycle.

The number of genes in a virus is very small. Let us remember that, as an example, the SARS-Cov2 has 15 of them.

Assuming that the comparison I raised with the present chapter is relevant, what can we infer from that?

The first conclusion is the unpredictability of the evolution of a virus. All reasons which were raised against the relevance of the comparison point at the same conclusion: a virus is much more complex than a Turing machine so that if the behavior of the latter is unpredictable, it is all the more the case for the former. Does it mean that there is nothing to do to limit as strongly as possible the damage a virus can do when it is virulent? Certainly not. The comparison I consider means that something which looks algorithmic has little chance to fight a virus. At that point, let us go back to the comparison. Computer scientists introduce sometimes a topology on data from which it can be proved that Turing machines viewed as operators on the data are continuous. The notion of continuity corresponding to that topology can be stated in a more concrete way: for a given Turing machine, if it is given long enough data and if there are no differences between the data within that length, the Turing machine gives the same output for those data. The larger that length with no differences, the finer, we say, the continuity. In other words, if different data entail the same behavior of the machine, it means that the considered data are big and that "close to the machine" we could say, they are not different. Every thing relies on the fineness of that continuity.

The consequences of that continuity theorem are that viruses try to cheat on the immune system of their hosts, it is the way they can survive. When I say that "the virus tries" it is a short-circuit for the sentence: among the "child"-virions produced by the "parent"-virions, those which are the more adapted to the present situation survive and only them. From the side of the immune system, its efficiency is a kind of measure of the fineness of its continuity. If it can be cheated only when data are very large and are little different, then it will be difficult for a virus to deceive it. Viruses and our immune systems have a long history of constant struggle, one against the other.

5. Conclusion

Homo sapiens are very proud of the knowledge that could be accumulated thanks to science. Let us remember that the basis of today's sciences stems from the late 16th century, mathematics being a bit older. Certainly, it is a little duration on the geological scale.

Indeed, that means less than 500 years, around 3,000 years for mathematics. Life on our planet has existed for over 3,000,000,000 years. For sure, life evolution requires much time to evolve from viruses and bacteria to plants, to animals, to man. Let us stress that human beings are newcomers at the geological scale. The way life appeared makes all its components interact, and that was apparently always the case. Science, of which homo sapiens are so proud, is the work of how many people? Certainly less than one million and a half persons at world scale at the present moment and we have to compare that with the more than 7,000,000,000 human beings at present. Moreover, the number of today's scientists is several times more than the number of dead scientists since the beginning of science, at some point 5,000 years ago, roughly speaking. That large estimate shows us that scientific activity, whatever the appraisal of its results, is a very partial and very recent activity of mankind. Can it allow us to play with our environment as if it were something given for our free use and at our free will? It seems to me that the answer is no. We should be careful when dealing with our environment. We should try to foresee as far as possible the consequences of our actions on the environment.

Let us be more modest and let us try to take benefit of what can we learn from the other fields of science, from each field of science. Theoretical computer scientists observe their environment and that careful look brought them valuable results. Perhaps Turing machines may be of help to better understand phenomena studied by other sciences. When computation and information are involved, we may have relevant points of view for other people. Theoretical computer science brings some light to philosophical problems, too, but that point goes far beyond the goal of this chapter.

I hope that the reader will find these comments of interest, whatever his/her point of view.

Acknowledgement

I express my thanks to Professor Andrew Adamatzky for his attention to my work.

References

1. A. M. Turing, On computable numbers, with an application to the Entscheidungsproblem. *Proc. London Math. Soc.*, ser. 2, 42:230–265, 1936; 43:544–546, 1937.
2. M. Margenstern, What can we learn from universal Turing machines? *Int. J. Parallel Emergent Distrib. Syst.* 37(2), 123–151, 2022.
3. L. M. Pavlotskaya, Solvability of the halting problem for certain classes of Turing machines. *Math. Notes (Springer).* 13(6):537–541, 1973 (translated from Razreshimost problemy ostanovki dlja nekotorykh klassov mashin Tjuringa, *Matematicheskie Zametki.* 13(6):899–909, June, 1973).
4. T. Neary and D. Woods, Four small universal Turing machines. *Fundam. Inform.* 91(1):123–144, 2009.
5. M. Margenstern, Frontier between decidability and undecidability: A survey. *Theor. Comput. Sci.* 231(2):217–251, 2000.
6. D. Woods and T. Neary, The complexity of small universal Turing machines: A survey. *Theor. Comput. Sci.* 410(4-5):443–450, 2009.
7. M. L. Minsky, *Computation: Finite and Infinite Machines.* Prentice Hall, Englewood Cliffs, NJ, 1967.

https://doi.org/10.1142/9789811257155_0009

Chapter 9

A Computable Piece of Uncomputable Art Whose Expansion May Explain the Universe in Software Space

Hector Zenil

Oxford Immune Algorithmics, Reading, United Kingdom
The Alan Turing Institute, British Library, London, United Kingdom
Algorithmic Dynamics Lab, Unit of Computational Medicine, Center for Molecular Medicine, Karolinska Institute, Stockholm, Sweden
Algorithmic Nature Group, LABORES for the Natural and Digital Sciences, Paris, France
hector.zenil@cs.ox.ac.uk

At the intersection of what I call uncomputable art and computational epistemology, a form of experimental philosophy, we find a most exciting and promising area of science related to causation with an alternative, possibly best possible, solution to the challenge of the inverse problem. That is, the problem of finding the possible causes, mechanistic origins, first principles, and generative models of a piece of data from a physical phenomenon. Here we explain how, by generating and exploring software space following the framework of Algorithmic Information Dynamics, it is possible to find small models and learn to navigate a sci-fi-looking space that can advance the field of scientific discovery with complementary tools to offer an opportunity to advance science itself.

1. Failure is at the Origin of Computer Science

At the beginning of the 20th century and through the end of the Second World War, computers were human, not electronic. The work of a computer consisted in solving tedious arithmetical operations with paper and pencil. This was looked upon as work of an inferior order.

At an international mathematics conference in 1928, David Hilbert and Wilhelm Ackermann suggested the possibility that a mechanical process could be devised that would be capable of proving all mathematical assertions without need of human intervention, given the tedious and almost mundane nature of such a step-by-step undertaking without much human creativity involved.

This proposition is referred to as *Entscheidungsproblem* in German, or "the decision problem" in English. If a human computer did no more than execute a mechanical process, it was not difficult to imagine that arithmetic would be amenable to a similar sort of mechanization. The origin of the *Entscheidungsproblem* dates back to Gottfried Leibniz, who, having (around 1672) succeeded in building a machine based on the ideas of Blaise Pascal that was capable of performing arithmetical operations (named *Staffelwalze* or the Step Reckoner), imagined a machine of the same kind that would be capable of manipulating symbols to determine the true value of mathematical principles. To this end, Leibniz devoted himself to conceiving a formal universal language, which he designated *characteristica universalis*, a language that would encompass, among other things, binary language and the definition of binary arithmetic.

In 1931, Kurt Gödel[1] concluded that Hilbert's intention (also referred to as "Hilbert's program") of proving all theorems by mechanizing mathematics was impossible to realize, except in the case of trivial, constrained versions of mathematical frameworks since the body of mathematics needed to contain basic axioms such as those in arithmetic or set theory. Gödel advanced formulas that codified arithmetical truths in arithmetical terms that could not be proved without arriving at a contradiction. Even worse, his formulas implied that there was no set of axioms that contained arithmetic free of unprovable true formulae, and that any extension to axioms would only lead to the same inconclusive statements in the theory.

In 1944, Emil Post, another key figure in the development of the concepts of computation and computability (focusing especially on the limits of computation) found[2] that this problem was intimately related to one of the 23 problems (the 10th) that Hilbert, speaking at the Sorbonne in Paris, had declared the most important challenges for 20th century mathematics.

Usually, Hilbert's program is considered a failure, but not only did it save human creativity from full mechanization it also led to computation as a science by its own virtue, perhaps even more than others that can be translated to finite or purely mechanistic procedures. Even though Gödel debunked[1] the notion that what was true could be proved, he presented a negative solution to the "decision problem" (Hilbert's tenth problem), followed by negative results based on Gödel's.[3–5]

1.1. *All-for-one machine, and one machine for all*

Not long after Gödel, Alan M. Turing made his appearance. Turing contemplated the problem of decision in much cruder terms. If the act of performing arithmetical operations is mechanical, why not substitute a mechanical device for the human computer? Turing's work represented the first abstract description of the digital general-purpose computer as we know it today. What Turing defined in his article as an "a" computer (for "automatic") is now known as a Turing machine.

A Turing machine is an abstract device which reads or writes symbols on a tape one at a time, and can change its operation according to what it reads, and move forwards or backwards through the tape. The machine stops when it reaches a certain configuration (a combination of what it reads and its internal state). It is said that a Turing machine produces an output if the Turing machine halts, while the locations on the tape the machine has visited represent the output produced.

The most remarkable idea advanced by Turing is his demonstration that there is an "a" machine that is able to read other "a" machines and behave as they would for an input s. In other words, Turing proved that it was not necessary to build a new machine for each different task; a single machine that could be reprogrammed sufficed. This erases the distinction between program and data, as well as between software and hardware, as one can always codify data as a program to be executed by another Turing machine and vice versa, just as one can always build a *universal machine* to execute any program and vice versa.

Turing also proved that there are Turing machines that never halt, and if a Turing machine is to be *universal*, and hence able to simulate any other Turing machine or computer program, it is actually imperative that it never halts for an infinite number of inputs of a certain type (while halting for an infinite number of inputs of some other type). Indeed, this is what Turing would have been led to expect, given Gödel's results and what he wanted to demonstrate: that Hilbert's mechanization of mathematics was impossible. This result is known as the undecidability of the halting problem.

In his seminal article, Turing defined not only the basis of what we today know as digital general-purpose computers, but also software, programming, and subroutines. And thus, without a doubt, it represents the best answer to date that we have to the question "What is an algorithm?"

In fact, in Alan Turing's work on his universal machine, he even introduced the concept of a subroutine that helped him in his machine construction. These notions are today the cornerstone of the field that Turing, more than anyone else, helped establish, *viz.* Computer Science.

2. A Slice of the Uncomputable Universe

Once we approach the problem of defining what an algorithm is and arrive at the concept of *universality* that Turing advanced, the question to be considered in greater detail concerns the nature of algorithms. Given that one now has a working definition of the algorithm, one can begin to think about classifying problems, algorithms, and computer programs by, for example, the time they take or the storage memory they may require to be executed. One may assume that the time required for an algorithm to run would depend on the type of machine, given that running a computer program on a Pentium PC is very different from executing it on a state-of-the-art super computer. This is why the concept of the Turing machine is so important — because any answers to questions about problems and algorithmic resources will only make sense if the computing device is always the same. And that device is none other than the universal Turing machine. So, for example, every step that a Turing machine performs while reading its tape is counted as a time step.

Many algorithms can arrive at the same conclusion by different paths, but some may be faster than others. This is now a matter that's carefully considered when fixing the framework for Turing's model of computation: one asks whether there is an algorithm that surpasses all others in economy as regards the resources required, when using exactly the same computing device. These are the questions that opened up an entire new field in the wake of Turing's work, the development of which Turing would certainly have been delighted to witness. This field is today referred to as the theory of Computational Complexity, which would not have been possible without a concept such as that of the universal Turing machine. The theory of Computational Complexity focuses on classifying problems and algorithms according to the time they take to compute when larger inputs are considered, and on how size of input and execution time are related to each other. This is all connected to two basic resources needed in any computation: space (storage) and time. For example, one obvious observation relating to this theory is that no algorithm will need more space than time to perform a computation. One can then quickly proceed to ask more difficult but also more interesting questions, such as whether a machine can execute a program faster if it is allowed to behave probabilistically instead of deterministically. What happens if one adds more than one tape to a universal Turing machine operation? Would that amount to implementing an algorithm to solve a problem much faster? Or one may even ask whether there is always an efficient alternative to every inefficient algorithm, a question that may lead us to a fascinating topic connecting computers and physics.

A full technical explanation of Figure 2 is provided in Ref. 6. My results were later used to validate and produce some other results, such as https://content.iospress.com/articles/computability/com 190250, https://content.iospress.com/articles/computability/com 073 these papers (that have cited mine! including Calude's!).

Each pixel represents a computer program with empty input running on an abstract computer (a Turing machine). The color indicates the runtime of the program upon completion of its computation. The darker a pixel, the longer the program took to produce its output. White pixels are computer programs that do not produce any output because they never halt (just as happens when your computer stalls forever, sometimes prompting you to reboot).

Red pixels are the computer programs that take the longest time relative to other programs of the same size before finishing (they are also called Busy Beavers). I call this color type coding the computability spectrum. Programs were sorted by size from smallest to largest (in terms of number of instructions) in such a way that one could be certain that any program would eventually be reached (including versions of MS Windows and Angry Birds that can be encoded as one-run computer programs).

Because there is no upper limit to the size of programs, the space of all computer programs is infinite but discrete, perhaps mirroring our own world. Because an enumeration of computer programs is linear (you start with program 1, then proceed to program 2, and so on), pixels depicted in this 2-dimensional figure are distributed along a "space-filling" Peano curve. This is a curve that guarantees that 2 contiguous elements in the enumeration remain as close as possible to each other in the 2D configuration.

The picture has the striking particularity that it is ultimately uncomputable. Imagine you point a telescope and take a picture of an object in the computational universe for which some pixel values cannot be determined, not only in practice but in principle — that is, no matter how good your camera, or how many "terapixels" some future version of it may have. This slide of the computational universe was only possible because it is a tiny region of small computer programs for which all pixels can be determined by zooming in deep enough.

However, Gödel and Turing proved that if you zoomed out to a larger view of the computational universe, the picture would begin to look completely blurry because pixel colors would not be determinable. In other words, this picture is the inverse equivalent of Hubble's Deep Field space — the larger the space field, the blurrier.

Just as our access to the visible universe is determined by the speed of light, due to similar limitations we can only see a small part of the computational universe, limitations both physical and computational (it seems we cannot build more powerful computers than the ones we already have to overcome these limitations). Hence, most of the computational space will remain unknowable, not only due to a limitation of resources but due to a fundamental barrier originating in the digital world and extending into our own reality.

Only incremental improvements to this picture are possible. While the picture is ultimately uncomputable and unattainable, one can make an informed guess as to what it would look like if we zoomed out. Theory predicts the average color of the picture at a greater scale, the theory in question being the theory of algorithmic probability, closely related to what is known as the Chaitin Ω number, or the halting probability. It turns out that the picture will be mostly white, as in an ever-expanding universe, finding computer programs that halt and produce something will be increasingly harder. Indeed, most programs will never halt, and those that do will do so relatively quickly in proportion to their size.

3. How to Navigate in Software Space

In his seminal book,[7] Wolfram introduced the concept of the software space in the context of mining it to find explanations of the physical world. The software space is, however, infinite, unreachable, and full of landmines, because of uncomputability and irreducibility,[8] and any attempt to traverse it will hugely depend on the enumeration chosen leading to biases in its exploration some of which, for example, can be seen in Figures 1 and 2 where some regions share more levels of gray compared to others.

First, it is important to recognize that a positive result of Turing universality is that there is only, and only one, software space, but many ways to generate it and navigate it. Indeed, to generate

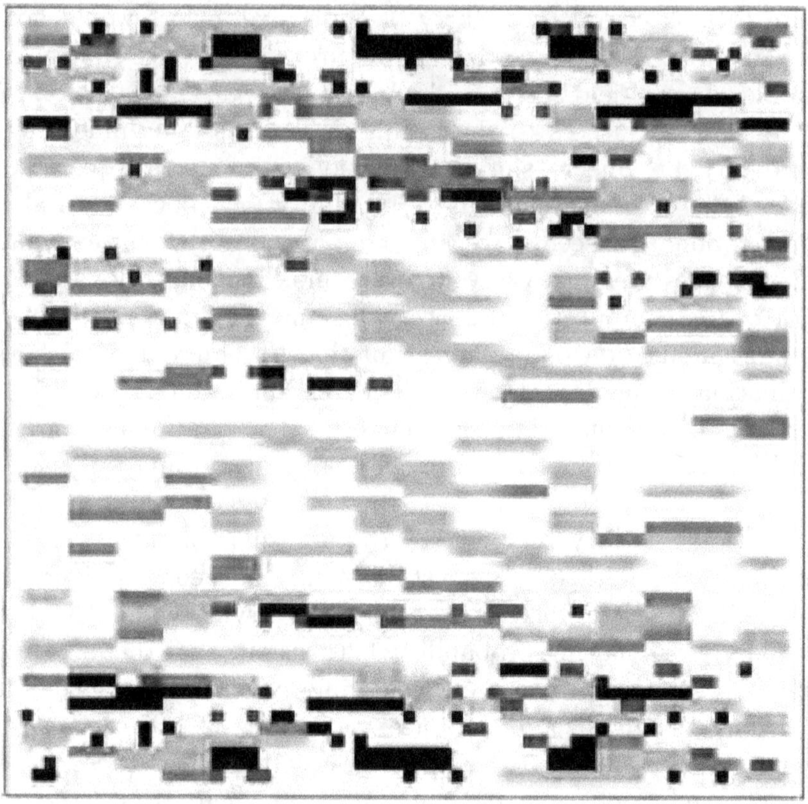

Fig. 1. Visualizing the Computational Universe: Runtime Space in a "Peano Curve"[6] by this author. 1st prize in the Computational Imagery category of the "2011 Kroto Institute Scientific Image Competition" held in the UK.

software space is to navigate it and the other way around and it all depends on the computing model and the enumeration of the computer programs that populate the universe, that is the order in which they are generated and navigated. Because of Turing universality, we know that no matter how we generate or navigate the space, the same programs will be found in any simulated space, and each space is just a different representation or simulation of another one and the same. However, this does not mean that one can generate and explore programs of interest at the same rate. In some models, this can be slower than others by a simple enumeration

Fig. 2. Runtime deep field of a segment of runtimes from 7,529,536 Turing machines in (3,2). The (3, 2) Busy Beavers are barely visible as isolated red points (online and color printed versions only).

reshuffling. Functional programming languages, for example, will tend to generate recursive functions, the type of functions that have some preference in mathematics and produce regular algorithmic or statistical patterns, more often and quicker than say, an enumeration based on rule-states (like Turing machines) where, for example, states that are not used, or rules from the product of machine symbols and states, will dominate certain local regions which, if they were not interesting, will remain non-interesting and the other way around.

So, how to navigate the space to find anything interesting in the sense of matching data and model to solve inverse problems? And why would we wish to navigate the space in the first place? Along the lines of Wolfram,[7] software space is an opportunity because computer programs in software space have the opportunity to represent an explanation of computable phenomena. This is not necessarily something completely new, in fact, all human scientific exploration so far has been based on the idea of finding computable explanations that we write down in the form of computable formulae or in computable simulations running on Turing-equivalent computers. What is new and introduced by Wolfram is its systematic exploration as an opportunity, and the discovery of its richness given that, under some strong but reasonable assumptions, such as the (physical version of the) Church–Turing thesis, one can expect to find many models,[9] if not all, to match any piece of empirical data. So, actually finding these small models offers an opportunity to look for discrete models of real-world phenomena to solve inverse problems. What had not been attempted and framed were strategies to navigate in such space to reduce the exploration, quantify, and reuse the findings in a clever way, given the (infinite) computational resources needed to computably approximate the uncomputable.

In Ref. 10, for example, we have found an exploration that explains some aspects of biological evolution, including modularization (e.g., genes), production of high-level structure, and pathways of mutation (e.g., leading to cancer). Other applications of ours have ranged from thermodynamics,[11] to finance,[12] cognitive sciences,[13] molecular biology,[14] and agriculture;[15] but a long list of uses from independent groups has also been recorded, ranging from psychometrics to chemistry, and cancer research.

3.1. *The beautiful theory of Algorithmic Probability*

Algorithmic probability tells us that most programs will produce an output following exponential decay ($-\log_2 Pr(s)$). Hence, most of them will produce the same strings and a very few will produce algorithmically random digital sequences. In other words, among the computer programs that do produce an output, most will be highly ordered, as if they were creating a structured digital universe of highly ordered patterns out of nothing (just like ours). Some of these programs that produce the same output may do so faster than others, but precisely how much faster a program can be is another open foundational question, and one of the greatest challenges in mathematics and computer science. If it were possible to find a program considerably faster than any other, then everything could be computed in "reasonable" time, regardless of the apparent difficulty of the computation.

In a later follow-up piece,[16] a connection between Turing machines and Turing patterns is made, explaining how symmetry-breaking creates structure — as opposed to randomness — out of nothing, with only computability being assumed.

At the center of my ideas for guiding searches of interesting regions and programs as models in the computational universe, in this deep field view of the software space, are some methods I have introduced based on or inspired by the seminal concept of Algorithmic Probability, itself introduced by Solomonoff[17] and Levin[18] and having affinities with Chaitin's Ω number or halting probability.[19] This concept of algorithmic probability is formally defined as follows:

$$AP(s) = \sum_{p:U(p)=s} 1/2^{|p|}.$$

That is, the sum over all the programs p for which a universal Turing machine U outputs s and halts.

The notion behind AP is very intuitive. If one wished to produce the digits of π randomly, one would have to try time after time until one managed to hit upon the first numbers corresponding to an initial segment of the decimal expansion of π. The probability of success is extremely small: 1/10 digits multiplied by the desired quantity of digits. For example, $1/10^{2400}$ for a segment of 2,400 digits of π. But if instead of shooting out random numbers one were to shoot out computer programs to be run on a digital computer, the result would be very different.

A program that produces the digits of π would have a higher probability of being produced by a computer. Concise and known formulas for π could be implemented as short computer programs that would generate any arbitrary number of digits of π.

3.2. *Simplicity is likeliness and likeliness, simplicity*

The length of the shortest program that produces a string is today the mathematical definition of randomness, as introduced by Kolmogorov,[20] Chaitin,[19] and Solomonoff,[17] and later expanded by Levin[18] and also called Kolmogorov–Chaitin complexity.

The idea is relatively simple. If a string s of length $|s|$ cannot be produced by a program $|p|$ shorter than $|s|$ in bits, then the string s is random because it cannot be effectively described in a shorter way than by s itself, there being no program p that generates s whose length is shorter than s. Formally,

$$C_U(s) = \min\{|p| : U(p) = s\},$$

where U is a universal optimal Turing machine. The invariance theorem[17,19,20] guarantees that such an optimal universal Turing machine exists for which the value of C under different Turing-complete language implementations over the same object is bounded by a "small" constant. Formally, if U_1 and U_2 are such implementations and $C_{U_1}(s)$ and $C_{U_2}(s)$ are the values of algorithmic complexity of s for U_1 and U_2, respectively, there exists an optimal U and constant c such that

$$|C_{U_1}(s) - C_{U_2}(s)| < c.$$

Clearly, the longer the string, the less relevant the c, and the more stable the algorithmic complexity values. One of the disadvantages is that, given the halting problem for Turing machines, C is not computable, which is to say that given a string, there is no algorithm that returns the length of the shortest computer program that produces it. Efforts have been taken before to approximate it through statistical compression algorithms[21] but these fail at characterizing algorithmic complexity in full and are rather implementations of variations of Shannon entropy.[22,23]

Algorithmic probability and algorithmic complexity K are inversely proportional, as established by the so-called algorithmic

Coding theorem:[24,25]

$$| - \log_2 AP(s) - C(s)| < c,$$

where c is a constant independent of s. The Coding theorem implies that the algorithmic complexity can be estimated from the frequency of a string. In other words, complexity is in inverse relation to algorithmic probability.

To illustrate the above, let us consider π. Under the assumption of Borel's absolute normality of π, whose digits appear randomly distributed, and with no knowledge of the deterministic source and nature of π as produced by short mathematical formulae, we ask how an entropy vs. an algorithmic metric performs. First, the Shannon entropy rate (thus assuming the uniform distribution along all integer sequences of N digits) of the first gN digits of π, in any base, would suggest maximum randomness at the limit. However, without access to or without making assumptions as regards the probability distribution, approximations to algorithmic probability would assign π high probability, and thus the lowest complexity, by the Coding theorem, as has been done in Refs. 26–29.

Just as with π but in application to graphs and networks, it has been proven how certain graphs can be artificially constructed to target any level of Shannon entropy,[23,30] preserving low algorithmic randomness.

But how relevant is the algorithmic Coding theorem in explaining, e.g., natural phenomena, if it only applies to Turing-universal systems? We know that the natural world allows and can carry out Turing-universal computation because we have been able to build computers that take elements from nature and make them perform as general-purpose computers. However, we don't know how much of the natural world is Turing-computable or how physical laws may naturally implement any form of computation.

So, in Ref. 31 we showed that up to 60% of the bias found in the output of systems that are not Turing universal may be explained by the algorithmic Coding theorem. This means that this theorem is far more relevant than expected, as it not only explains the way data and patterns distribute for a very particular computational model, but does so, to a considerable extent, for basically any computing model. It is not difficult to imagine that nature operates at some of these levels, perhaps not even on a fixed one but at multiple scales, with the algorithmic Coding theorem being relevant to all of them.

3.3. *My own navigation through software space*

The work of Greg Chaitin has been at the center of my interests. Not only was he one of the examiners for my PhD in computer science in 2011, but my work would have not been possible without some of his theoretical work. I met Greg in 2006, having been invited to his home and to his office at IBM's legendary Thomas J. Watson Research Center (see Figure 3), both in Yorktown Heights, NY, US.

Fig. 3. (Top) A picture of Greg Chaitin I took outside his longtime office at IBM Research headquarters, the Thomas J. Watson Research Center. (Bottom) The two sides of the medal I helped Wolfram design, featuring material relating to Chaitin's life achievements on the obverse and Leibniz's own medal celebrating binary arithmetic on the reverse.

His house was typical of an intellectual, full of objects and books piled one on top of the other. But Greg also had exotic pieces of sculpture scattered around, and an over-representation of Leibniz- and Gödel-related books, the two major influences on his work and thinking.

In 2008, I organized a two-part panel discussion (Figure 4) during the Wolfram Science conference at the University of Vermont in Burlington, an event that, together with a second meeting that I

Panel Part I: Is the universe random?

Panel Part II: What is computation and How does nature compute?

Fig. 4. Part I and Part II panel discussion pictures. (Top) From left to right: Hector Zenil, Stephen Wolfram, Paul Davies, Ugo Pagallo, Greg Chaitin, Cris Calude, Karl Svozil, Gordana Dodig Crnkovic, and John Casti. (Bottom) From left to right: Hector Zenil, Greg Chaitin, Ed Fredkin, Rob de Ruyter, Tony Legett, Cris Calude, Tom Toffoli, and Stephen Wolfram. Transcriptions in Refs. 32 and 33.

organized (with Adrian German) in 2008 at Indiana University Bloomington, will, I believe, come to be regarded as events of significant historical importance in the discussion of the ideas of the late 20th and early 21st centuries around questions of the analog and the digital. The first part of the panel discussion addressed the question "Is the Universe random?" Participating were Cris Calude, John Casti, Greg Chaitin, Paul Davies, Karl Svozil, and Stephen Wolfram. The second part addressed the question "What is computation and how does nature compute?". The participants were Ed Fredkin, Greg Chaitin, Tony Legett, Cris Calude, Rob de Ruyter, Tom Toffoli, and Stephen Wolfram, with George Johnson, Gerardo Ortiz, and myself serving as moderators. Transcriptions of both parts were published in my volume.[32,33]

In 2007, I helped Stephen Wolfram design a commemorative medal to celebrate Gregory Chitin's 60th birthday, which also involved minting for the first time a 300 year old medal originally designed by Gottfried Leibniz to celebrate the invention or discovery of binary arithmetic (see Figure 3). I published a blog post soon after we came up with the idea for the medal, explaining the pre- and post-minting story of the medal.[34]

At the center of my research is Greg Chaitin's work on what is known as algorithmic probability, which in turn is related to Chaitin's Omega (Ω) number, also called the halting probability. My very first research paper[35] was published in Chaitin's 60th birthday Festchrift.[24] This paper provided the first numerical evidence of a bias present in nature, a claim that I later advanced in a contribution that won an FQXi prize[36] in a competition on the subject of the digital or analog nature of the universe. Such claims have more recently been picked up and expanded upon[37] — efforts in which my own methods have again played an important role — and have also been further explored by myself in the context of scientific discovery.[14,38,39]

3.4. *The Coding Theorem Method (CTM)*

The algorithmic *Coding Theorem Method* (CTM)[26,27] provides the means for approximation via the frequency of a string. Now, why is this so? The underlying mathematics originates in the relation specified by algorithmic probability between frequency of production of a string from a random program and its algorithmic complexity. It is also therefore designated the algorithmic *Coding theorem*, in contrast to another well-known coding theorem in classical information theory. Essentially, the numerical approximation hinges on the fact that the more frequently a string (or object) occurs, the lower its algorithmic complexity. Conversely, strings with a lower frequency have higher algorithmic complexity. Otherwise stated,

$$CTM(s) = -\log_2 AP(s) + c.$$

The way to implement a compression algorithm at the level of Turing machines, unlike popular compression algorithms which are heavily based on Shannon entropy, is to go through all possible compression schemes. This is equivalent to traversing all possible programs that are a compressed version of a piece of data, which is exactly what the CTM algorithm does.

3.5. *The Block Decomposition Method (BDM)*

Our approach to Chaitin's halting probability and Solomonoff–Levin's algorithmic probability consists in asking after the probability of a matrix being generated by a random Turing machine on a 2-dimensional array, also called a *termite* or *Langton's ant*.[40] Hence, an accounting procedure is performed using Turing machines that aims to approximate the algorithmic complexity of the structures identified. This technique is referred to as the *Block Decomposition Method* (BDM), as introduced in Ref. 29. The BDM technique requires a partition of the adjacency matrix corresponding to the graph into smaller matrices. With these building blocks at hand we numerically calculate the corresponding algorithmic probability by running a large set of small 2-dimensional deterministic Turing machines, and then — by applying the algorithmic Coding theorem, as discussed above — its algorithmic complexity.

Following such a divide-and-conquer approach we can then approximate the overall complexity of the original adjacency matrix by the sum of the complexity of its parts. Note that we have to take into account a logarithmic penalization for repetition, given that n repetitions of the same object only add $\log n$ to its overall complexity, as one can simply describe a repetition in terms of the multiplicity of the first occurrence. Technically, this translates into the algorithmic complexity of a labelled graph G by means of BDM, which is defined as follows:

$$BDM(G,d) = \sum_{(r_u,n_u)\in A(G)_{d\times d}} \log_2(n_u) + K_m(r_u), \qquad (1)$$

where $K_m(r_u)$ is the approximation of the algorithmic complexity of the sub-arrays r_u arrived at by using the algorithmic Coding theorem, while $A(G)_{d\times d}$ represents the set with elements (r_u, n_u), obtained by decomposing the adjacency matrix of G into non-overlapping squares, i.e., the block matrix, of size d by d. In each (r_u, n_u) pair, r_u is one such square and n_u its multiplicity (number of occurrences). From now on $BDM(G, d = 4)$ will be denoted only by $BDM(G)$, but it should be taken as an approximation to $C(G)$ unless otherwise stated (e.g., when taking the theoretical true $C(G)$ value). Once CTM is calculated, BDM can be implemented as a look-up table, and hence runs efficiently in linear time for non-overlapping fixed size submatrices.

4. Algorithmic Information Dynamics, a Software Calculus

Unlike most complexity measures[21] that are designed for static objects, except those related to dynamical systems (e.g., inverse problems with differential equations, Lyapunov exponents, etc), the measure of algorithmic complexity we introduced before can be adapted to characterize how dynamical systems move in software space by looking at the rate of change of algorithmic complexity of an object evolving over time. This measure is universal in the sense that it can deal with any computable feature that a system may display over time, either spontaneously or as a result of an external perturbation/intervention/interaction. Indeed, Algorithmic Information Dynamics is the combination of algorithmic probability and perturbation analysis in software space.

At the core of Algorithmic Information Dynamics,[39,41] the algorithmic causal calculus that we have introduced is the quantification of the change of complexity of a system under natural or induced perturbations, particularly the direction (sign) and magnitude of the difference of algorithmic information approximations denoted by C between an object G, such as a cellular automaton or a graph, and its mutated version G', e.g., the flip of a cell bit (or a set of bits) or the removal of an edge e from G (denoted by $G \backslash e = G'$). The difference $|C(G) - C(G \backslash e)|$ is an estimation of the shared algorithmic mutual information of G and $G \backslash e$. If e does not contribute to the description of G, then $|C(G) - C(G \backslash e)| \leq \log_2 |G|$, where $|G|$ is the uncompressed size of G, i.e., the difference will be very small and at most a function of the graph size, and thus $C(G)$ and $C(G \backslash e)$ have almost the same complexity. If, however, $|C(G) - C(G \backslash e)| \leq \log_2 |G|$ bits, then G and $G \backslash e$ share at least n bits of algorithmic information in element e, and the removal of e results in a loss of information. In contrast, if $C(G) - C(G \backslash e) > n$, then e cannot be explained by G alone, nor is it algorithmically not contained/derived from G, and it is therefore a fundamental part of the description of G, with e as a generative causal mechanism in G, or else it is not part of G but has to be explained independently, e.g., as noise. Whether it is noise or part of the generating mechanism of G depends on the relative magnitude of n with respect to $C(G)$ and to the original causal content of G itself. If G is random, then the effect of e will be small in either case, but if G is richly causal and has a very small generating program, then e as noise will have a greater impact on G than would removing e from an already short description of G. However, if $|C(G) - C(G \backslash e)| \leq \log_2 |G|$, where $|G|$ is, e.g., the vertex count of a graph, or the runtime of a cellular automaton, G, then e is contained in the algorithmic description of G and can be recovered from G itself (e.g., by running the program from a previous step until it produces G with e from $G \backslash e$).

We have shown, for example, how we can infer and reconstruct space-time evolutions via quantification of the disruption of a perturbation[39] and even de novo.[42] From there, it can then be extracted as a generating mechanism from the ordered time indices, from least to most disruptive, and produce candidate generating models. Simpler rules have simpler hypotheses, with an almost perfect correspondence in row order. Some systems may appear more disordered than others, but locally the relationship between single rows is for the most part preserved (indicating local reversibility).

We have also shown that the later in time a perturbation is injected into a dynamical system, the less it contributes to the algorithmic information content of the overall space-time evolution.[39] We then move from discrete 2D systems to the reconstruction of phase spaces and space-time evolutions of N-dimensional, continuous, chaotic, incomplete and even noisy dynamical systems.

Algorithmic Information Dynamics suggests that by performing perturbations to data and systems and quantifying the changes that these produce on the data models in software space, we can navigate software space to better understand a system or data in physical space.

4.1. *What the Universe may be made of*

Could it be that the physical universe is made or comes from a computational process and is therefore within software space? This would make our work even more relevant if computable model-driven explanations are approachable with the methods we have advanced allowing some degree of navigation in the boundless infinite software space. Among the many interesting questions is how common computation universality is in software space? If it is very common, one can expect both forward and inverse problems to be more difficult than expected. First, because the forward problem means that a universal model would be capable of arbitrary output making thus initial configurations and interactions with other systems more important. For the inverse problem, that also means that matching model to data is more difficult because the relationship is between even larger sets than possibly imagined, as many more models will be able to match many more data, again making specific parameters all the more important.

What we have found,[43] is that indeed, universality seems to be quite common. For this, we came up with an approach[44] able to produce statistical evidence toward universality, which is rather strong, to some extent even stronger than simple computation universality, a stronger version called intrinsic universality consisting on not only being able to emulate any computable function but also any computable function of the type and scale of the emulator itself. And this is what we found, that most computer programs were capable of emulating both quantitatively and qualitatively most, or all, other computer programs in their own model space. We did this by looking at the set of ever growing compilers (computable translations) that coupled with the original computer program would allow the program to compute quantitatively and qualitatively any other computer program in its own rule space and thus beyond its rule space assuming such evidence amounted to intrinsic universality.

Our compilers (and compilers in general) are nothing but small finite computer programs that cannot be themselves universal because they only operate once a computation has started and do not interact with the computation itself.

A positive result (if pervasive universality is read as a negative one, depending on the angle) is that we found also a counterbalance in the equation. Not all compilers for all rules where easily found to be equally versatile, there was a strong inequality indicating that some programs are more naturally programmable than others, thus reintroducing an asymmetry compatible with other measures of complexity, namely resource (time) complexity. Pervasive universality also means that software space is richer than possibly first imagined and may indicate that the real world is equally rich (and thus, difficult to match) but also fascinating and with the possibility to explain the universe itself.[45,46]

The methods and framework that we have introduced, Algorithmic Information Dynamics (AID),[47] provides tools to attempt to find those matches,[38,39,48] even between sets of data and sets of models, together with a ranking rule to gage such models based on algorithmic likeliness. For this reason, we believe AID is a first step in the right direction.

References

1. K. Gödel, Über formal unentscheidbare Sätze der Principia Mathematica und verwandter Systeme, I. *Monatshefte für Mathematik und Physik* 38:173–198, 1931.
2. E. L. Post, Finite combinatory processes — Formulation 1. *J. Symb. Log.* 1:103–105, 1936.
3. M. Davis, Hilbert's tenth problem is unsolvable. *Am. Math. Mon.* 80:233–269, 1973.
4. M. Davis, Y. V. Matiyasevich, and J. Robinson, Hilbert's tenth problem: Diophantine equations: Positive aspects of a negative solution. In *Mathematical Developments Arising from Hilbert Problems.* Felix E. Browder, editor. *Proceedings of Symposia in Pure Mathematics*, XXVIII.2. American Mathematical Society, 1976, 323–378.
5. Y. V. Matiyasevich, Enumerable sets are Diophantine. *Doklady Akademii Nauk. SSSR* (in Russian) 191:279–282. MR 0258744. English translation in *Sov. Math.* 11(2):354–357, 1970.
6. H. Zenil, From computer runtimes to the length of proofs: With an algorithmic probabilistic application to waiting times in automatic theorem proving. In *Computation, Physics and Beyond International Workshop on Theoretical Computer Science*, M. J. Dinneen, B. Khousainov, and A. Nies, editors, WTCS 2012, LNCS 7160, 223–240, Springer, 2012.
7. S. Wolfram, *A New Kind of Science.* Wolfram Media, Champaign, IL, 2002.
8. H. Zenil, F. Soler-Toscano, and J. J. Joosten, Empirical encounters with computational irreducibility and unpredictability. *Minds Mach.* 22(3):149–165, 2012.
9. H. Zenil, Algorithmic data analytics, small data matters and correlation versus causation. *Berechenbarkeit der Welt?* 453–475, 2017.
10. S. Hernández-Orozco, N. A. Kiani, and H. Zenil, Algorithmically probable mutations reproduce aspects of evolution, such as convergence rate, genetic memory and modularity. *R. Soc. Open Sci.* 5(8):180399.
11. H. Zenil, C. Gershenson, J. A. R. Marshall, and D. A. Rosenblueth, Life as thermodynamic evidence of algorithmic structure in natural environments. *Entropy.* 14(11):2173–2191, 2012.
12. H. Zenil and J. P. Delahaye, An algorithmic information theoretic approach to the behaviour of financial markets. *J. Econ. Surv.* 25(3):431–463, 2011.
13. N. Gauvrit, H. Zenil, F. Soler-Toscano, J.-P. Delahaye, and P. Brugger, Human behavioral complexity peaks at age 25. *PLoS Comput. Biol.* 13(4):e1005408, 2017.
14. H. Zenil and P. Minary, Training-free measures based on algorithmic probability identify high nucleosome occupancy in DNA sequences. *Nucleic Acids Res.* gkz750, 2019.

15. H. Zenil, J. P. Delahaye, and C. Gaucherel, Image information content characterization and classification by physical complexity. *Complexity.* 26–42, 2010.
16. H. Zenil, Turing patterns with Turing machines: Emergence and low-level structure formation. *Nat. Comput.* 12(4):499–515, 2013.
17. R. J. Solomonoff, A formal theory of inductive inference. Parts i and ii. *Inf. Control.* 7(1):1–22, 224–254, 1964.
18. L. A. Levin, Laws of information conservation (nongrowth) and aspects of the foundation of probability theory. *Prob. Peredachi Infor.* 10(3):30–35, 1974.
19. G. J. Chaitin, On the length of programs for computing finite binary sequences: Statistical considerations. *JACM.* 16(1):145–159, 1969.
20. A. N. Kolmogorov, Three approaches to the quantitative definition of information. *Int. J. Comput. Math.* 2(1-4):157–168, 1968.
21. H. Zenil, A review of methods for estimating algorithmic complexity: Options, challenges, and new directions. *Entropy.* 22:612, 2020.
22. H. Zenil, Towards demystifying shannon entropy, lossless compression, and approaches to statistical machine learning. In *Proceedings of the International Society for Information Studies 2019 summit.* University of California, Berkeley, 2020.
23. H. Zenil, N. A. Kiani, and J. Tegnér, Low Algorithmic Complexity entropy-deceiving Graphs. *Physical Review E.* 96:012308, 2017.
24. C. Calude (editor), *Randomness and Complexity: From Leibniz to Chaitin.* World Scientific Pub Co, Singapore, 2007.
25. T. M. Cover and J. A. Thomas, *Elements of Information Theory*, 2nd edition. Wiley-Blackwell, 2009.
26. J.-P. Delahaye and H. Zenil, Numerical evaluation of the complexity of short strings: A glance into the innermost structure of algorithmic randomness. *Appl. Math. Comput.* 219:63–77, 2012.
27. F. Soler-Toscano, H. Zenil, J.-P. Delahaye, and N. Gauvrit, Calculating Kolmogorov complexity from the frequency output distributions of small turing machines. *PLoS ONE* 9(5):e96223, 2014.
28. H. Zenil, F. Soler-Toscano, J.-P. Delahaye, and N. Gauvrit, *Two-Dimensional Kolmogorov Complexity and Validation of the Coding Theorem Method by Compressibility*, 1:e23, 2015.
29. H. Zenil, F. Soler-Toscano, N. A. Kiani, S. Hernández-Orozco, and A. Rueda-Toicen, A decomposition method for global evaluation of Shannon entropy and local estimations of algorithmic complexity. *arXiv:1609.00110 [cs.IT]*, 2016.
30. M. Morzy, T. Kajdanowicz, and P. Kazienko, On measuring the complexity of networks: Kolmogorov complexity versus entropy. *Complexity.* 2017:3250301:1–3250301:12, 2017.
31. H. Zenil, L. Badillo, S. Hernández-Orozco, and F. Hernández-Quiroz, Coding-theorem like behaviour and emergence of the universal distribution from resource-bounded algorithmic probability. *Int. J. Parallel Emergent Distrib. Syst.* 2018.

32. H. Zenil, *Randomness Through Computation: Some Answers, More Questions.* World Scientific Pub Co, Singapore, 2011.
33. H. Zenil, *A Computable Universe: Understanding and Exploring Nature as Computation.* World Scientific Pub Co, Singapore, 2012.
34. H. Zenil, Blog of http://www.mathrix.org/liquid/archives/the-history-of-the-chaitin-leibniz-medallion Accessed on February 26, 2019.
35. J.-P. Delahaye and H. Zenil, On the Kolmogorov-Chaitin complexity for short sequences. In *Randomness and Complexity: From Leibniz to Chaitin*, C. Calude, editor. World Scientific Publishing Company, 2007.
36. H. Zenil, The World is Either Algorithmic or Mostly Random. Third Prise Winning Essay — Foundational Questions Institute (FQXi) Contest "Is Reality Digital or Analog?" 2011.
37. K. Dingle, C. Camargo, and A. A. Louis, Input-output maps are strongly biased towards simple outputs. *Nat. Commun.* 9(761), 2018.
38. H. Zenil, N. A. Kiani, A. Zea, and J. Tegnér, Causal deconvolution by algorithmic generative models. *Nat. Mach. Intell.* 1:58–66, 2019.
39. H. Zenil, N. A. Kiani, F. Marabita, Y. Deng, S. Elias, A. Schmidt, G. Ball, and J. Tegnér, An algorithmic information calculus for causal discovery and reprogramming systems. *iScience.* 19:1160–1172, 2019.
40. C. Langton, Computation at the edge of chaos: Phase transitions and emergent computation. In *Emergent computation*, S. Forest, editor. The MIT Press, 1991, 12–37.
41. H. Zenil, N. A. Kiani, and J. Tegnér, Algorithmic information dynamics of emergent, persistent, and colliding particles in the game of life. In *From Parallel to Emergent Computing (book)*, A. Adamatzky, editor. Taylor & Francis/CRC Press, 2019, 367–383.
42. S. Hernández-Orozco, H. Zenil, J. Riedel, A. Uccello, N. A. Kiani, and J. Tegnér, Algorithmic probability-guided machine learning on non-differentiable spaces, *Front. Artif. Intell.* 25, 2021.
43. J. Riedel and H. Zenil, Cross-boundary behavioural reprogrammability reveals evidence of pervasive universality. *Int. J. Unconv. Comput.* 13:14–15, 309–357, 2018.
44. H. Zenil, On the dynamic qualitative behaviour of universal computation. *Complex Syst.* 20(3):265–278, 2012.
45. S. Wolfram, A class of models with the potential to represent fundamental physics, *Complex Systems.* 29(2), 107–536, 2020.
46. H. Zenil, Compression is comprehension, and the unreasonable effectiveness of digital computation in the natural world. In *Unravelling Complexity* (Gregory Chaitin's 70 festschrift), S. Wuppuluri and F. Doria, editors, World Scientific Publishing, 2019.
47. H. Zenil, N. A. Kiani, F. S. Abrahão, and J. Tegnér, Algorithmic information dynamics. *Scholarpedia.* 15(7):53143, 2020.
48. J. Riedel and H. Zenil, Rule primality, Minimal generating sets and Turing-universality in the causal decomposition of elementary cellular automata. *J. Cell. Autom.* 13:479–497, 2018.

https://doi.org/10.1142/9789811257155_0010

Chapter 10

Brain-inspired Computing: Achievements and Challenges

Filippo Romani[*,‡] and Victor Erokhin[†,§]

[*] *Nalco Europe B.V. Oude Rhijnhofweg 17, 2342 BB Oegstgeest, The Netherlands*

[†] *Institute of Materials for Electronics and Magnetism, Italian National Research Council, IMEM-CNR, Parco Area delle Scienze 37A, 43124 Parma, Italy*

[‡] *filippo.romani@gmail.com*

[§] *victor.erokhin@imem.cnr.it*

We discuss a possibility of the realization of brain-inspired computational systems. We present a comparison of essential features of the brain with those of modern computers. In particular, we discuss the difference in the architecture of computers, nervous systems, and learning algorithms. We also review the effect of noise, cross-talk of elements, and emotions. Where possible, we refer to developed technical approaches. However, as the area is a rather new one, there are more questions than solutions. Thus, we make our general considerations and present drawings for illustrating some concepts.

1. Introduction

The term unconventional computing currently implies a wide range of systems, based on different working principles including chemical reactions,[1] biological activity of simple living organisms,[2] optical treatment of data,[3] quantum computing,[4] etc. The only common feature among them is that their architecture and data treatment algorithms are different from those used in current computers. Of course, it is impossible to make an overview of all possible unconventional computers within this chapter. Therefore, we will consider here the so-called neuromorphic, or brain-inspired computing.[5] In particular, we will think about some essential features of the brain that are different from those of modern computers. We will try to analyze how the presence of these features can change the performance of the computing system. Where possible, we will provide references to research articles, reporting scientific and technological findings, allowing approaching neuromorphic computers. However, here we will concentrate our attention on challenges of brain-inspired computing rather than on possible technical solutions. It is to note that at the beginning of the computer epoch, all activity was dedicated to the realization of systems that should allow to understand better how the brain works.[6] However, as it often happens, the results of these activities were very exciting and useful but had nothing to do with the initial goal. Modern computers are used in practically all aspects of our current daily life, but they do not allow to reconstruct directly basic features of the brain organization and function. Therefore, the main efforts in mimicking the nervous system are now done at the software level. As an example, neural networks were considered only as the hardware implementations at the beginning,[6] while now they are mainly realized as software programs.

Even if traditional computers can do individual operations much faster than the brain does, they have several drawbacks. The most obvious ones are connected to the impossibility of parallel information processes and in energy efficiency. The processor of a single nucleus computer can perform only one operation in a defined moment. In the case of a multi-nuclei processor, the number of operations that can be treated simultaneously is equal to the number of nuclei. Such processors are widely used in GPUs (Graphic Processing Units), for processing graphic information, which requires high energy consumption. Thus, even if a single operation will be treated very fast in a single nucleolus processor, the processing of the whole set of data, obtained from different sensoristic systems (visual, acoustic, tactile, etc.) will require much more time with respect to the brain, because in the brain all these data are processed in parallel. The concept of the energy consumption is also very important. Considering that the brain contains about 10^{15} synapses[7] and the characteristic time of their weight variation is in ms range, we can estimate that a single data treatment event must be not more than 10^{-18} J — the value is absolutely not possible to reach using current electronic technologies for the processor fabrication. However, if the energy consumption (and, therefore, dissipation) will be higher than the mentioned value with the same level of integration, it will result in the significant increase of the system temperature, which will result in the generation of errors or in the destruction of some elements or even the whole system. Just to illustrate this statement: high potential video cards (GPUs) of modern computers allow them to work with big data, even if the integration degree of elements is much lower than with that in the brain, but require the use of external cooling systems, otherwise they will stop working.

Of course, the difference of the brain with respect to computers is not limited by the possibility of parallel information processing and energy consumption. In this chapter, we will consider four important differences, namely: architecture, learning algorithms, internal activity, and effect of noise, and emotions. We will critically examine the importance of these principal items and how they can influence the characteristics of computational systems.

The chapter contains graphic illustrations: of course, these drawings include the personality of the artist, which underlines once more the fact that even researches in the field of neuromorphic computing are still very far from understanding and reproducing brain functions.

2. Architecture

The difference in the architecture of the nervous system (and brain) and computers, probably, is a key parameter responsible for their different performances during data processing. In computers, processor and memory are separate systems. Properties of the processor are fixed and determined by their predisposed design and current status of the electronics technology processes, responsible for the degree of integration of elements (level of complexity) and frequency of the operation (number of operations, that can be executed in a fixed period of time, that directly depends on the sizes of realized electronic elements). Memorized information makes no influence on the processor properties. Instead, elements of the brain (neurons and synapses) are responsible for both storage and processing of the information. The brain is a very complicated network of neurons, connected with each other. The neuron body (soma) is a threshold element, allowing the income of all stimuli, but permitting the further propagation of the signal only when a certain threshold is reached (temporal integration of all input signals during a fixed time interval is higher than a threshold value). Each neuron has several dendrites, allowing the signal from other neurons to come to this neuron, and one axon, that is used for the further propagation of the signal, when the threshold value is reached. Contact of the dendrite of one neuron with the axon of the other one is called synapse. This is a very important element: it can vary the weight function (transmission of the signal amplitude) according to the duration and/or frequency of its involvement into the signal transmission process. We will consider the rules responsible for the synaptic weight variation in the successive section, dedicated to the learning algorithms. Here it is important only to mention that these elements are responsible for both recording and processing of the information. It means that the map of the weights memorizes the accumulated information, on the one hand, and varies the connection of nonlinear elements of the "processor", on the other hand. In other words, when we remember something, the "digested" information varies also the "processing" capabilities of the brain by reinforcing or inhibiting possible signal pathways. Thus, this consideration destroys a "sweet dream" of many students (and not only students): to connect a memory device to the brain and to know immediately all the recorded information without a hard and time-consuming study process. If we want to remain humans and not machines (at least in the modern understanding of the term "machine"), we must learn in a traditional way. In this respect, other "brain improving" algorithms, implying taking of special pills, widely described in science fiction, also will not work well. If the pills contain a defined set of neurotransmitters, it can accelerate the learning process, switching the brain in a mode allowing faster learning. Mainly, it will be connected to the emotional state of the person (as all of us know from experience, in some situations we can study much faster and efficiently than in others). We will consider the role of emotions in more detail in the following section. The approach suggested by Douglas Noel Adams in "The Hitchhiker's Guide to the Galaxy"[8]: to insert a special Babel fish into the ear for understanding languages seems more adequate: "The Babel fish," said The Hitchhiker's Guide to the Galaxy quietly, "is small, yellow and leech-like, and probably the oddest thing in the Universe. It feeds on brainwave energy not from its carrier but from those around it. It absorbs all unconscious mental frequencies from this brainwave energy to nourish itself with. It then excretes into the mind of its carrier a telepathic matrix formed by combining the conscious thought frequencies with nerve signals picked up from the speech centers of the brain which has supplied them. The practical upshot of all this is that if you stick a Babel fish in your ear you can instantly understand anything said to you in any form of language. The speech patterns you actually hear decode the brainwave matrix which has been fed into your mind by your Babel fish".[8]

Therefore, it is possible to imagine that the human brain has obtained a new intrinsic transducing system (decoding the information, that was considered initially as noise, into codes, already present in the brain of a particular person). Therefore, the new symbiotic system cannot be considered as a simple combination of a human (more generally, sapience) and the Babel fish, but a new being with new properties and capabilities, such as fast learning of foreign languages.

As it was stated in the Introduction, another important item, coming from the difference in the architecture, is the fact that in a computer processor only one operation is executed in each time moment. Therefore, as the operation frequency of computers is successively increasing, the computer is able to resolve arithmetical operations faster and faster than the humans do. Moreover, a decade ago it was considered that a computer will never be able to win a chess game playing against a grandmaster. Instead, as we know now, it can. In fact, in order to win the game, it is not necessary to perform a simultaneous analysis of all available information (the position is fixed and will not be able to change before the next step), but it is necessary to perform just deep calculations of possible results of one motion or the other. If the time is not limited, the computer will always win. In this respect, it seems more "neuromorphic" if the computer will win a game of poker playing against several players and keeping in consideration their individual features (not in general, but in a particular moment): face, mood, risk preferences, history of playing manner (long (as usual) and short (today) strategies). Of course, even in this case the characteristic features of the computer (recording of all displayed cards and calculation of the most probable layout of partners cards) will help to avoid silly mistakes, but it will not be a key factor for winning.

Regarding the physical implementation of brain-inspired information processing systems, most of the reported works were dedicated to mimicing properties, rather than the architecture.[9,10] According to our knowledge, there is only one work where the architecture of the part of the nervous system of a simple animal (pond snail) was reproduced according to the developed model and its function (conditioning) was demonstrated.[11] Regarding the construction of the 3D stochastic systems with the possibility of distant connections of threshold elements, several works have been reported. However, the only one,[12] based on the self-organization of organic molecules, have revealed interesting learning capabilities that will be considered in the successive section, with features intrinsic for living beings' nervous system, allowing also unsupervised learning.

The other interesting approach is connected to the so-called "processor in memory", where the time of the communication of the processor with memory should be minimized.[13] It is to note that the communication between the processor and memory is the slowest process in modern computers. Therefore, successful realization of this paradigm is expected to guarantee a significant increase of the operation speed. Nevertheless, it seems that it can result not only in the speed increase, but can also open new perspectives of learning at the hardware level.

3. Learning Algorithms

When we talk about learning of traditional computers, we usually mean its programming (the worth case of supervised learning: once done it cannot be varied without changing the program) or learning of artificial neuron networks, realized at the software level. The theory of neuron networks is well developed and the main algorithm of learning is connected to the back propagation.[14] In simplest cases, we deal with perceptrons.[15] In the case of elementary single layer perceptron,[16] where there is only one adjustable layer of connections (synapses), learning is rather simple and it is necessary to know only the configuration of input stimuli and the output signals. Training of the system (supervised learning) is connected to the variation of synapse weight function in order to make classification of input stimuli according to the pre-determined algorithm. When the number of layers is increased, learning is more complicated. The algorithm in this case implies the possibility of having access to all elements of the network even if they are in the intermediate position, not in the input or output layer.[17] To make a parallel to the brain, we must know the weight function (and its temporal variation) of each synapse analog and have the possibility to vary them in a desirable way. Of course, in the brain we do not have such possibility. Only terminal "inputs", such as sensor systems and motor neurons with adequate feedback, can be activated. Intermediate (numerous) layers of neurons with associated synapses cannot be directly accessed. Therefore, the brain uses other algorithms for learning, different from those currently used for artificial neuron networks' training.

The learning of the brain is directly connected to the phenomenon of so-called synaptic plasticity: the possibility of changing the strength of connections between neurons. It is somehow similar to the adjustment of weight functions of connections in artificial neuron networks. Nevertheless, there is a significant difference: in the brain there are rules, allowing to vary the synaptic weights automatically according to the sequence of the signal propagation.

The main mechanism responsible for the synaptic strength variation is called Spike Timing Dependent plasticity (STDP).[18,19] Let us briefly consider this mechanism. The signal in the brain is propagated in the form of spikes. Consider we have two neurons, A and B, connected by a synapse. Cell A is a presynaptic neuron, while B is a postsynaptic one. Each cell can generate spikes. STDP rule establishes the relationship between the strength of the synaptic connection between these cells and to the time interval between spikes produced by them. If the time interval between spikes is rather long (tens of ms), nothing happens with the synaptic strength. When the interval is shorter, we can see the variation of the synaptic connection strength. If the spike on the cell A was before that on the cell B, the strength will be reinforced, while in the case that cell B produces the spike before the cell A, it will be suppressed. The level of the variation depends on the duration of the interval between these spikes: the shorter it is, the more pronounced the variation of the synaptic connection. It is possible to claim that this rule automatically establishes the causal relationship. If we consider the spike on cell A as the event "1" and the spike on cell B as the event "2", we can see that short delayed interval of the spiking of cell B after spiking of cell A will increase the synaptic strength, indicating that event "1" is the cause and event "2" is the effect. Repetition of the situation or shortening the interval between spikes will increase gradually the possibility of spiking of the cell B as the result of cell A spiking. Instead, when spiking of the cell B occurs before spiking of the cell A, the synaptic strength will be depressed, indicating that the event "2" cannot be the cause of the event "1".

As you can see, the brain does not require the access to intermediate neurons, the connections between them will be adjusted automatically according to the causality principles. In addition, it seems that mono-synaptic junctions in the brain are responsible for short-term associations, while hetero-synaptic junctions provide long-term memory.[20]

In artificial neuromorphic systems, the most studied learning algorithm was connected to the reproduction of classic conditioning of Pavlov's dog.[21] In this case, learning was connected to the association of neutral stimulus (bell ring) with the presence of food. If these stimuli were repeated together several times, the dog would begin to salivate even when the bell ring alone was present. Such kind of conditioning was realized using different artificial electronic elements, which is not very difficult.[22] However, we want to attract your attention to the work, Ref. 23 where the realized system used STDP as a learning algorithm.

There is another important difference between the brain and machine learning that is directly connected to the difference in their architecture, discussed in the previous section. The architecture of the computers of a certain type is absolutely the same and it cannot be changed. If the feature of a certain computer does not correspond to the initial design, it will be considered as a defective one. Loading of the operation system and software as well as its programming can be hardly called "learning". Thus, as it was mentioned in the previous section, learning can be done only at the software level, using artificial neuron network algorithms.

The situation is different in the case of animals and humans. In general, brains of animals of the same type have similarities in architecture, which is due to genetic factors. However, the brain of each individual has its own characteristic features. These features will develop during the growth phase. It is possible to distinguish "baby" and "adult" learning. The brain of a just born baby can be considered as a complicated system with numerous connections between different neurons. However, the strength of these connections is randomly distributed and must be systematized during learning. In the early stages of life, a "baby" learning (or imprinting) takes place. In the brain the mechanism of "indelible memory"[24] is connected to the elimination of several connections and the potentiation of others. In different parts of the brain, maxima of the connections' formation occurs in different ages, followed by the cutting of some branches of axons and synapses. The degree of cutting depends on the richness of the surrounding world: maximum of this cutting occurs during the sensorial lack in the critical age.[25] Therefore, learning during the critical age plays a fundamental role in the formation of characteristic features, determining the individuum, as they make contributions in the formation of the adult anatomic network. It can be defined as a formation of a person. The stronger the formed connections correspond to the genetic features, the deeper the imprinting. In order to illustrate the correlation of the imprinting depth, we can consider a well-known example, when ducklings were nurtured by a hen. They followed the hen and did not try to go into the water. However, when that happened, they began to swim, because the obtained imprinting did not correspond to the genetically imposed features. However, in most of the cases when there are no such contradictions, characteristic features, established during learning at the early stage of life will be maintained during the whole life period, because imprinting results not only in the variation of synaptic strengths, but also in the modification of the brain architecture at the physiological level. Thus, even the impression that the surrounding world makes on a growing child will be varied in time according to the acquired experience. In the case of the "adult" learning, the situation is different. In our everyday life, learning can result in short- and long-term connections (depending on the frequency and duration of the external stimuli), necessary for resolving of currently available problems according to the configuration of the external stimuli and accumulated experience. However, the established connections will be in a dynamic equilibrium: they can be reconfigured in the case of the appearance of new strong external stimuli.

Regarding the practical realization of brain-inspired learning algorithms, it is worth mention two strongly developing ways: spiking neuron networks[26] and reservoir computing.[27]

In the case of spiking neural networks, brain-like algorithms of learning, such as STDP, are used. It is expected that such systems must allow unsupervised learning as it establishes the causality of events without the necessity of the external supervisor (automatically).

Another very interesting system for considering in neuromorphic applications is reservoir computing. In a few words, the system must contain a set of input and outputs, connected to a "reservoir" — black box with nonlinear properties and possibilities of numerous variable signal pathways. It can be realized both at the software[28] and hardware[29] levels. At the software level, reservoirs are neuron networks with randomly generated connections. Important difference with respect to other neuron networks (perceptrons, for example) is that these connections remain randomly distributed and not all of them can be accessed and varied. One can only provide the actions on input and output terminals according to the correctness of the established connections. In the case of hardware realization of reservoir computers, one must have materials with nonlinear properties (optical, electrical) also allowing memorization.

Learning of this system implies only adequate feedback systems between inputs and outputs. Likewise to the brain, there is no possibility to access nodes in the "reservoir". We can only apply correcting stimuli between different inputs and outputs, controlling that the system begins to perform recognition and classification of different configurations of the external stimuli.

4. Internal Activity and Effect of Noise

In the case of traditional computers, internal activity, implying spontaneous cross-talk of elements and components, is an absolutely negative event that can result in errors in the performed calculations or even the damage of the whole system. On the contrary, it is impossible to imagine cognitive processes in the brain without this cross-talk of neurons. When we think, we even do not need to have new information from sensory systems (it is enough to have already obtained recorded information, that, as we have discussed in Section II, have resulted in the variation of synaptic strengths in different signal pathways): cross-talk of elements itself allows to recognize and classify objects, forecast future events, and create new concepts. As a conclusion, an impression of the surrounding reality will depend not only on the actual signals from sensory systems, but also as a result of previous internal activity (thinking) of the brain. As it is well known, the wisdom can allow to reconstruct the ocean just looking at a drop of water. It means that if the brain is already trained to make generalizations of objects and phenomena, it will be able to imagine an ocean as a qualitative step of the quantitative increase of the water content in a confined space. Thus, it is not necessary to have new stimuli from the sensor systems. In fact, reading an impressive book or listening to an exciting lecture can change our way of thinking. Moreover, the influence on our mentality can also be indirect: basic ideas distributed in society also affect the mentality of individual members.

Let us make a comparison of art and science. During the Middle Ages the main concept of the universe's organization was connected to the fact that the Earth is in its center and all stars and planets are rotating around it. The same situation can also be seen in art: the observer is in the center and "parallel lines" become more distant from each other the more distant they are from the observer (the situation is called "reversed perspective").

During the Renaissance period, the understanding of the universe's organization completely changed: the Earth is not in the center people realized. In parallel, the artists began to make their drawings using a direct perspective, where parallel lines meet each other at the infinity point.

Important optical findings came to light in the 18th century: diffraction has shown that the shape of shadows can be not straight and they can be colored. In parallel, a new style in art, "impressionism", was born.

Finally, starting from the 20th century the physical understanding of the world's organization was significantly changed. We will not list all of the new features here. However, it is possible to say that also the art was drastically varied, assuming the possibility of the influence of the observer on the surrounding reality, coexistence of multiple (parallel) realities, coexistence of consciousness and subconsciousness, and complex organization of the space–time continuum. It is to note, that most of the artists did not read scientific works of Schrödinger, Bohr, Einstein, Freud, etc., but just expressed general ideas "flying" in society. Probably, surrealism was the first attempt for expressing these new ideas. However, now all of these ideas are widely used in literature and movies. Summarizing, the internal brain activity of individual persons is very strongly (even if not directly) affected by the basic ideas distributed in society. For us, it is very easy to understand scientific concepts, diffused in the past, but it seems that modern art would be not clear (and even awful) for the average person of the Middle Ages. We even do not consider the acceptance of scientific ideas: their authors would have been burned at the stake of the Inquisition.

The other important difference between the computer and the brain is the effect of noise (Figure 1). In the case of computers, noise always plays a negative role and can result in computational errors. In the case of the brain, it is different. In some cases, the effect of noise is positive and in some cases it can, for example, facilitate object recognition.[30] Just to draw a parallel, we can consider the

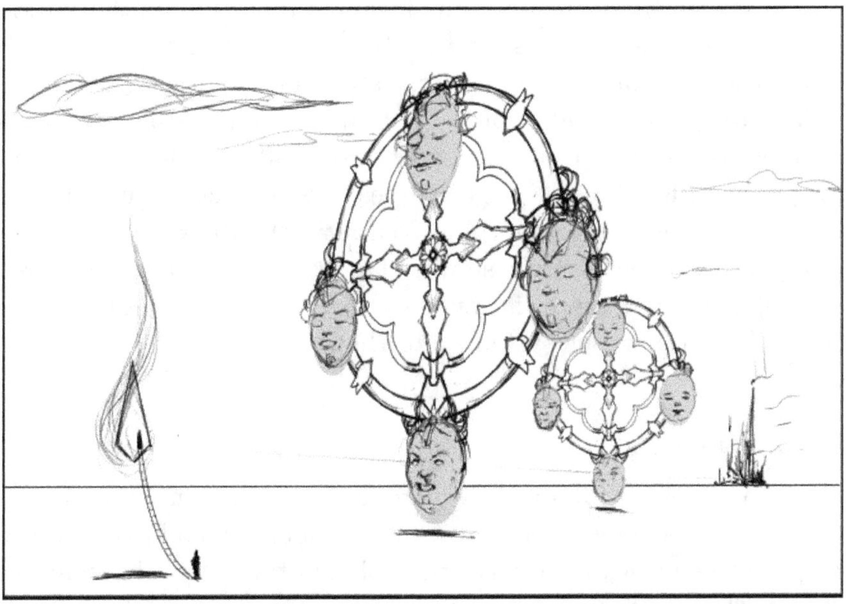

Fig. 1. Illustration of the creative process. Think, for example, of an artist who enters his studio with the purpose of painting a subject that should represent a vague concept like a feeling. He will use some memories and perhaps some visual references too, but the longer the process goes on, the more he will be influenced by the work he is doing. Emotions will start to play a role in the interpretation of what he is painting, plus there is a good amount of noise in the process, perhaps a brush stroke will look like something totally unplanned. Also, why cannot I assume that connections in his brain will change as well? So, at the end of the session the artist will slightly different from when he started.

brain as a very complicated system with a lot of local energy minima. The presence of noise can facilitate the transfer of the system to a deeper minimum, allowing to make a better decision. In principle, even a speech in a foreign language (especially if its phonetics is very different from one's own) can be considered as noise. However, applied learning algorithms will transfer this "noise" into a meaningful information. The nature of noise can be different: disfunction of the physiological activity of the organism or "confusional" identification of the external stimuli. We think that creative processes can also be considered as the superposition of the noise action to internal processes in the brain, whose architecture and connections are established during previous "baby" and "adult" stage learning. In this respect, all of us have reflections of the external world, but only some of us are able to make "ordering" of this noisy information and produce a picture, a story, music, or a scientific idea.

Regarding technical implementations of the ideas of this section, it is worth noting that currently cross-talk of elements is not always considered as a negative feature, especially in the field of unconventional neuromorphic computing.[31] Regarding the noise effect, it has been already shown that it can play a positive role, allowing faster and more efficient switching of the network properties, as it occurs in nervous systems and the brain.[32,33]

5. Emotions

This is also a very important difference between computers and the brain. The behavior of animals and humans depends not only on the performed learning and configuration of the external stimuli, but also on the emotional state of the particular being. The behavior and decisions of the person are strongly affected by emotions. In the brain, different emotions are connected to the production of neurotransmitters. Relative concentrations of these neurotransmitters can vary significantly the features of the nervous system. It can result, for example, in switching of learning algorithms from one to the other. In particular, we have discussed STDP learning in the appropriate section. Even if this algorithm is considered now as the

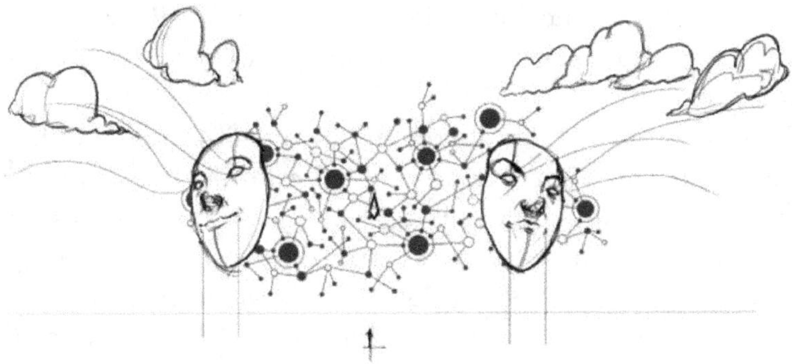

Fig. 2. The perception of the surrounding world depends not only on the age, education, and experience but also on the emotional state of a person.

most important (and studied), it is not the only one.[34] There are alternative learning algorithms, such as anti-STDP and Sombrero. The presence of neurotransmitters (connected to the emotional state of the person) can switch the learning algorithm that the system uses.

The perception of the surrounding world is, of course, strongly dependent on the emotional state (Figure 2). The same event, or image, can even produce an opposite effect on the person if she/he is angry or satisfied. In principle, it is possible to say that emotions play a role on ethic and aesthetic factors, influencing our decisions and behavior. Of course, ethical and aesthetical criteria are different for different persons. However, the state of the community is a very important factor, influencing each individual person. In the case of aesthetics, variations occur rather often. We can, for example, consider the influence of fashion on the dressing and behavior of people. Of course, there are a lot of people who ignore the currently existing fashion. However, even they cannot completely ignore it, participating in events where "dress code" is strictly obligatory. Aesthetic criteria can be different for different countries or even different groups of people, what is connected to their education, the work they are doing, etc. In the case of ethics, the changes are more slow. However, if we consider a rather long historical period, we can see these changes: the ethical criteria in the Middle Ages were significantly different from ones we are using now. Similarly to the case of aesthetics, ethical criteria are different for different groups of people, connected to the country where they live, educational level, and religion.

Of course, in the case of computers the situation is different. It is rather difficult to imagine aesthetical criteria that computers can use. Regarding ethics, there is at least one example, described in science fiction. We mean three laws of Robotics, introduced by Isaac Asimov in his books. It is interesting, that, as it has been shown by the author, even small weakening of any one of these laws can result in serious and unpredictable consequences. Similar situation is also valid for humankind: the 10 commandments are well known but not always executed.

Regarding the technical implementation of emotions for unconventional computing, there are few works where the possibility of the realization of an emotional computer were considered at the level of modeling. A cube of emotions[35] was considered as a starting point of these works. Within this consideration, emotions were attributed to the presence of different neurotransmitters, which was connected to the redistribution of computer resources (memory, processor power, switching between different computer subunits).[36] However, all these works are at the level of modeling, and we have no clear idea how it can be realized at the hardware level. Therefore, the area of emotions in the case of unconventional computing is the least studied one.

6. Conclusion

In this chapter, we tried to make an overview of the existing problems and approaches in the field of the realization of brain-inspired computers. It is obvious that there are much more questions than answers, which makes the problem even more interesting. Artistic illustration of the vision of the problem is shown in Figure 3.

The artist decided to develop the idea of a sort of human-like computer in a typical retro science-fiction fashion. Having in mind the concept of the internal noise and its creative potential, the artist came up in his drawing with those large constructs which are creating and running imaginary cities (the skyscrapers that you can see in the

Fig. 3. Harvesting the huge potential of the human mind: Large brain inspired computers stand tall in the emptiness of a desert landscape. Faraway in time and space superhuman computers are quietly running imaginary cities, their sensory system is fed with the appropriate stimuli and the emotional status is carefully monitored. When needed, neurotransmitters are adequately dosed. Those are the remains of an amazing species that spread across the galaxies and that now rests by those giant human-like constructs that just like the heads on Easter island are now watching over their slumber. Or perhaps they are not resting at all, but rather tirelessly working and ready to conquer yet another frontier!

sphere). The whole block is a sort of an enhanced brain provided with energy, visual stimulations, neurotransmitters, and all the stimuli that a brain needs. The large displays sow the emotional state of it, which we know is a key component. These buildings are linked together in a sort of network and they are standing in a desolate landscape, we like to think of them as the remains of a civilization, like the heads on Easter island. One of the ways to think about it is that most of the humans have uploaded themselves in those cities, another is that the buildings themselves are the last remaining evidence of consciousness on the planet. The little figure at the bottom is something added for a bit of mystery, could be a human or a robot, we don't know.

We have not presented technical details of strategies of neuromorphic systems. Brief description of approaches can be found in Ref. 37, while a more detailed description is presented in Ref. 38.

Acknowledgments

We want to thank Svetlana and Konstantin for useful discussions.

References

1. A. Adamatzky and B. De Lacy Costello, Experimental logical gates in a reaction-diffusion medium: The XOR gate and beyond. *Phys. Rev. E.* 66:046112, 2002.
2. T. Shirakawa, A. Adamatzky, Y.-P. Gunji, and Y. Miyake, On simultaneous construction of Voronoi diagram and Delaunay triangulation by Physarum polycephalum. *Int. J. Bifurcation Chaos.* 19:3109–3117, 2009.
3. A. D. McAulay, *Optical Computer Architectures: The Application of Optical Concepts to Next Generation Computers.* John Wiley and Sons, New York, 1991.
4. B. E. Kane, A silicon-based nuclear spin quantum computer. *Nature.* 393:133–137, 1998.
5. F. Akopyan, J. Sawada, A. Cassidy, R. Alvarez-Icaza, J. Arthur, P. Merolla, N. Imam, Y. Nakamura, P. Datta, G. J. Nam, B. Taba, M. Beakes, B. Brezzo, J. B. Kuang, R. Mahohar, W. P. Risk, B. Jackson, and D. S. Modha, True north: Design and tool flow of a 65 mW 1 million neuron programmable neurosynaptic chip. *IEEE Trans. Comput. Aided Des. Integr. Circuits Syst.* 34:1537–1557, 2015.

6. F. Rosenblatt, The perception: A probabilistic model for information storage and organization in the brain. *Physiological Rev.* 65:386–408, 1958.
7. V. Erokhin, A. Schuz, and M.P. Fontana, Organic memristor and bio-inspired information processing. *Int. J. Unconv. Comput.* 6:15–32, 2010.
8. D. Adams, *The Hitchhiker's Guide to the Galaxy.* Harmony Books, New York, 1980.
9. S. H. Jo, T. Chang, I. Ebong, B. b. Bhadviya, P. Mazumder, and W. Lu, Nanoscale memristor device as synapse in neuromorphic systems. *Nano Lett.* 10:1297–1301, 2010.
10. J. Wang and F. Zhuge, Memristive synapses for brain-inspired computing. *Adv. Mater. Technol.* 4:1800544, 2019.
11. V. Erokhin, T. Berzina, P. Camorani, A. Smerieri, D. Vavoulis, J. Feng, and M. P. Fontana, Material memristive device circuits with synaptic plasticity: Learning and memory. *BioNanoScience.* 1:24–30, 2011.
12. V. Erokhin, T. Berzina, K. Gorshkov, P. Camorani, A. Pucci, L. Ricci, G. Ruggeri, R. Sigala, and A. Schuz, Stochastic hybrid 3D matrix: Learning and adaptation of electrical properties. *J. Mater. Chem.* 22:22881–22887, 2012.
13. M. Gokhale, B. Holms, and K. Iobst, Processing in memory — The terasys massively-parallel PIM array. *Computer.* 28:23–31, 1995.
14. F. Rosenblatt, *Principles of Neurodynamics: Perceptrons and the Theory of Brain Mechanism.* Spartan Books, Washington DC, 1961.
15. P. D. Wasserman, *Neural Computing: Theory and Practice.* Van Nostrand Reinhold, New York, 1989.
16. V. A. Demin, V. Erokhin, A. V. Emelyanov, S. Battistoni, G. Baldi, S. Iannotta, P. K. Kashkarov, and M. V. Kovalchuk, Hardware elementary perceptron based on polyaniline memristive devices. *Org. Electronics* 25:16–20, 2015.
17. P. J. Werbos, *The Roots of Backpropagation. From Ordered Derivatives to Neural Networks and Political Forecasting.* John Wiley and Sons, Inc., New York, NY, 1994.
18. S. Song, K. D. Miller, and L. F. Abbott, Competitive Hebbian learning through spike-timing-dependent plasticity. *Nat. Neurosci.* 3:919–926, 2000.
19. N. Caporale and Y. Dan, Spike timing-dependent plasticity: A Hebbian learning rule. *Annual Rev. Neurosci.* 31:25–46, 2008.
20. C. H. Bailey, M. Giustetto, Y. Y. Huang, R. D. Hawkins, and E. R. Kandel, Is heterosynaptic modulation essential for stabilizing Hebbian plasticity and memory? *Nature Rev. Neurosci.* 1:11–20, 2000.
21. I. P. Pavlov, Experimental psychology and psychopathology in animals. In *Lectures on Conditioned Reflexes*, International Publishers, vol. 1, 1928, 47–60.
22. M. Ziegler, R. Soni, T. Patelczyk, M. Ignatov, T. Bartsch, P. Meuffels, and H. Kohlstedt, An electronic version of Pavlov's dog. *Adv. Func. Mater.* 22:2744–2749, 2021.

23. A. A. Minnekhanov, A. V. Emelyanov, D. A. Lapkin, K. E. Nikiruy, B. S. Shvetsov, A. A. Nesmelov, V. V. Rylkov, V. A. Demin, and V. V. Erokhin, Parylene based memristive devices with multilevel resistive switching for neuromorphic applications. *Sci. Rep.* 9:10800, 2019.
24. J. W. Lichtman and H. Colman, Synapse elimination and indelible memory. *Neuron.* 25:269–278, 2000.
25. D. Purves and J. W. Lichtman, Elimination of synapses in the developing nervous system. *Science.* 210:153–157, 1980.
26. S. Ghosh-Dasidar and H. Adeli, Spiking neural networks. *Int. J. Neural Syst.* 19:295–308, 2009.
27. G. Tanaka, T. Yamane, J. B. Heroux, R. Nakane, N. Kanazawa, S. Takeda, H. Numata, D. Nakano, and A. Hirose, Recent advances in physical reservoir computing: A review. *Neural Netw.* 115:100–123, 2019.
28. S. Ortin, M. C. Soriano, L. Pesquera, D. Brunner, D. San-Martin, I. Fischer, C. R. Mirasso, and J. M. Gutierrez, A unified framework for reservoir computing and extreme learning machines based on a single time-delayed neuron. *Sci. Rep.* 5:14945, 2015.
29. C. Du, M. A. Zidan, W. Ma, S. H. Lee, and W. D. Lu, Reservoir computing using dynamic memristors for temporal information processing. *Nat. Commun.* 8:2204, 2017.
30. A. Fiasconaro and B. Spagnolo, Stability measures in metastable states with Gaussian colored noise. *Phys. Rev. E.* 80:041110, 2009.
31. A. Dimonte, T. Berzina, M. Pavesi, and V. Erokhin, Hysteresis loop and cross-talk of organic memristive devices. *Microelectron. J.* 45:1396–1400, 2014.
32. G. A. Patterson, P. I. Fierens, and D. F. Grosz, On the beneficial role of noise in resistive switching. *Appl. Phys. Lett.* 103:074102, 2013.
33. S. Battistoni, R. Sajapin, V. Erokhin, A. Verna, M. Cocuzza, S. L. Marasso, and S. Iannotta, Effect of noise sourcing on organic memristive devices. *Chaos Solit. Fractals.* 141:110319, 2020.
34. M. Talanov, E. Zykov, V. Erokhin, E. Magid, S. Distefano, Y. Gerasimov, and J. Vallverdu, Modeling inhibitory and excitatory synapse learning in the memristive neuron model. In *ICINCO Proc. 14th Int. Conf. Informations on Control, Auromation and Robotics*, vol. 2, 2017, 514–521.
35. H. Lovheim, A new three-dimensional model for emotions and monoamine neuro-transmitters. *Med. Hypotheses.* 78:341–348, 2012.
36. J. Vallverdu, M. Talanov, S. Distefano, M. Mazzara, A. Tchitchigin, and I. Nurgaliev, A cognitive architecture for the implementation of emotions in computing systems. *Biol. Inspired Cogn. Archit.* 15:34–40, 2016.
37. V. Erokhin, Memristive devices for neuromorphic applications: Comparative analysis. *BioNanoScience.* 10:834–847, 2020.
38. V. Erokhin, *Fundamentals of Organic Neuromorphic Systems.* Springer Nature, Switzerland, 2022.

https://doi.org/10.1142/9789811257155_0011

Chapter 11

Novel Reversible Logic Elements for Unconventional Computing

Kenichi Morita
Hiroshima University,
Higashi-Hiroshima 739-8527, Japan
km@hiroshima-u.ac.jp

Logic elements used in conventional computers are mostly logic gates such as AND, OR, NOT, NAND, etc. Their history is quite long, since logical operations of AND, OR and NOT were already known more than 2,000 years ago.[1] They were obtained by analyzing human thinking and reasoning, and thus it is easy for us to understand them. However, when we investigate future computing systems, we should not be tied to such old traditions. Reversible computing[2] is a paradigm of computing that reflects physical reversibility, one of the fundamental microscopic physical laws of nature. It searches for novel and unconventional methodologies that are directly related to reversible microscopic phenomena. We consider here a reversible logic element with 1-bit memory (RLEM), and examine its possibilities as a novel logical device. Though its physical realizability in the nano-level is not known at present, it gives new vistas on unconventional computing devices. We shall see that it has very different features that cannot be seen in conventional logic gates. In particular, reversible computing systems such as reversible Turing machines can be constructed out of RLEMs in a very unique way. We also see that an RLEM can be implemented reversible environments such as a billiard ball model of computation, and a very simple reversible cellular automaton. In the following, these features are explained using many illustrations without giving technical details.

1. Reversible Logic Element with 1-bit Memory

A *reversible logic element with 1-bit memory* (RLEM) is a kind of reversible finite automaton. Figure 1 shows a typical example of an

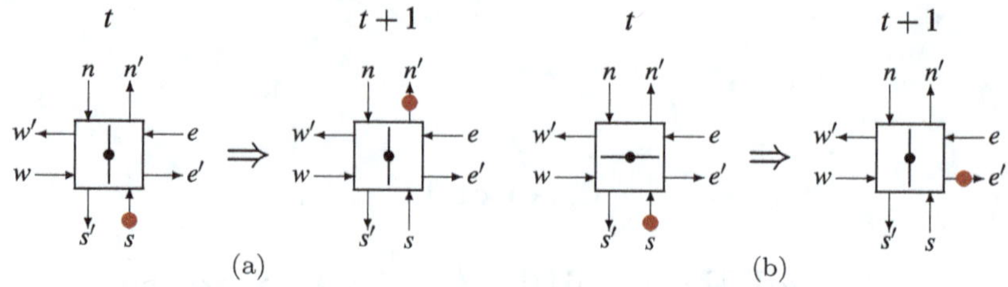

Fig. 1. Rotary element (RE) and its operations. (a) The parallel case, and (b) the orthogonal case.

RLEM with four input ports and four output ports called a *rotary element* (RE).[3] Conceptually, it has a rotatable bar that controls the move direction of an incoming signal (or a particle). It takes one of the two states H and V depending on the direction of the bar (i.e., horizontal or vertical). If a signal comes from the direction parallel to the bar, it goes straight ahead and the state does not change. If a signal comes from the direction orthogonal to the bar, it turns rightward and the state changes. It is reversible in the following sense: From the state at $t+1$ and the output, the state at t and the input are uniquely determined.

2. Composing Reversible Computers out of Rotary Elements

A *reversible Turing machine* is an abstract model of a reversible computer where every computational state has at most one predecessor. Hence, we can trace back its computing process uniquely. It is known that any (irreversible) Turing machine can be simulated by a reversible one without leaving garbage information on the tape, and thus reversible Turing machines are computationally universal.[4] We can construct any reversible Turing machine out of rotary elements. Figure 2 shows a circuit that simulates a simple reversible Turing machine that checks if a unary number given on its tape is even.[5] In this figure, a finite-state control of the reversible Turing machine is in the left part, and a tape unit is in the right part. If a particle is given to the Begin port, it starts to compute.

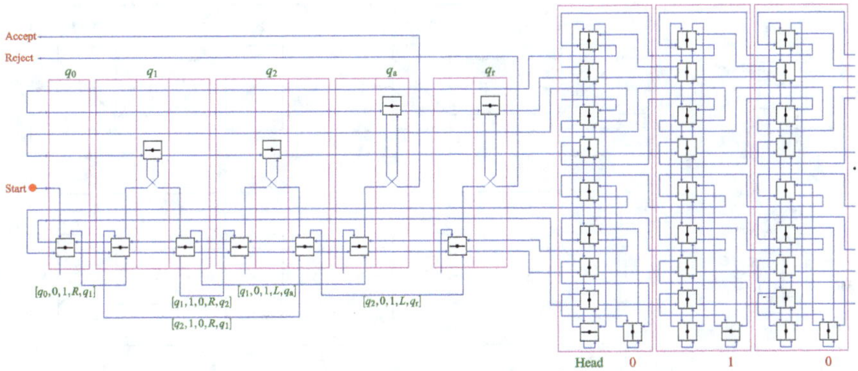

Fig. 2. An example of a reversible Turing machine composed of rotary elements.

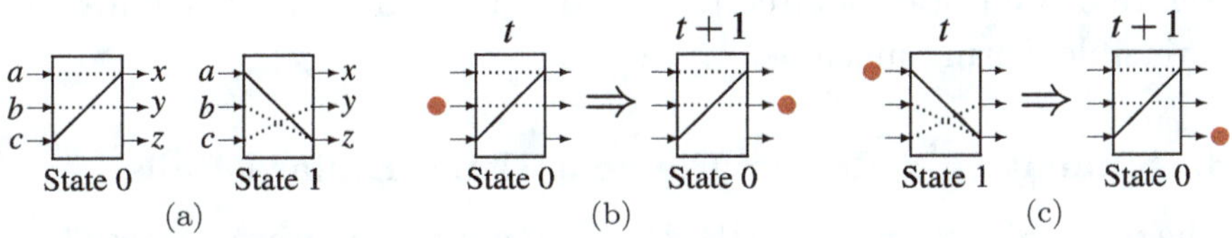

Fig. 3. (a) States of RLEM 3-7. (b) The case where the state does not change. (c) The case where the state changes.

3. Universal RLEMs

There are infinitely many RLEMs if we do not restrict the number of input/output ports. Figure 3 shows RLEM No. 3-7, where "3" stands for 3-input and 3-output, and "7" is its serial number. Two boxes in Figure 3(a) indicate its two states. The dotted and solid lines give input–output relation in each state. If an input signal goes through a dotted line, the state does not change (Figure 3(b)). If it goes through a solid line, the state changes (Figure 3(c)). Note that RE can also be represented by such a figure, but we employ Figure 1 for ease in understanding. An RLEM is called *universal* if it can simulate any other RLEM. Remarkably, it has been proved that *every* RLEM (except degenerate ones) is universal if it has three or more input/output ports.[6] Therefore, RLEM 3-7 and RE are of course universal. Figure 4 shows how to simulate an RE by RLEM 3-7. Replacing each occurrence of REs in Figure 2 by the circuit of

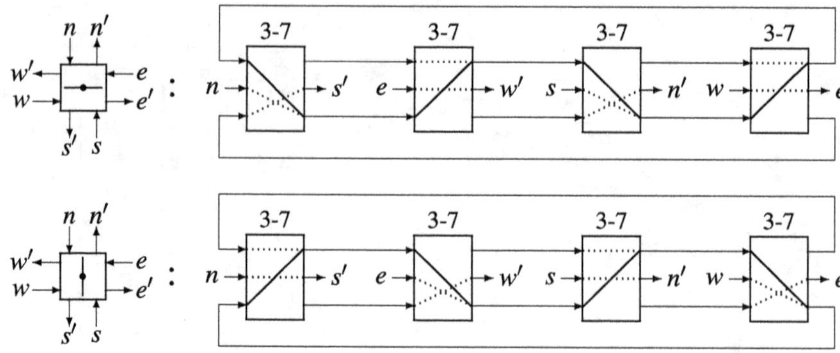

Fig. 4. Simulating a rotary element by a circuit composed of RLEM 3-7.

Figure 4, we obtain a circuit made of RLEM 3-7 that simulates the reversible Turing machine.

4. Simulating a Rotary Element Using Billiard Balls

The *billiard ball model* (BBM) of computation was proposed by Fredkin and Toffoli[7] to show that the Fredkin gate, a universal reversible logic gate, is realizable in the BBM. It is a kind of reversible mechanical system consisting of ideal elastic balls and reflectors. We can see that an RE is also realized in the BBM as shown in Figure 5 where small rectangles are reflectors.[2] Here, two kinds of balls, i.e., a *state ball* and a *signal ball*, are used. One of the key points of the construction is that the state ball (yellow) is put stationarily at the position H or V in a resting mode. Figure 5 shows the cases where the directions of the bar and the incoming signal are orthogonal as in Figure 1(b). Consider the case that the state ball is put at the position H, and the signal ball (green) comes from the input port s. The signal ball collides with the state ball at the position H. Then, the state ball and the signal ball move along the paths p_0 and p_1, respectively. When the state ball comes to the position V, these balls collide again. Then, the state ball stops at V, while the signal ball moves eastward and goes out from the output port e'. By this the operation of Figure 1(b) is realized. The case that the state ball is at the position V, and the signal ball comes from s, is trivial. In this

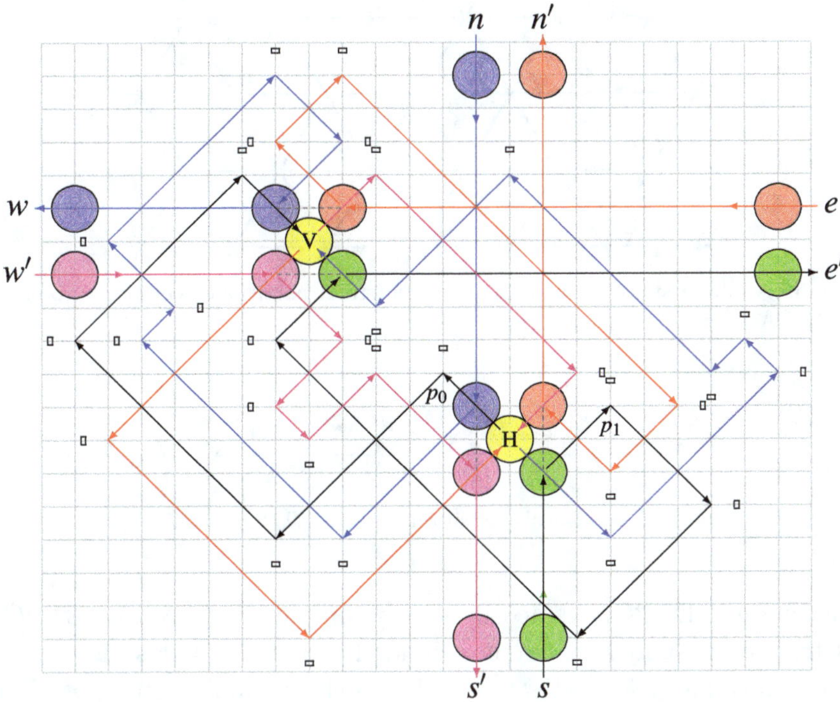

Fig. 5. A rotary element realized in the billiard ball model, an idealized mechanical computing model.

case, the signal ball simply moves northward without interacting with the state ball and the reflectors, and thus Figure 1(a) is realized. In this way, the whole circuit that simulates a reversible Turing machine given in Figure 2 is also embeddable in the BBM.

5. Realizing RLEMs in a Simple Reversible Cellular Space

There is yet another spatiotemporal model of a reversible environment called a *reversible cellular automaton.* A cellular automaton (CA) consists of an infinite number of finite automata called *cells* that are placed and connected uniformly in a space. A cell changes its state depending on the states of its neighboring cells. We use a special type of a CA called an *elementary triangular partitioned cellular*

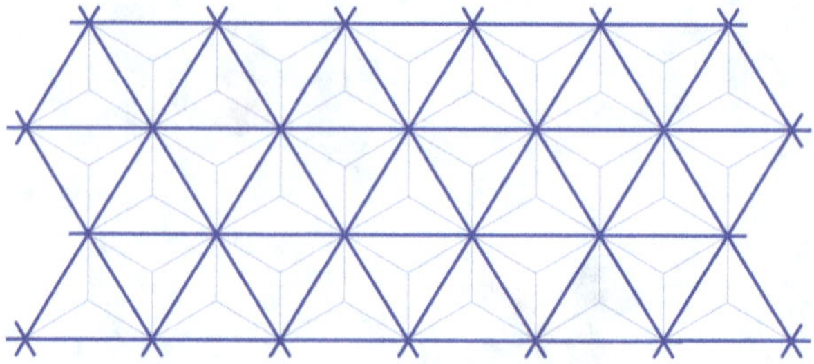

Fig. 6. Cellular space of an elementary triangular partitioned cellular automaton (ETPCA).

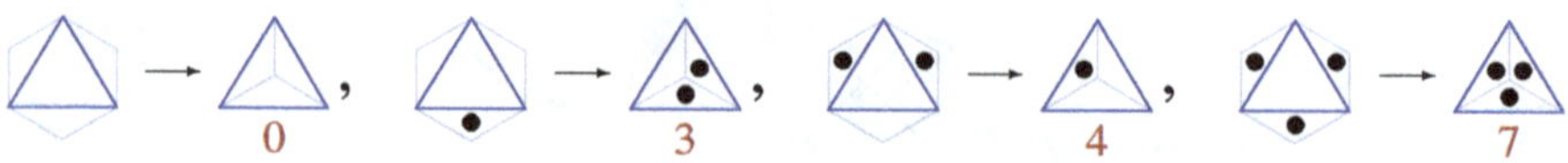

Fig. 7. Local transition rules of ETPCA 0347.

automaton (ETPCA) since it is very simple. Each cell of ETPCA is triangular, and it is further divided into three parts (Figure 6). Each part has two states 0 and 1, which are indicated by a blank and a dot. Here, we consider a particular local transition function defined by the four local transition rules shown in Figure 7, which is identified by the number 0347.[8] We assume that these rules are rotation-symmetric, i.e., for each rule in Figure 7 there exist rules obtained by rotating both sides of it by a multiple of 60 degrees. As seen from Figure 7, the next state of a cell is determined by the three parts of its neighboring cells.

Consider the configuration given at time $t = 0$ in Figure 8. Applying the local function to all the cells in parallel, we have a configuration at $t = 1$. Configurations at $t = 2, 3, \ldots$ are obtained likewise. Note that the dot patterns at $t = 0$ and $t = 6$ are the same except that the latter is shifted rightward. Hence, it is a space-moving pattern called a *glider*, which can be used as a signal.

The local transition function defined by the rules in Figure 7 is injective, since there is no pair of distinct rules whose right-hand sides are the same. By this, for each configuration we can find a previous configuration uniquely. Such an ETPCA is called *reversible*.

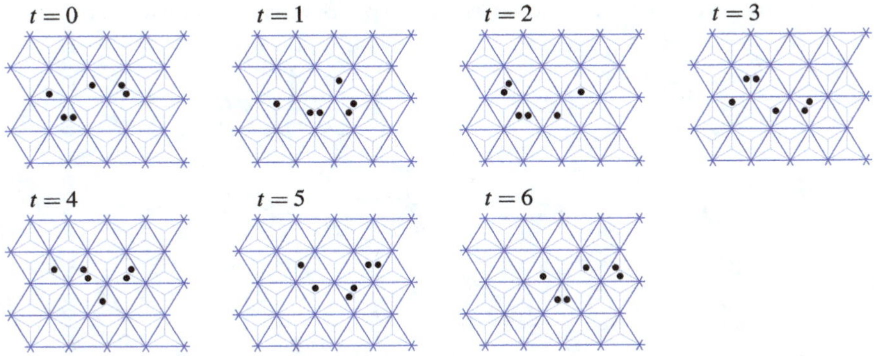

Fig. 8. A space-moving pattern called a *glider* in ETPCA 0347.

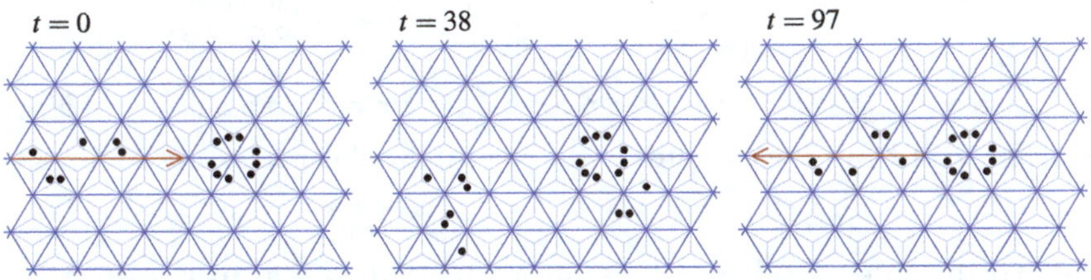

Fig. 9. Backward turn of a glider is realized by a collision with a block.

The moving direction of a glider is controlled by a stable pattern called a *block*. Figure 9 shows the process of the backward turn by a single block. At $t = 0$ a glider (left) is about to collide a block (right). At $t = 38$ the glider is split into a *rotator* (left) and a *fin* (right). The fin moves around the block. At $t = 97$ the rotator and the fin meet, and a glider is reconstructed. Then the resulting glider moves leftward. Appropriately placing several blocks, we can also realize the right turns by 60 and 120 degrees, the left turn by 120 degrees, and the U-turn of a glider.

Combining several useful phenomena found in the reversible cellular space of ETPCA 0347, we can construct a pattern that simulates RLEM 3-7 (Figure 3) as shown in Figure 10. A glider is given to one of the three input use ports as a signal. After changing the state, the glider comes out from one of the output ports. The pattern contains many backward turn, right turn, and

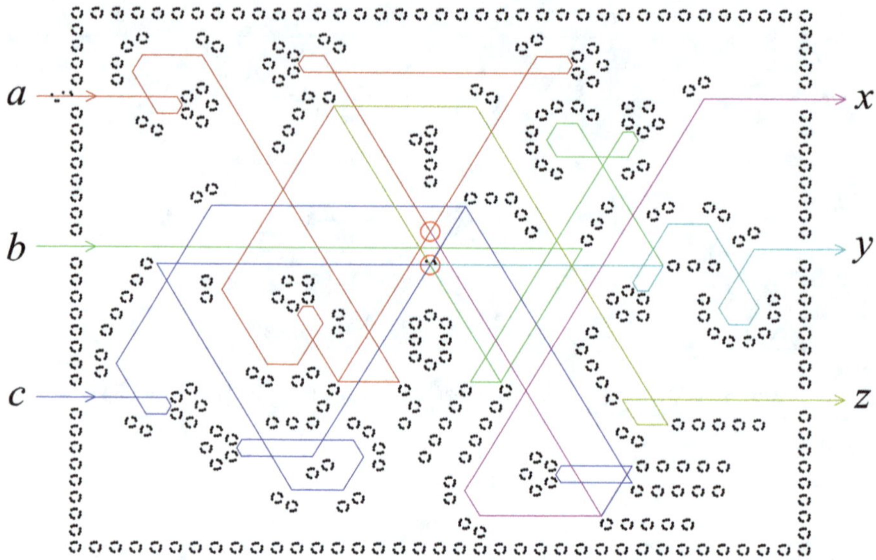

Fig. 10. RLEM 3-7 implemented in the reversible cellular space of ETPCA 0347.

U-turn modules. To keep the state of RLEM 3-7, a fin is put around the center of the pattern. A fin is a periodic pattern of period 6 consisting of three dots. Two small circles in the figure show possible positions of a fin. If a fin is at the lower (upper, respectively) position, then we consider the state is 0 (1). Testing whether the fin is at the position 0 or 1, and shifting it between these two positions are performed by collisions of the glider. In this way, RLEM 3-7 is simulated in the cellular space of ETPCA 0347 having an extremely simple local function. Since a rotary element can be composed of RLEM 3-7, circuits that simulate reversible Turing machines like the one in Figure 2 are also simulated in this cellular space. An emulator for ETPCA 0347 has been given,[9] which works on the general purpose cellular automaton simulator *Golly*.[10] There, computing processes of reversible Turing machines composed of RLEM 4-31 can be seen.

References

1. J.-M. Bocheński, *Ancient Formal Logic*. North-Holland Publishing Company, Amsterdam, 1951.

2. K. Morita, *Theory of Reversible Computing*. Springer, Tokyo, 2017. Doi: 10.1007/978-4-431-56606-9.
3. K. Morita, A simple reversible logic element and cellular automata for reversible computing. In *Proc. MCU 2001*, M. Margenstern, Y. Rogozhin, editors. LNCS 2055, 2001, 102–113. Doi: 10.1007/3-540-45132-3_6.
4. C.-H. Bennett, Logical reversibility of computation. *IBM J. Res. Dev.* 17:525–532, 1973. Doi: 10.1147/rd.176.0525.
5. K. Morita, Constructing reversible Turing machines by reversible logic element with memory. In *Automata, Universality, Computation*, A. Adamatzky, editor. Springer, 2015, 127–138. Doi: 10.1007/978-3-319-09039-9_6. Slides with figures of computer simulation: Hiroshima University Institutional Repository. http://ir.lib.hiroshima-u.ac.jp/00029224.
6. K. Morita, T. Ogiro, A. Alhazov, and T. Tanizawa, Non-degenerate 2-state reversible logic elements with three or more symbols are all universal. *J. Mult.-Valued Log. Soft Comput.* 18:37–54, 2012.
7. E. Fredkin and T. Toffoli, Conservative logic. *Int. J. Theoret. Phys.* 21:219–253, 1982. Doi: 10.1007/BF01857727.
8. K. Morita, A universal non-conservative reversible elementary triangular partitioned cellular automaton that shows complex behavior. *Nat. Comput.* 18(3):413–428, 2019. Doi: 10.1007/s11047-017-9655-9.
9. K. Morita, Reversible world: Data set for simulating a reversible elementary triangular partitioned cellular automaton on Golly. Hiroshima University Institutional Repository. http://ir.lib.hiroshima-u.ac.jp/00042655, 2017.
10. A. Trevorrow, T. Rokicki, T. Hutton, *et al.*, Golly: An open source, cross-platform application for exploring Conway's Game of Life and other cellular automata. http://golly.sourceforge.net/ 2005.

https://doi.org/10.1142/9789811257155_0012

Chapter 12

Logic, Stress, and Satisfaction

Andrew Schumann

Department of Cognitive Science and Mathematical Modelling, University of Information Technology and Management in Rzeszow, Sucharskiego 2, 35-225 Rzeszow, Poland

andrew.schumann@gmail.com

In this chapter, it is claimed that (at least classical) logic is not a natural process of thinking. It is always biased by two moods: stress and satisfaction. Some biased forms of conjunction (under stress) and disjunction (under satisfaction) are observed from the level of actin filament networks — proteins responsible in perceiving signals by cells. Furthermore, these forms are scale-invariant and they can be observed then on swarm patterns and making decisions by humans.

1. Introduction

In unconventional computing any natural system, such as physical, chemical or biological one, is regarded as a medium which can implement some logical (or, more generally, arithmetic) functions. It meas that an appropriate hypothesis is put forward according to which logic is implemented by nature itself. For instance, it was shown that *Physarum polycephalum*, the unicellular organism behaving as a giant amoeba, can approximate some simple algorithms of Kolmogorov–Uspensky machines[1] or algorithms of cellular automata.[2] This amoeba can be examined even as some biological logic gates.[3–7] Then these results were generalized to demonstrate that some proteins of amoeboid organisms are responsible for their intelligent behavior. Therefore these proteins can be considered a medium implementing different logical and arithmetic functions. So, some networks of these proteins such as actin filament networks can be presented as a logic or computation medium, too.[8–15]

Hence, there are many evidences from unconventional computing that logic is implemented in some biological processes, indeed — at least on the activity of *Physarum polycephalum* and within actin filament networks.

In this chapter, I am going to show that logic is implemented biologically only in some biased forms. A pure logic does not exist. In the philosophy of the 19th century, the idea that logic is a natural way of thinking for all humans was dominant. Nevertheless, in the 20th century it was proved experimentally that humans do not think logically in its strict sense. They make gross blunders even at the primary level of logic. For example, the logical conjunction $A\&B$ is often understood not correctly — some referents do not know that its probability $P(A\&B)$ must be always smaller than the individual probability $P(A)$ or $P(B)$.[16] It is an insensitive blunder in understanding conjunction as one of the basic logical operations.

Let $\vdash$ be a sign of logical deducibility. So, $A \vdash B$ means that from A we can deduce B, i.e., A is a premise for inferring B, and $\vdash B$ means that B is an axiom, i.e., it holds true everywhere without premises. Then any inference of A from B in classical logic: $A \vdash B$, assumes that an appropriate implication $A \Rightarrow B$ holds true. This statement is called the deduction theorem: if $A \vdash B$, then $\vdash A \Rightarrow B$.

In classical logic there are the following two basic inference rules:

modus ponens If $\vdash A \Rightarrow B$ and $\vdash A$, then $\vdash B$ (it means that from the true implication $A \Rightarrow B$ and the true statement A we can infer that B is true, also);

modus tollens If $\vdash A \Rightarrow B$ and $\vdash \neg B$, then $\vdash \neg A$ (it means that from the true implication $A \Rightarrow B$ and the true negation of B we can infer the true negation of A, as well).

These inference rules are always valid. There are similar deductions which are invalid in fact:

affirmation of the consequent If $\vdash A \Rightarrow B$ and $\vdash B$, then $\vdash A$ (but it is invalid that from the true implication $A \Rightarrow B$ and the confirmation of its consequent B we can infer that A is true);

denial of the antecedent If $\vdash A \Rightarrow B$ and $\vdash \neg A$, then $\vdash \neg B$ (but it is invalid that from the true implication $A \Rightarrow B$ and the denial of its antecedent A we can infer the denial of its consequent A).

Understanding that *modus ponens* and *modus tollens* are valid, while the affirmation of the consequent and the denial of the antecedent are invalid, constitute the nuts and bolts of elementary propositional logic. Without this understanding any reasoning will contain fallacies. But it turned out that sometimes even scientists cannot draw elementary correct conclusions from conditionals presented by true implications. So, seventy-two scientists including psychologists, biologists, and physicists from a large US university completed a questionnaire[17] to check their abilities to rate the validity of four forms of propositional inference: *modus ponens*, *modus tollens*, affirmation of the consequent, and denial of the antecedent. As a result, some of them made serious fallacies, see Table 1. In the meanwhile, there was detected the failure of nearly half of the scientists to recognize the logical validity of *modus tollens*, one of the two fundamental inference rules of propositional logic. This experiment shows that even scientists do not think purely logically with avoiding unintentional fallacies.

As it was shown,[18] the fallacy of affirming the consequent is very common for biologists. Sometimes they recognize the following reasoning as logically valid: "If bipolar patients have reduced cognitive processing speed, then we will observe slower reaction time on a test. We observe a slower reaction time, and therefore bipolar patients have reduced cognitive processing speed." Furthermore, they often use the false analogy to extrapolate from sharing some characteristics between two groups of animals to sharing their other characteristics.[18] Thus, logic is not natural even for humans and even for scientists! We can then wonder if we can expect to see a pure logic

Table 1. This table shows that scientists make gross fallacies in understanding *modus ponens*, *modus tollens*, affirmation of the consequent, denial of the antecedent. Even the logic training for them is not so helpful.

Referents	Psychologists (%)	Biologists (%)	Physicists (%)	Overall percentage across disciplines (no logic training) (%)	Some logic training ($n = 26$) (%)
1. Recognizing the logical validity of *modus ponens*,					
abstract scheme:	83	92	92	89	100
concrete example:	100	100	100	100	100
2. Recognizing the logical invalidity of denial of the antecedent,					
abstract scheme:	83	58	75	72	67
concrete example:	75	75	100	83	100
3. Recognizing the logical invalidity of affirmation of the consequent,					
abstract scheme:	83	75	75	73	87
concrete example:	75	50	92	72	82
4. Recognizing the logical validity of *modus tollens*,					
abstract scheme:	33	33	58	41	53
concrete example:	50	83	75	69	91

Note: The table is taken from the work of Kern *et al.*[17]

in nature — how far biological processes of *Physarum polycephalum* or actin filament networks can implement the pure logic.

In this chapter, I'll show that logic is ever biased in nature and it can be biased more or less, but its pure (unbiased) forms are almost impossible. In Section 2, this thesis is considered on the material of intra- and inter-cellular effects caused by a lateral activation and a lateral inhibition. Let us note that the lateral activation can be treated as a state of satisfaction on the microlevel, while the lateral inhibition as a state of stress. In Section 3, we regard some biased forms of intelligent reactions performed by different animals to show that these reactions repeat the lateral activation and the lateral inhibition, but now on the macrolevel.

2. Conjunctive Meaning of Lateral Inhibition and Disjunctive Meaning of Lateral Activation

We may observe the following fundamental duality in all active reactions of different organisms to external stimuli: each stimulus either attracts the organism or repels it. So, each stimulus which in fact changes the behavior of individuals, can be considered either an attractant (such as food pieces or sex pheromone) or a repellent (such as predators or poisons). This duality is topological: the attractant can be occupied **from** different directions, but from the repellent each individual moves **in** different directions. On the basis of this topological duality, it is possible to present (i) the attractants $|p|$, $|q|, \ldots, |r|$, occupied by some individuals, as true propositional variables p, $q, \ldots, r$; (ii) and the repellents $|\dot{p}|$, $|\dot{q}|, \ldots, |\dot{r}|$, avoided by some individuals, as true negations $\neg p$, $\neg q, \ldots, \neg r$.

Evidently, different attractants (repellents) have a different intensity of their influence on individual behaviors: they can more or less attract (repel) organisms. This property can be expressed by a degree of truth (falsehood) presented by the interval of real numbers $[0, 1]$, where it is possible to be closer to 0 (it means, do more to repel) or to 1 (do more to attract).

Now, let us consider several attractants $|p|$, $|q|, \ldots, |r|$ and several repellents $|\dot{p}|$, $|\dot{q}|, \ldots, |\dot{r}|$. Surely that all the repellents $|\dot{p}|$, $|\dot{q}|, \ldots, |\dot{r}|$ must be avoided. Then this reaction of individuals can be treated as the following true conjunction: $\neg p \wedge \neg q \wedge \cdots \wedge \neg r$ (it means, $|\dot{p}|$, $|\dot{q}|, \ldots, |\dot{r}|$ are avoided simultaneously). However, in respect to the attractants $|p|$, $|q|, \ldots, |r|$, reactions of individuals can be very different. First of all, there are the following two typical reactions: (i) fusion as an attempt of focusing only on one attractant (or on a small number of attractants) to occupy it (them) first and foremost; and (ii) splitting as an attempt of occupying several attractants at once, the more the better.

Both possible reactions to the attractants are explained by the following two complementary mechanisms in perceiving signals on the intra- and inter-cellular levels: (i) a lateral inhibition corresponding to a fusion, it expresses some stress conditions which increase the contrast of one signal to react more directly and efficiently and to be better focused; and (ii) a lateral activation corresponding to a splitting, it shows some comfortable conditions which decrease the contrast of signals to perform concurrent actions. These mechanisms in perceiving stimuli are scale-invariant. First, they occur during assembling and disassembling of actin filament waves to cause a fusion of protoplasmic tubes of *Physarum polycephalum* or splitting them. Second, they occur between groups of neurons within the visual cortex to process visual cues more or less focused. Third, they occur between active zones of swarms to react to their environments to occupy attractants more or less focused, too.

Repellents $|\dot{p}|$, $|\dot{q}|, \ldots, |\dot{r}|$ cause always a stress in individuals, therefore each individual tries to be concentrated on the repellents due to the lateral inhibition to avoid all of them in advance. But for attractants $|p|$, $|q|, \ldots, |r|$, each individual can be more laterally inhibited or more laterally activated. As a consequence, it starts to occupy all attractants or only some of them. The attractants in focus are connected by conjunction. The attractants which are not in focus are connected by disjunction. Hence, the lateral inhibition for signals $|p|$, $|q|, \ldots, |r|$ gives their increasing $p \wedge q \wedge \cdots \wedge r$, while the lateral activation for signals $|p|$, $|q|, \ldots, |r|$ gives their decreasing $p \vee q \vee \cdots \vee r$. The problem is that there is a smooth transformation between lateral inhibition and lateral activation. Therefore, there is the same smooth transformation between conjunction and disjunction which is difficult to be presented logically. In other words, conjunction as well as disjunction can be more or less biased (activated).

Many optical illusions occur, because our brain is trained to switch between lateral inhibition and lateral activation automatically under different conditions. For example, in the Müller–Lyer illusion, see Figure 1, we perceive a length of a line depending on whether the line has the inward wings, see Figure 1(a) (when we follow the lateral inhibition effect), or the outward wings, see Figure 1(b) (the

(a) Line with inward wings. (b) Line with outward wings. (c) Line without wings.

Fig. 1. The Müller–Lyer illusion. The lines of (a), (b), (c) are of the same length. Nevertheless, it seems to us that the line with the inward wings (i.e., (a)) is shorter than line (b), and the line with the outward wings (i.e., (b)) is longer than line (a).

lateral activation effect). It is worth noting that T. Sakiyama and Y.-P. Gunji showed[19] that the Müller–Lyer illusion holds for the group behavior of foraging ants. They explain this phenomenon by that each swarm of ants has the following two main logistic tasks: (i) to build a global route system connecting the nest with food sources to monopolize all reachable food sources (it corresponds to the lateral activation, that is, to the colony's ability to discover new food sources through exploration); and (ii) to exploit effectively and efficiently each found food source (it corresponds to the lateral inhibition, i.e., to the colony's ability to concentrate on some food sources).

In the Ebbinghaus illusion, see Figure 2, as well as in the Delboeuf illusion, see Figure 3, perceiving the disc surrounded by bigger figures is laterally inhibited, therefore this disc seems to be smaller, while perceiving the disc surrounded by smaller figures is laterally activated, therefore this disc seems to be larger. It was shown[20] that a kind of Ebbinghaus or Delboeuf illusions can be observed in the behavioral patterns of honeybees. So, when the viewing distance was unrestricted, an appropriate illusion occurred. However, when the viewing distance was restricted, bees did not demonstrate perceiving the Ebbinghaus or Delboeuf illusions.

Note that at least one Grey parrot (*Psittacus erithacus*) reports the Müller–Lyer illusion under certain conditions.[21] The Ebbinghaus and Delboeuf illusions are well traced in perceptions of macaque monkeys and some other species of monkeys.[22,23] The point is that the perception of brightness in macaques and humans is quite similar because of common underlying neural computations in the two species. The same illusions under some conditions are well traced

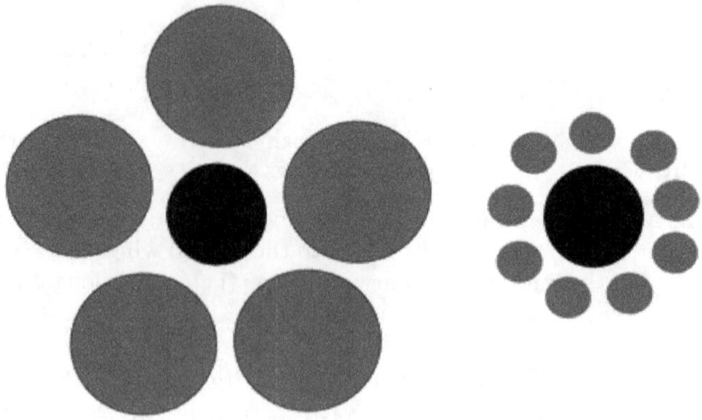

Fig. 2. The Ebbinghaus illusion. The two discs in both centers are of the same size, but the right one seems to be larger than the left one because of the surrounded discs of different diameters.

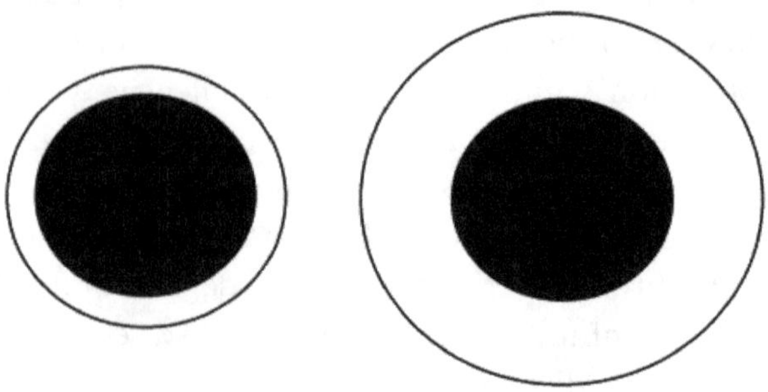

Fig. 3. The Delboeuf illusion. The two discs in both centers are of identical size, but the left one seems to be larger than the right one because of the surrounded rings of different diameters.

in perceptions of red-footed tortoises (*Chelonoidis carbonaria*) and bearded dragons (*Pogona vitticeps*).[24] Nevertheless, they are not so well reported for domestic dogs.[25] Thus, different mechanisms of lateral inhibition and activation may alter perceiving or not perceiving different illusions in a particular species. This data could improve our understanding of animal perception. In any case this perception is always biased. So, a reaction of any organism to stimuli can be more conjunctive (laterally inhibited) or more disjunctive (laterally activated) than it is in fact.

3. Cognitive Biases of Humans and Animals

Behavioral scenarios of human and animals may be more or less rational. If they are irrational, they are called cognitive biases. There are the same two main reasons of many biases: stress as psychic reaction (analogous to the lateral inhibition with one neural network) and satisfaction (analogous to the lateral activation of neurons). For example, managers usually overestimate the probability of positive forecasting[26] — they are not focused enough and pay more attention to information that supports their opinions. Among cognitive biases of managers there are the so-called framing (when they may frame the risky outcome differently, either considering the worst case or the best case) and confidence (their optimistic or pessimistic assessment of one's ability to perform, e.g., in a period of economic recession, people are less confident and more pessimistic about future economic conditions and as a consequence they avoid risky decisions).[27] There is also a "herd behavior", when they copy actions of a larger group due to the social pressure of conformity, e.g., some investors may follow others to buy the same stocks.[28] The study of different cognitive biases and their influences on making financial decisions is presented in behavioral finance,[29] a new approach in finance. So, managers are not enough logical and their decisions are at least partly always biased.

Now, cognitive biases are studied for animals, too, first of all for animals of zoo and aquaria to check their affective states to know whether their welfare is acceptable or it should be improved.[30] It is stated[31] that a balance of positive and negative affective states is always preferable for managing zoo or aquaria. For instance, cognitive biases are analyzed for bottlenose dolphins (*Tursiops truncatus*)[32] and western lowland gorillas (*Gorilla gorilla gorilla*).[33] It is learnt that under stress conditions animals follow more pessimistic judgements about ambiguous stimuli, pay more attention to negative stimuli, and remember more often negative memories.[30,34,35] As a consequence, it was detected[36] that, similar to humans, animals in their negative affective states exhibit negative affective biases on decision-making and memory tasks. Their emotional responses to a stimulus can be quick, but at the same time there are affective states which are longer (such as anxiety or depression) and they are not caused by a single stimulus but are the results of their experiences.[36] There was developed a special system of judgment bias tasks[37] to assess emotional states of animals by a cognitive measure of their optimism and/or pessimism by recording behavioral responses to ambiguous stimuli.

4. Conclusion

Hence, cognitive biases are normal mechanisms of thinking not only of humans, but also of animals. Humans as well as animals do not think purely logically. There are the following two main moods of thinking: stress and satisfaction. On the microlevel both moods are expressed by the two mechanisms in perceiving signals: lateral inhibition and lateral activation.

References

1. A. Adamatzky, Physarum machine: Implementation of a Kolmogorov-Uspensky machine on a biological substrate. *Parallel Process. Lett.* 17(4):455–467, 2007.
2. T. Shirakawa, H. Sato, and S. Ishiguro, Construction of living cellular automata using the Physarum polycephalum. *Int. J. Gen. Syst.* 44:292–304, 2015.
3. A. Adamatzky and T. Schubert, Slime mold microfluidic logical gates. *Mat. Today.* 17(2):86–91, 2014.
4. J. D. Jones and A. Adamatzky, Towards Physarum binary adders. *Biosystems.* 101(1):51–58, 2010.
5. R. Mayne and A. Adamatzky, Slime mould foraging behaviour as optically coupled logical operations. *Int. J. Gen. Syst.* 44(3):305–313, 2015. Doi: 10.1080/03081079.2014.997528.
6. A. Schumann, *p*-adic valued logical calculi in simulations of the slime mould behaviour. *J. Appl. Non-Class. Log.* 25(2):125–139, 2015. Doi: 10.1080/11663081.2015.1049099.
7. J. G. Whiting, B. P. de Lacy Costello, and A. Adamatzky, Slime mould logic gates based on frequency changes of electrical potential oscillation. *Biosystems.* 124:21–25, 2014.
8. A. Adamatzky and R. Mayne, Actin automata: Phenomenology and localizations. *Int. J. Bifurc. Chaos* 25(2): 2015. Doi: 10.1142/S0218127415500303.

9. R. Alonso-Sanz and A. Adamatzky, Actin automata with memory. *Int. J. Bifurc. Chaos.* 26(1): 2016.
10. A. Schumann, Toward a computational model of actin filament networks. In *Proceedings of the 9th International Joint Conference on Biomedical Engineering Systems and Technologies (BIOSTEC 2016) — Volume 4: BIOSIGNALS, Rome, Italy, February 21–23, 2016*, 290–297. Doi: 10.5220/0005828902900297.
11. A. Schumann, On arithmetic functions in actin filament networks. In *10th EAI International Conference on Bio-inspired Information and Communications Technologies (formerly BIONETICS)*, ACM, vol. 3, 2017. Doi: 10.4108/eai.22-3-2017.152402.
12. A. Schumann, Decidable and undecidable arithmetic functions in actin filament networks. *J. Phys. D: Appl. Phys.* 2017. https://doi.org/10.1088/1361-6463/aa9d7b.
13. S. Siccardi and A. Adamatzky, Actin quantum automata: Communication and computation in molecular networks. *Nano Commun. Netw.* 6(1):15–27, 2015. Doi: 10.1016/j.nancom.2015.01.002.
14. S. Siccardi and A. Adamatzky, Quantum actin automata and three-valued logics. *IEEE J. Emerg. Sel. Topics Circuits Syst.* 6(1):53–61, 2016.
15. S. Siccardi, J. A. Tuszynski, and A. Adamatzky, Boolean gates on actin filaments. *Phys. Lett. A.* 380(1):88–97, 2016.
16. A. Tversky and D. Kahneman, Extensional Versus Intuitive Reasoning: The Conjunction Fallacy in Probability Judgment. In *Reasoning: Studies of Human Inference and its Foundations*, J. E. Adler and L. J. Rips, editors. Cambridge University Press, 2008, 114–135. Doi: 10.1017/CBO9780511814273.008.
17. L. H. Kern, H. L. Mirels, and V. G. Hinshaw, Scientists' understanding of propositional logic: An experimental investigation. *Soc. Stud. Sci.* 13(1):131–146, 1983. Doi: 10.1177/030631283013001007.
18. E. Sjoberg, Logical fallacies in animal model research. *Behav. Brain Funct.* 13(1):3, 2017. Doi: 10.1186/s12993-017-0121-8.
19. T. Sakiyama and Y.-P. Gunji, The müller-lyer illusion in ant foraging. *PLOS ONE.* 12(8):1–12, 2013.
20. S. R. Howard, A. Avarguès-Weber, J. E. Garcia, D. Stuart-Fox, and A. G. Dyer, Perception of contextual size illusions by honeybees in restricted and unrestricted viewing conditions. *Proc. R. Soc. B: Biol. Sci.* 284(1867):20172278, 2017. Doi: 10.1098/rspb.2017.2278.
21. I. M. Pepperberg, J. Vicinay, and P. Cavanagh, Processing of the müller-lyer illusion by a grey parrot (psittacus erithacus). *Perception.* 37(5): 765–781, 2008. Doi: 10.1068/p5898.
22. X. Huang, S. P. MacEvoy, and M. A. Paradiso, Perception of brightness and brightness illusions in the macaque monkey. *J. Neurosci.* 22(21):9618–9625, 2002. ISSN 0270-6474. Doi: 10.1523/JNEUROSCI.22-21-09618.2002.
23. A. Parrish, S. Brosnan, and M. Beran, Do you see what i see? A comparative investigation of the delboeuf illusion in humans (homo sapiens), rhesus

monkeys (macaca mulatta), and capuchin monkeys (cebus apella). *J. Exp. Psychol.* 41(08): 2015. Doi: 10.1037/xan0000078.

24. M. Santacà, M. E. Miletto Petrazzini, C. Agrillo, and A. Wilkinson, Can reptiles perceive visual illusions? Delboeuf illusion in red-footed tortoise (chelonoidis carbonaria) and bearded dragon (pogona vitticeps). *J. Comp. Psychol.* 133(03): 2019. Doi: 10.1037/com0000176.
25. S.-E. Byosiere, P. A. Chouinard, T. J. Howell, and P. C. Bennett, Illusion susceptibility in domestic dogs. *Ethology.* 126(10): 949–965, 2020. https://doi.org/10.1111/eth.13083.
26. J. B. Heaton, Managerial optimism and corporate finance. *Financ. Manage.* 31(2):33–45, 2002.
27. J. Chen, H. Hong, and J. C. Stein, Forecasting crashes: Trading volume, past returns, and conditional skewness in stock prices. *J. Financ. Econ.* 61(3):345–381, 2001. https://doi.org/10.1016/S0304-405X(01)00066-6.
28. T. Gilovich, D. Griffin, and D. Kahneman (editors), *Heuristics and Biases: The Psychology of Intuitive Judgment.* Cambridge University Press, 2002.
29. J. Goldberg and R. V. Nitzch, *Behavioural Finance.* John Wiley & Sons, Hoboken, New Jersey, 2001.
30. I. L. K. Clegg, Cognitive bias in zoo animals: An optimistic outlook for welfare assessment. *Animals (Basel).* 8(7):104, 2018. ISSN 2076-2615. Doi: 10.3390/ani8070104.
31. B. M. Spruijt, R. van den Bos, and F. T. Pijlman, A concept of welfare based on reward evaluating mechanisms in the brain: Anticipatory behaviour as an indicator for the state of reward systems. *Appl. Anim. Behav. Sci.* 72(2):145–171, 2001. ISSN 0168-1591. https://doi.org/10.1016/S0168-1591(00)00204-5.
32. I. L. Clegg, H. G. Rödel, and F. Delfour, Bottlenose dolphins engaging in more social affiliative behaviour judge ambiguous cues more optimistically. *Behav. Brain Res.* 322:115–122, 2017. ISSN 0166-4328. https://doi.org/10.1016/j.bbr.2017.01.026.
33. S. Daoudi, G. Badihi, and H. M. Buchanan-Smith, Is mixed-species living cognitively enriching? Enclosure use and welfare in two captive groups of tufted capuchins (sapajus apella) and squirrel monkeys (saimiri sciureus). *Anim. Behav. Cogn.* 4(1):72–90, 2017. ISSN 0166-4328. https://doi.org/10.12966/abc.06.02.2017.
34. M. Eysenck, K. Mogg, J. May, A. Richards, and A. Mathews, Bias in interpretation of ambiguous sentences related to threat in anxiety. *J. Abnormal Psychol.* 100(2):144–150, 1991. ISSN 0021-843X. Doi: 10.1037//0021-843x.100.2.144.
35. W. F. Wright and G. H. Bower, Mood effects on subjective probability assessment. *Organ. Behav. Hum. Decis. Process.* 52(2):276–291, 1992. ISSN 0749-5978. https://doi.org/10.1016/0749-5978(92)90039-A.
36. E. Robinson and J. Roiser, Affective biases in humans and animals. In *Translational Neuropsychopharmacology*, T. W. Robbins and B. J. Sahakian, editors. Springer International Publishing, Cham, 2016, 263–286. ISBN 978-3-319-33913-9. Doi: 10.1007/7854_2015_5011.

37. S. Roelofs, H. Boleij, R. E. Nordquist, and F. J. van der Staay, Making decisions under ambiguity: Judgment bias tasks for assessing emotional state in animals. *Front. Behav. Neurosci.* 10:119, 2016. ISSN 1662-5153. Doi: 10.3389/fnbeh.2016.00119.

https://doi.org/10.1142/9789811257155_0013

Chapter 13

Mu-Psi and Music Space: Insights into a 21st Century Music Practice

Cristian Vogel
NeverEngineLabs, Copenhagen, Denmark
info@cristianvogel.com
info@neverenginelabs

In this chapter, I take as a starting point a manifesto called Mu-Psi, published by Carla Scaletti in 2006. I will discuss how the ramifications of such a manifesto, be they new music experiences, methods, or creative directions, can be paths for expanding the innate ability we all have to *imagine music* — which applies to composer and listener alike. I explore how an expanded understanding of the activity and variability inherent in music has the potential to inspire in other mediums and to melt away imposed restraints or controlling forces which may be acting on the human imagination at large. I touch on how the contribution of musical imagination has shaped our development as individuals and societies and conclude with examples from my own practice, which involve unconventional computation for defining and quantizing musical variability from the realms of music framed as an *n-dimensional* abstract phase space.

1. Mu-Psi

In 2006, composer and software developer Carla Scaletti posted a manifesto on her website. The first part of it reads:

> Mu-Psi is to music what science fiction is to narrative art. It is sound-art that seeks to transcend the personal and to express universal concepts, patterns, and processes. Just as a science fiction

> begins with the premise of some possible future universe and proceeds to fill in the details and consequences of that premise, a Mu-Psi sound work begins with a hypothesis, a *"what-if"* premise, and proceeds to explore the ramifications of that premise.[1]

This essay takes the Mu-Psi manifesto as a starting point, framing it as an accessible example of a curious and egalitarian spirit that stems from a truly modernist music practice. An accomplished musician and sound designer, Scaletti is also one of the two computer scientists behind the invention and development of the Kyma sound design language.[2] Kyma is a 30-year-old, specialized software and hardware environment intended for the manifestation of "what-if" sonics. It excels at uninhibited creative exploration in sound with an elegant, flowing feel. For me, the Mu-Psi manifesto which underpins Scaletti's work is representative of the spirit of modernist musical thinking — beyond genre, beyond aesthetics, beyond any kind of defining technique. It echoes an intellectual attitude of "what-if" which accompanied many of the more radical visions of 20th Century modernism.[3] Scaletti's manifesto avoids some of the biases and pitfalls of the old modernism, by addressing today's more egalitarian cultural production — the manifesto comes presented with an invitingly imaginative entry point. The invitation to playfully imagine a hypothesis and then seriously undertake the technical exploration of its ramifications in the form of transcendental art, evidences an advanced understanding of the systemic impact of music — in the case of Kyma, in particular real-time performance, synthesis and music computation, such music can be sonified using the most up to date levels of technical precision, thereby making it truly modern. Mu-Psi positions itself as a means to reckon with modernity through a primarily sonic imagination and audible exploration. This is in contrast to scientific, philosophical, or literary methodologies — musicians don't only read words, we also read sound[1,4]

I shall now briefly outline — including a sprinkling of references for further reading — a glimpse into the high levels of complexity in music, both as an abstract system and as a human cultural activity. The following section fills in some of the background for Mu-Psi and 21st Century musical modernism.

2. Many Rivers to Cross

One cannot step twice into the same river.

Heraclitus

In nature, there is never one identical music experience. Music is affective, situated, and embodied; listening to the sound waves of music changes us[5] — and the setting or the focal quality of our hearing changes music.[6] There is no "real music" because music as a whole will never be reducible to one set of particular techniques, styles, instruments, or sounds. We need to start thinking about music as something much bigger and much more abstract than recordings or popular songs.

How about a multidimensional space of continuous activity and malleable variability, a musical *ontology*?[7] We are talking about **Music** with a capital "M" — some kind of vast, abstract, limitless dynamical system of all possible music objects, something along the lines of Borges's Labyrinths.[8] In the normal course of life, it is often easier to talk about a piece of music, than to talk about "Music" in all its mind-boggling abstraction. Let it be clear that when we refer to a piece of music — a recording or a song — we are referring to a music object — a relatively fixed arrangement of specified sounds, mined and polished from the raw flux of a music ontology. That extraction and mediation is accomplished by a musician, composer, or active listener. State-of-the-art research suggests we could possibly include algorithms as mediators for music discovery, although arguably software can only *process* music as digital data — this is far from the act of listening. Given the astounding parametric diversity and timescales of music, it can also be surprisingly egalitarian. For example, a composer or designer might equate seemingly unrelated attributes. Some examples; pitch and rhythm,[9] spectra and tuning,[10] tonality and emotion[11] or — as in an example from the field of data sonification — duration and threat detection.[12] It is interesting to note that in the context of music and auditory design thinking, such experimental couplings are refined through the act of listening — perhaps the results of the research are as narrowly focused as a scientific experiment or as unbridled as a contemporary art form.

The diversity, unity, and inclusiveness of the culture of music manifests most prominently on the level of human engagement and this, for me, is one of the most compelling reasons why music should not be trapped inside the hustle of an entertainment industry existing solely on the political economy of music objects. I would argue that the experience of music and musical meaning are by nature free to all who listen. From a political perspective, it has been argued that music as an abstract system *is* capable of defending itself from any form of absolute control[13] — music activity is difficult to suppress because of the reality that modernity affords; music objects are in great abundance, each one imbued with immeasurable levels of nuance and variability in meaning.

Music taps directly into our most profound notions of being, delivering universal information mediated through interwoven scales of vibrational energy. Think of how the tiniest vibrations of a speaker membrane can evoke memories or move bodies in rhythm. A sonic musical experience can unfold with or without audience, with or without the conventions of spoken language, and with or without technology. Music and sound have empathic, cultural, political, and ecstatic affects.[14] Sound and music technology have contributed directly to the development of mathematics, computers, medical and scientific thought, and vice versa.[15] All of this is Promethean fire for a Mu-Psi premise.

Being exposed to the musical imagination of individuals and groups is a very common experience — and has accompanied humans for thousands of years. Auditory phenomena can alter the way we situate ourselves in space and in time. Indeed, the physical form of sound waves literally vibrates and alters our neurons and cells, and has done so over the entirety of human history. In 2008, a vulture bone flute was discovered at *Hohle Fels* cave in Germany, dated to 40,000 years, and considered the world's earliest recognizable music instrument.[16] In the modern day, when music and sound are proposed in a methodological role beyond conventional ideas of entertainment or aesthetics, they are generally considered unscientific by the mainstream and are often marginalized. This seems at odds with the rich legacy of sonic creativity and life. Some examples might be how the fields of data-sonification or archeo-acoustics remain niche. The latter emphasizes how ancient places and artifacts might have *sounded*, rather than how they might have looked. Sounding out the relationships between the location of archeological sites and special acoustic properties of those locations can help us infer just how important sound and music might have been for our ancient ancestors.[17] Another rather niche technical example from my own work is the "WireFrames" library for Kyma which implements real-time sample rate computation using synchronized audio signals acting upon each other. Using only audio signals at high sample rates, it can compute the dynamics of flocking systems, Lorenz systems, or Nearest Neighbor algorithms, the results of which can be sonified in real time as digital oscillators or spectral resynthesis.[18] I believe the applications of high speed computation in the audio signal domain have enormous potential in fields beyond those of sound design, yet I struggle to convince non-sonic practitioners.

Let us get back to the Mu-Psi manifesto. I would like now to compose some Mu-Psi premises, and invite the reader to consider the ramifications. Are they purely speculative or musical folly? My premises build on the idea that sound and music formed a big part of the original augmented realities in our previously un-electrified world. Evidence of prehistoric instruments, carved from bone or rock, suggest that the ecstatic experience of music (in the original sense of ekstasis — to be outside or beyond one's self) and its transformative affects, has been with us for millennia. Here are three Mu-Psi premises which explore the ramifications of a human development where music, acoustic, and sound phenomena have maintained their sacred status from antiquity.

> *What if....* music and sonic meaning were not marginalized today as little more than entertainment? We need all the help we can get as we face catastrophic threats to life on Earth. How can Music contribute to improving our relationship with nature?

> *What if....* Music contains within it all of humanity's prayers, gods, imagining and dreams as a vast mandala, parts of which we can play back as vibrations, perceivable in the air around us?
>
> *What if....* sonic communities exist who worship ecstatic sound objects. What if they consider the creation of music as sacred as the invention of prayers, as useful as the most empowering technology?
>
> *What if....* an abundance of ecstatic sound objects and reproduction technology rapidly empowers the auditory imagination of individuals and societies triggering an unprecedented feedback of *ekstasis* inspiring new ways of relating to nature and each other?

What kind of crazy music could stem from these premises? To me, and surely others from the music community, such premises do not seem so crazy, fictional, or futuristic. They seem like a description of the *sonic-now*. It is already happening, but needs more amplification!

3. The Art of Flux

As we have implied, the Mu-Psi manifesto is both current and futuristic — the contemporary modernity. It is a calling to be radically creational in a medium that does not necessarily rely on any kind of transaction, decoding, or interpretation. Music is sensorially immersive, like a flux, like noise — it flows through us and expresses many possible states at once. People tend to take it for granted for example, that music and sound are both in the airwaves around us *and*, somehow, inside our bodies at the same time. This ubiquitous freedom grants a significant purpose and responsibility to the creation of music, and it reminds us that the artistry and craft around music can be a demanding and unconventional pursuit. The second half of the Mu-Psi manifesto contains a mission statement.

> Art has a purpose and plays a powerful role in society. The artist's mission is to:
>
> - Stimulate imaginative thinking, discussion, and creation of alternative worlds, solutions, possibilities
> - Entertain by stimulating the intellect
> - Guide audiences along paths through abstract conceptual spaces

- Invite the audience on an intellectually challenging journey from which they will return refreshed, rejuvenated, and inspired
- Create sonic metaphors for powerful universal concepts
- Create an ecstatic experience for the audience
- Play within a set of constraints; experimentally violate one or more constraints to explore the ramifications.

This mission statement along with the whole Mu-Psi attitude aligns well with another source of encouragement I would like to share here — Adam Harper's *Infinite Music*.[19] Some of the ideas packed into this dense little book significantly widened the scope of my own 21st Century music practice. It is a fast-paced investigation into the structural and philosophical complexity of a Mu-Psi-like premise — *What if.... the variability of music is infinite?* The book — with its mind-bending footnotes — attempts to explore the ramifications of that idea. An exciting concept that he welcomes us to early on, and one I have also been introducing gently to the reader in this chapter, is the collected theories about the fabric of an *n-dimensional* "Music Space" where infinite musical variability exists as a quantizable space of possible musical attributes. Against this backdrop, Harper blows the conventional idea of music wide open. For example, he postulates that experienced practitioners working in Music Space easily integrate and specify non-musical forms such as architecture, images, text, bodies, or history as musical variables. In fact, everything can be material to be appropriated, synthesized, recombined, saturated, repeated, or precisely specified in the neverending pursuit of extending a universal sonic imagination. I would add that when lived, this mind frame is often undirected and unfocused yet devoted to innovating in sound. Why? As it says in the Mu-Psi manifesto — extending the musical imagination is an ecstatic process of continuous spiritual and embodied transformation for all who engage with the process or the production.

> The widening of an imagination to accommodate a new and unusual idea or possibility can be a rewarding experience in itself, but this process is also the engine of our development and betterment as individuals and as societies.
>
> Harper, 2011, p. 9

4. Music Space

At this point, I would like to go through some examples from my own practice which involve some unconventional ideas and in the praxis, some unconventional music computing. I consider them as works informed by the ramifications of thinking about music as a (near infinite) Music Space. Firstly, I refer the reader to a video piece,[a] part of a collaboration with composer S⊘S Gunver Ryberg[20,21] and film maker Marta Bala.[22]

> Poise. Intention. Action. Precision. Here we are on our knees, mining the Sugar Cave. It is our environment of being, it is a beautiful place. It's made of music. It dreams of us and our dreams flow back into it. It's resourceful, sometimes we've even seen it feed upon itself...Inside, Music is everything but sound can be absent. In fact, some of our favourite parts are silent. There's lots of silence in there.[23]

The spoken text I wrote for the video piece expresses metaphorically what it feels like to live and work in the flux of *our* Music Space — which we called the "Sugar Cave". As well as being the speakable name of our esoteric moniker `SGR^CAV`, it stands as a metaphor for the abstract vastness of Music Space with its resonant caves, audible dust, spectral trails, and voids. In the video piece, we come across as Kerouac-style adventurers, on our knees, mining Music Space for mythical sonic bounty using special technologies and following secret trails before packing our discoveries down into encapsulated and diffusible forms for public consumption. At least that's the way I read it. The piece is hinting that a Mu-Psi/Music Space life can be directional but also confusing — like a drunken walk. The main example in this piece, is the attempt to describe how it is that composers and sound artists might not always work with sound when they are deriving musical meaning. This notion could be construed as equivalent to thinking about computation without computers, and equally as path widening.

[a]Viewing link: https://vimeo.com/126520100.

5. Soundless Music

Let us break down that idea a bit more. John Cages's piece of silent music *4'33"* is a modernist milestone,[24] performed with famously disruptive impact by David Tudor in 1952. Seventy years later — demonstrative of the power of true modernism — to think about the paradox of equating silence with music, still serves as a thought exercise. Consider the following example. Have you ever observed a fellow passenger at rush hour, reading a musical manuscript in preparation for a meeting perhaps? Have you ever wondered if that person is hearing sounds in their head? If you dwell on it — how is it even possible to hear imaginary sounds in such a noisy place? *What if... we all have a Mind's Ear?* Of course, sound is as important to musicians as words are to a writer, yet paradoxically, composers are regularly considering non-sonic parameters in the normal course of music composition.

Turning to post-modern pop, Bernard Sumner of New Order once sang *A sound form in a vacuum may seem a waste of time*.[25] Maybe. But then again, maybe the line was hinting that imagining a sound form in a vacuum isn't a waste of time at all. These words about silence are embedded inside layers of Sequential Circuits synths and Oberheim drum machines.[26] Therefore, the medium which surrounds the poetic notion of a silent sound in a vacuum is not a void, but of course music in all its void-filling richness. Music — when it is not sounding — exists as potential sound form in a vacuum waiting to be released, waiting to be realized in time and space. From a personal perspective, I have witnessed many times how a precisely specified mix, adapted and mutated in a listener's presence. My recording apparently contained layers of unspecified behavior which only emerged when listening to it with others. This amazes me. To maximize this effect, I try to design it in — sound with combinatorial potential — and that often means imagining and specifying a range of sonic possibilities. This is achieved through applied non-sonic activities, such as tweaking in the code editor or configuring rows of modules or guitar pedals for performative potential. This is modern composing. It is often a virtuoso act of sonic imagination.

6. Crawling Around in the Mud

Let us now get further into what it is like to compose something like a Mu-Psi premise, and techniques for extracting from the fabric of Music Space. In my studio — sometimes in silence, sometimes out loud — I spend substantial hours designing the tooling and explorative systems that will help in the endless process of innovating sound and music, the process of mapping out Music Space. On a day-to-day basis, these systems might form networks of unpredictable signal modulations and algorithms acting on many musical attributes at once. Examples of stochastic algorithms might be the zig-zag of the "Drunken Walk" or the recombinance of the "Markov Chain" — when applied to notes, frequencies, rhythms, and timbres, these time-based processes always evoke ranges of expressive data, to listen to and consider as musical potential. Shepherding the myriad of possibilities toward a functional goal such as an audio recording, an improvization, or a written score is an activity that musicians all do, one way or another. I would argue it is at the heart of the practice for many. *Crawling* and *Shephärding* over a long history of iterations.

At some point during a creation, I will begin to structure the sonic and musical architecture through a process of interactive listening. This stage of the process usually unfolds when situated in the studio, but not always. It involves a lot of parameter adjustments while listening to results — volume is one of the most important parameters, hence the critical importance of studio acoustics and good isolation. Through interactive listening, the idea is to understand what might be evolving musically and how to sieve that emergence through a universal, predetermined, and/or subjective set of decisions. It can be exciting, it can be repetitive, it can be exhausting. Hence, the `SGR^CAV` mantra — *Poise. Intention. Action. Precision.* — designed to guide safe critical decision-making while in the flow.[b]

A notion of safety and responsibility in the electronic music studio is important. The studio should be understood as an experimental

[b]Perhaps also somewhat inspired by SØS Gunver Ryberg's former stunt training.

laboratory containing technologies that have been solely designed for sonic *ekstasis* and the production thereof. A transformative sonic temple if you will, optimized for listening to what Conny Plank described as the "potential of noise".[27]

The intensity levels in the studio can be, and need to be, overwhelming at times. I encourage getting away from the machines every now and then, going out on sound explorations through the city, or going out to performances and music events. In fact, one of the cool things about being immersed into a compositional process is how the act of listening to everything else is a continuation of the work of composition and decision-making on your own project — just a little offset in time. For example, I may be inspired by something I've heard and later sit down to compose a remembered sound in the studio. Starting from remembered sounds is another example of musical imagination taking a role in production. To clarify — I am rarely successful at recreating something exactly from memory — that isn't the point of the exercise. That would be better served by field recording. Instead, this exercise is intended to open paths, when the process might have become blocked or stagnated.

7. Play. Create. Guide

To conclude, I would like to return to focus on three points of the Mu-Psi Artists' Mission statement.

(1) Play within a set of constraints; experimentally violate one or more constraints to explore the ramifications
(2) Create sonic metaphors for powerful universal concepts
(3) Guide audiences along paths through abstract conceptual spaces

This concluding section will outline examples of these three, applied to my own practice and with suggestions for further reading.

7.1. *Play*

When discovering a particular terrain of musical imagination, put down some constraints which might help mark it out. In other words, lay down some tracks. In the case of my 2013 album —

Eselsbrücke[28] — I introduced intervallic constraints from the very beginning, before sounding anything. The works on the album were derived from rows of constant integers and an interest in the musical potential of the intervals between the elements. Four row transformations were allowed, as a means of evolving higher order sets. The rows all expressed distinct intervallic patterns, and the algorithms for generating them were straightforward, although it became challenging to find a non-subjective way of making choices about them. The gritty details of how I applied sieve theory[29] to create the sets, and how I solved the artistic challenge of struggling against intuitive decisions, are beyond the scope of this chapter. Nevertheless, I encourage the reader to take a look at the creative and technical essay published by *Fylkingens Journal* online, which goes over the project in detail.[30]

7.2. *Create*

This subsection, relates to the metaphors part of "Create Metaphors". After the recording process and concerts were all exhausted for the *Eselsbrücke* album, I still felt compelled to document a little more about the universal concepts — and following the Mu-Psi manifesto — do that as metaphor. As I had already taken all the decisions to compose the sonic metaphors, I moved into the non-sonic music territory. Painting might have been adequate, but I am not so good at that. Instead I decided to write short stories. Three short stories emerged while thinking about the music and the musical processes on *Eselsbrücke*. The stories are set in a space that, for me at least, seemed to resonate with the synthetic, time computed, and intervallic dialectics of the music processes on that record. In the frame of "Music Space", we could see these stories as music objects specified to be text based, but containing musical meaning. Perhaps they could be interpreted as scores or librettos, for improvising another *Eselsbrücke* music object, perhaps not a solo this time, but instead an opera with a group of musicians. Perhaps we could imagine a new literary genre of *Mu-Psi stories* — creative writing about abstract conceptual spaces that were originally musical narratives. To sum up; with the music and the words, a corner of Music Space was rendered in the form of recorded music and in the form of literary non-literal texts — parallel examples of encapsulated musical meaning.

7.3. *Guide*

I leave the reader with an extract from one of those Mu-Psi short stories. The whole chapter can be downloaded at https://github.com/cristianvogel/mupsi-musicspace-writing-music-spectra. The piece of music it is based on is also available in that repository, alongside full color spectrograms kindly prepared by Guy Birkin.[31] A short conclusion to this essay follows the excerpt.

8. Mount The 137 (excerpt)

> *Every strategy we invent is in the hope that it will unfold unexpectedly. We invent the real in the hope of seeing it unfold as a great ruse. From strategy we expect control, but from seduction we hope for surprise.*

When the bridge emerged, the solution crossing solution, it was fast and vivid and its sudden sound woke him instantly. Hope had finally matrixed out all the mapping decisions and executed the patterns. The synthetic nucleotide had alloyed into an instant shimmering vertical stack. The tall structure stretched out across the river. It was constructed from crystal clock state machines, its edges honed by the strong resonances the clocks traced in the fluxscape. Lucky Connor was particularly impressed by the way Hope had used a sieve to make structural decisions about the clocks that made up the bridge. Connor always carried around a light green sieve in his stuff, and thats the one Hope had used. He always kept it, because this particular sieve had a great pattern, one that expressed the space between all prime numbers. He had often wondered what it was really for — it was never much good at draining rice. He could see that Hope had used the sieve to sift out and frame fundamental decisions for the bridge structure — the height at which each clock was suspended, for example, had been chosen through the sieve, as had the speed of each clock being used. The lowest and slowest clock ticked — changed states — just once between one bank and the other. Its on-state cast a wide fresh crystal alloy across the fluxscape, stretching out exactly half way across the river. The highest and fastest clock ticked 137 times to reach the other side, leaving staccato fragments of crystal, high in the fluxlight. When a clock tocked itself into the off-state, it resonated an outline of its previous state. These empty outlines, transparent shadows or memories of stability, filled the gaps between each solid tick. In these intervals, the background of the river scene was pulled taut to become foreground. Unlike the reflective crystal alloy links of the on-states, the off-states of each clock refracted the view of the over bank and magnified Lucky Connor's projections of nextlevel flow — like mirages manifesting out of distant hot air into something so desirable, so close.

9. Conclusion

The prose metaphorises the stacked oscillator and pulse structures from which a spectral "bridge" (Figure 1) emerged. The spectral shape was a characteristic of almost all the pieces on the album stemming from the intervallic constraints, particularly relating to the stacking of prime number tempos. Interestingly, even though the music was derived from a relatively "cold" starting point — applying prime number sieves for constrained decision-making — the album ended up having a surprisingly powerful emotional and spectral signature, notably different to my previous work. For me, the structure of the story has a sense of the On–Off switches and State Machines, as it splices between my sometimes generative

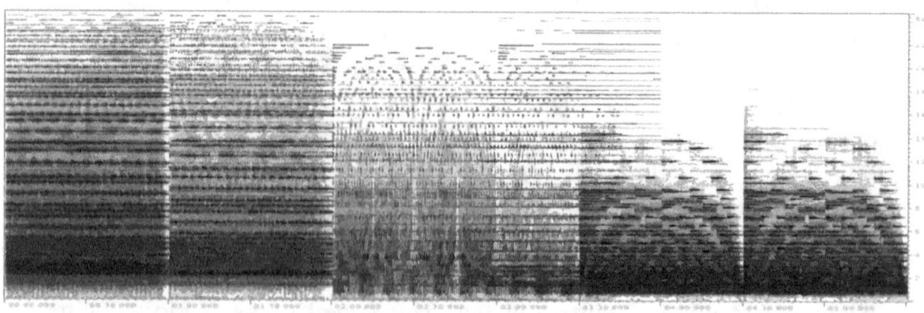

Fig. 1. Approximate spectrogram of Mount The 137 showing the spectral "bridge".

Mu-Psi prose and that of another text, a book by Jean Baudrillard. The spliced passages were appropriated and processed, much as I do with sampled sounds. My original writing and the appropriated texts both have intervallic, ecstatic, and emotionally additive qualities, which contrasted well together and structurally with the sound of the music. I feel *Eselsbrücke* adhered to many aspects of the Mu-Psi manifesto, placing it as a concrete example in my catalog of how Mu-Psi and Music Space directed my musical imagination toward unconventional means and methods for the discovery and traversal of Music Space.

References

1. C. Scaletti, Mu-psi manifesto. Accessed: May 2021 https://carlascaletti.com/.
2. C. Scaletti, Computer music languages, kyma, and the future. *Comput. Music J.* 26:69–82, 2002.
3. W. Duckworth, *Talking Music: Conversations with John Cage, Philip Glass, Laurie Anderson, and Five Generations of American Experimental Composers.* Da Capo Press, New York, 1995.
4. G. Goodale, *Sonic persuasion: Reading Sound in the Recorded Age.* University of Illinois Press, Champaign, IL, 2011.
5. J. M. Shapiro, B. W. Drinkwater, A. W. Perriman, and M. Fraser, Sonolithography: In-air ultrasonic particulate and droplet manipulation for multiscale surface patterning, *Adv. Mat. Technol.* 6(3):2000689, 2021.
6. E. Clarke, Ways of listening: An ecological approach to the perception of musical meaning, Oxford University Press, Oxford, 2011.
7. A. Harper, Composition, change and musical objects in infinite music. Darmstädter Beiträge zur Neuen Musik, Mainz, pp. 15–28, 2014.
8. J. Borges, *Labyrinths: Short Stories and Other Writings.* New Directions, New York, 1964.
9. J. Harvey, *The Music of Stockhausen: An Introduction*, Faber, London, 1975.
10. W. Sethares, *Tuning, Timbre, Spectrum, Scale*, Springer, 1998.
11. T. Eerola and J. Vuoskoski, A review of music and emotion studies: Approaches, emotion models, and stimuli. *Music Percept.* 30:307–340, 2013.
12. S. Lenzi, G. Terenghi, R. Taormina, S. Galelli, and P. Ciuccarelli, Disclosing cyber attacks on water distribution systems. An experimental approach to the sonification of threats and anomalous data. ICAD Conference Paper, Northumbria University, 2019.
13. J. Attali, *Noise: The Political Economy of Music*, Vol. 16. University of Minnesota Press, Minneapolis/London, 1985.
14. B. Ehrenreich, *Dancing in the Streets: A History of Collective Joy.* Metropolitan Books, New York, 2006.

15. L. Gaston-Bird, *Women in Audio*. Routledge, New York/London, 2019.
16. Bone flute is oldest instrument Published June 24, 2009 https://www.nationalgeographic.com/culture/article/bone-flute-is-oldest-instrument–study-says.
17. S. Miyagawa, C. Lesure, and V. A. Nóbrega, Cross-modality information transfer: A hypothesis about the relationship among prehistoric cave paintings, symbolic thinking, and the emergence of language. *Front. Psychol.* 9: 2018.
18. C. Vogel, Wire frames library, *NeverEngineLabs*, 2015–2021. https://github.com/cristianvogel/wireframes.
19. A. Harper, *Infinite Music: Imagining the Next Millennium of Human Music-Making.* Winchester: Zero Books, 2011.
20. S. G. Ryberg, www.sosgunverryberg.com. Accessed: May 2021.
21. C. Vogel and S. G. Ryberg, `sgr^cav` https://sgrcav.bandcamp.com/.
22. M. Bala, www.martabala.de. Accessed: May 2021.
23. C. Vogel and S. G. Ryberg, `sgr^cav`: Moved by magnets, 2015. Cassette album, Video, Performance.
24. J. Pritchett, What silence taught john cage: The story of 4'33. In *The Anarchy of Silence: John Cage and Experimental Art*, Museu D'Art Contemporani de Barcelona (MACBA), 2009, 166–177.
25. N. Order, Your silent face, 1983. FACT75.
26. Music magazine archive Archival article accessed May 2021.
27. C. Plank, The potential of noise, 2017. https://www.imdb.com/title/tt4979826/.
28. C. Vogel, Eselsbrücke, 2013. Limited edition CD and Digital Download.
29. C. Ariza, An object-oriented model of the xenakis sieve for algorithmic pitch, rhythm, and parameter generation. In *ICMC*, 2004.
30. C. Vogel, Donkey bridges: On creative and technical process behind eselsbrücke. *Fylkingens J.* 2014. Accessed: May 2021, https://hz-journal.org/n18/vogel.html.
31. G. Birkin, Spectrograms of cristian vogel's album eselsbrücke, 2013. Accessed: May 2021 https://aestheticcomplexity.wordpress.com/2021/05/24/spectrograms-of-cristian-vogels-album-eselsbrucke-2013/.

https://doi.org/10.1142/9789811257155_0014

Chapter 14

In Search of Textile Roots of Computing: The Genesis of the First *MaSo* Carpet in India*

Tim Otto Roth
imachination projects,
Oppenau, Germany
tor@imachination.net

Generally, computation is associated as calculus with arithmetic. But looking back into cultural history there are techniques that in principle allow a computation even without processing numbers. The *MaSo* carpet project demonstrates, with a team of weavers in a traditional Indian loom, that a tessellated structure such as a hand-knotted carpet can be used as a binary computer. As a carpet evolves like a computer screen row-wise over time, it is predestined to apply an iterative algorithmic principle to deduce the color of a knot in the next row from the local color constellation in the previous row. The creation of such a carpet turned out not only to be an exploration in the history of automata. It also manifested a social-cultural and anthropological experiment revealing a specific human error management but also certain limits of imagining complexity.

1. In Search of Historic Automata Patterns

There is one page in Steven Wolfram's seminal book *A New Kind of Science* deviating stylistically but also conceptually from his

*This chapter is dedicated to Bhola Baranwal. This *gem of carpet weaving* as he was called by his Indian fellows passed away in January 2021 when I wrote this chapter.

automata theory stimulating further speculations: On page 43, the physicist presents a set of depictions of historic patterns from various epochs from Palaeolithic times, from 22,000 year BC to Islamic art in the 14th century. Here Wolfram is missing patterns based on the self-organizing principles of cellular automata he discusses in his book:

> Repetitive patterns are common and some nested patterns are seen, but the more complicated kinds of patterns discussed in this chapter do not ever appear to have been used.[1]

The illustrations do show a mineral affinity with mosaics, pottery, and stone architecture. But one important pattern medium is missing: textiles. This is quite remarkable as most textile technologies create tissues applying a 'historic' procedure. For instance, in weavings or knotted carpets a row, respectively, a weft needs to be finished to start a new row. This inherent teleology in textiles is missing in mosaics [Fig. 1]. Above all woven textiles or knotted

Fig. 1. A 'computed' floor pattern as it could have been created in principal thousands of years ago by a self-assembling principle: Temporary mosaic made of circles of retroflecting foil applied to the existing floor tiles in occasion of the exhibition *Logical Fantasies*[2] at Kunsthalle Jesuitenkirche Aschaffenburg (Germany) in 2020. The row-wise evolving pattern illustrates the dynamics of the automaton rule (Wolfram code 30) driving the activity of the 18 pipes of the self-organizing water organ *aura calculata* installed in the apsis of the Baroque church. Image: Stefan Stark Photoproduction.

carpets are discrete systems with a regular cell space defined by warps and wefts. So why not use a textile system iteratively generating the color composition of the next row as a result of the constellation in the previous row? Interestingly, you can find a kind of algorithmic technique yet in Andean and Ancient[3] weaving, which possibly also inspired early arithmetic concepts. But these early pattern generating algorithms in fabrics are based on combinatorial variations of differently colored warps and wefts,[4] but do not rely on an iterative principle. Having examined textile artifacts from various cultures, I couldn't spot self-organizing patterns going beyond the simple growth patterns of alternating colors you find for instance in woven Navaho rugs.[5] The English mathematician J. C. P. Miller made a connection between automata and textiles: By the end of the 1960s he used automata generated patterns as a template for *Touch Tapestry*, a leisure kit to create pictorial tapestry. Interestingly, he ignored the inherent potential of tapestry to compute the patterns by its own textile means.[6]

2. Textile Automata — The *MaSo* Carpets as Historic Enactment

To answer the question why an automaton principle was never applied for the pattern generation of knotted carpets — a cultural technique existing for thousands of years, the best thing is to make an experiment in a traditional loom. This was the starting point to the most adventurous endeavor in my artistic career examining the riddle of self-organization: In winter 2019, I traveled together with my filming partner Miriam Seidler to one of the centers of traditional Indian carpet industry in the Badohi region near the Holy City of Varanasi by the river Ganges.[a]

For our Western minds, the idea was a quite simple heuristic approach: Let's give a workshop to some Indian weavers instructing them on a simple rule on how to tie a two color carpet and then see how it goes at the loom. Our local project partner was Bhola Baranwal who founded recently the label *Organic Weave*

[a]See clip on www.maso.gallery.

promoting and producing fully organic hand-knotted carpets. Bhola Baranwal organized the workshop and made an arrangement with a correspondent loom in Hossainipur, a small weaver village somewhere in the remote country side 1.5 hours car drive away from Varanasi.

The traffic jams on the way to our workshop were the perfect introduction: The 15 female and male participants immediately got the general idea of self-organization thinking on the riddle of the Indian traffic, where nobody obeys rules and just reacts to the local situation. It was more tricky to communicate the instructions on how to weave a self-organizing carpet. Therefore, we had to radically break down the rule set to a minimum set of conditions. Finally, for our carpet to be made of white cotton and red wool we simplified a one-dimensional automaton rule (Wolfram code 90) to a single condition: The weavers tie a red knot if the knot in the previous line had exactly one red neighbor, otherwise they tie a white knot. As it turned out, the *Four Connect* games we brought with us from Germany were perfect learning tools to simulate with the two differently colored chips this self-organization rule [Fig. 2]. The participants, all people without higher education and speaking no word of English, had a steep learning curve and most of them seemed to be more or less familiar with the rule set after just half an hour.

Fig. 2. Automaton workshop in Town Khamaria (Uttar Pradesh, India) using *Four Connect* games as simulation tool to train the weavers. Image: Tim Otto Roth/imachination projects.

3. Handknotted Computers — A Social Experiment

It was encouraging how smoothly the workshop was accomplished, but the situation was much more challenging implementing the rules afterwards at the loom in the neighboring village. The three weavers [Fig. 3] selected to start the carpet felt confused and seemed to be somehow blocked. These difficulties were related to the history of hand-knotted carpets, which used to be always a reproduction: Traditionally, weavers were used to copy a drawn map or to translate a notation, which in some cases was even sung by a supervisor. This is different from the *MaSo* carpet: For the first time, the pattern of the carpet emerges integrally alongside the process of weaving without any of the pre-conceived images of patterns created in map or talim systems of weaving. So the experienced weavers who for decades have only reproduced patterns are now asked to reflect on which color is to be knotted in the next step without aid of a map. Instead of just reproducing a precomputed map, the weavers are getting to be computers in the original sense of the word.[7] It's hard to imagine how challenging this new role for the weavers was: For the first time, they

Fig. 3. The three Indian weavers presenting their computation progress on the loom after 4 days. Image: Tim Otto Roth/imachination projects.

had to decide on their own on the pattern — a situation which made them feel in the beginning quite uncomfortable as they were also now responsible for possible mistakes.[8] So the *MaSo* carpet project turned out to be a social experiment. Thought barriers are overcome; the level of confidence raised. As Indian society is still marked by the caste system and its clear hierarchies, this procedure resembles a minor revolution.[b]

Here also the name *MaSo* comes into play — an acronym for **Ma**thematical **So**cialism, which was used for the first time in a title for an exhibition at the Goethe Institute in Hanoi, where I showed in 2017 plenty of automata-related works. *MaSo* is an ironic reaction on the political conditions of stone-age communism in Vietnam, as it parodies the concept of equality as a core principal of socialism: The automata works including the sound installation *aura calculata* reveal that even if every agent is 100% equal, behaving according to the same rules as its neighbor, this does not necessarily result in an uniform activity in lockstep, where every unit shows the same activity.[9] On the contrary, despite its determinstic character, cellular automata can reveal an emergent behavior that we can't predict.[10] With the Indian weavers *MaSo* gets again another twist.

4. Creative Errors — Aberrations of "Natural Automata"

Besides the peculiar social conditions in India, which we documented also in the 15 minute movie *In Search of Gandhi's Spinning Wheel* accompanying the project, another interfering human factor comes into play. While a computer can make calculations without making mistakes, this new mode of carpet production depends on the weavers' concentration: a miscalculated knot has an effect on all subsequent knots. Consequently, this results in a change of pattern, yet not affecting the fascinating look. The first *MaSo* carpet [Fig. 4] four square meters in size is made of around 270,000 individual

[b]There also comes a gender and age component into play: The most recent collection of *MaSo* carpets (Figure 5) was knotted by young ladies, who mastered the weaving with ease.

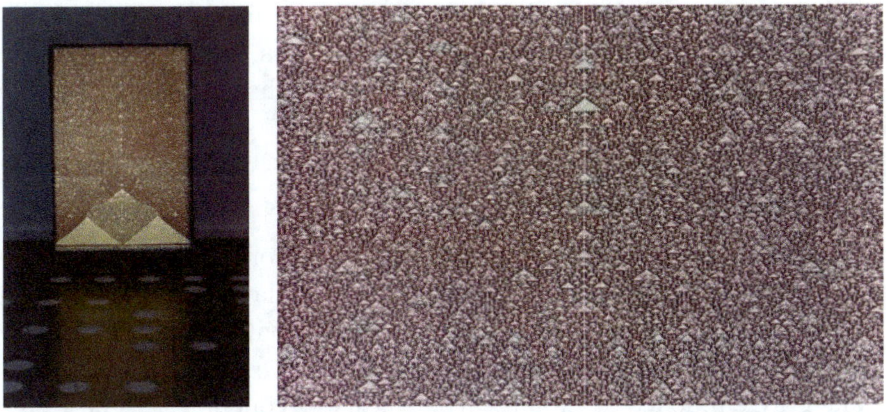

Fig. 4. The *MaSo* carpet N°1 premiered in the exhibition *Logical Fantasies*. It was presented in the former Jesuit church in Aschaffenburg pending about 2 meters off the wall, so the visitors can also see the much clearer articulated back side (right), which is normally hidden. Red organic wool, white organic cotton, ca. 240*170 cm, 2019. Image: Stefan Stark Photoproduction, Tim Otto Roth/imachination projects.

knots — so mistakes are inevitable and become an essential part of its creation. Working with human and therefore implicitly fallible agents, in fact, contributes to the unique character of this 'computing carpet'.

This recalls the mathematician John von Neumann differentiating the failure management of artificial and natural automata: Machines are quite error-prone as a single processing flaw can result in a faulty result. Here natural automata react in a much more robust way as "rigid, all-or-none concepts" of formal logic inherent in every computing device.[11] For instance, natural automata as biological neuronal systems react more reliably on potential failures in the system and recover even from severe lesions.

Consequently, these mistakes made by the weaving "natural automata" differ significantly from machine-based errors: These aberrations don't appear arbitrarily, but they seem to follow the weaver's aesthetic preferences and in some cases reflect a recurrence of conventional pattern thinking. These individual creative faults are hard to anticipate and even simulate as the occurrence of these human bugs do not correspond to computational random patterns.

This human touch is clearly visible also in the first carpet. It starts with clearly articulated macroscopic structures and suddenly turns into an amorphous forest of little trees — this happened right at the moment when we left the village after 4 days.

5. The *MaSo* Carpet Project Goes Back to the Roots

When the carpet was evolving row by row, not only the weavers were surprised with the emerging pattern, even Bhola Baranwal, who had seen so many carpets around the globe, felt bewildered: "When I saw the first 10 rows, I was so surprised about the emerging pattern." The gray eminence of Indian carpet industry concludes: "The whole carpet is so beautiful. I have never seen a pattern like that before — it's magic!" Bhola's empathetic reaction touches the anthropological question why automaton rules haven't been used much earlier for

Fig. 5. Studio impression with the almost perfectly knotted *MaSo* carpets N°53 & 54 from a more recent collection: The different edge conditions show a significant effect. Blue organic wool, orange organic cotton, each 100*75 cm, 2020. Image: Tim Otto Roth/imachination projects.

pattern creation? As the Indian experiment revealed the weavers couldn't imagine that such simple conditions could result in such an emergent behavior. It finally did afford the practical comprehension to open the door to this discovery of complex interactions. In the end, it's a radical rethinking of complexity, an idea which took in mathematics also thousands of years until the advent of computers triggered the first automaton concepts.

References

1. S. Wolfram, *A new kind of science.* Wolfram Media, Champaign, IL, 2002.
2. Kunsthalle Jesuitenkirche Aschaffenburg (editor), *Logische Phantasien [Logical Fantasies].* Kehrer, Heidelberg, 2020.
3. C. Brezine, Algorithms and automation — The production of mathematics and textiles. In *The Oxford Handbook of the History of Mathematics,* E. Robson and J. Stedall, editors. Oxford University Press, Oxford, 2009, 469–492.
4. B. Grunbaum and G. C. Shephard, Satins and twills — An introduction to the geometry of fabrics. *Math. Mag.* 53(3):139–161, 1980.
5. T. O. Roth and A. Deutsch, Universal synthesizer and window — Cellular automata as a new kind of cybernetic images. In *Imagery in the 21st Century,* O. Grau and T. Veigl, editors. MIT Press, Cambridge, MA, 2011, 269–288.
6. J. P. C. Miller, Periodic forests of stunted trees. *Phil. Trans. R. Soc. Lond. A.* 266(11):63–111, 1970.
7. D. A. Grier, When computers were human, Princeton University Press, Princeton, 2005.
8. S. Talukdar, Weaving mathematical principles into Indian carpets, *NewsClick India* (April, 2021). https://www.newsclick.in/Indian-Carpets-Mathematics-Weaving.
9. R. Barry, Mathematical Socialism — Tim Otto Roth lights up Hanoi. *The Quietus,* November, 2017. https://thequietus.com/articles/23568.
10. T. Roth, About the robustness of 1d cellular automata revising their temporal entropy. *Physica D: Nonlinear Phenomena.* 425 (November, 2021). Doi: 10.1016/j.physd.2021.132953.
11. J. v. Neumann, *The general and logical theory of automata,* In *Collected Works — Design of Computers, Theory of Automata and Numerical Analysis,* A. H. Taub, editor. Pergamon Press, Oxford, 1961, 288–326.

https://doi.org/10.1142/9789811257155_0015

Chapter 15

Unconventional Sensing: Doing it in an Unusual Way in Unusual Settings

Zoran Konkoli

Department of Microtechnology and Nanoscience - MC2, Chalmers University of Technology, Goteborg, Sweden

zorank@chalmers.se

The standard way of sensing implies a linear flow of information, from the object we wish to analyze toward the observer. The object of interest will be referred to as "the environment". The environment can be many things, a temperature in a room or the amount of cracks in the material. The usual sensor is like a factory line where the flow of information is well understood and every step is carefully engineered. Normally, a sensing instance is a one-time event. It can be, and often is, repeated leaving the impression of continuity. We know exactly how every part of the sensor is supposed to behave. But it is a one-time event nevertheless and the execution of every sensing event is always nearly the same. But there are other ways of sensing, and over the course of the last few years we have explored a novel, somewhat unconventional way of sensing, referred to as the SWEET sensing approach. SWEET is more a template than it is an approach. The SWEET is an algorithmic template for building intelligent sensing substrates. The user is supposed to engineer different instances of it depending on the situation. The typical SWEET sensing process is very flexible, and is essentially a dynamic process, where the single time instance is irrelevant, but rather the behavior over an extended time interval is of paramount importance. It is a sort of indirect sensing which happens through a proxy. There is an ongoing dialogue between the observer and the proxy, in which the observer accumulates small clues proxy provides about the system of interest. In turn, the proxy also accumulates information through a dialogue. This chapter summarizes how such an unconventional sensing setup can be realized and what its essential ingredients are. The main objective is to present the material in a pedagogical way so that colleagues from other disciplines can understand the method and use it to solve their problems.

The standard way of sensing implies a linear flow of information, from the object we wish to analyze, the environment, toward the observer. In sensing applications, the "environment" can be many things, a temperature in the room, the amount of cracks in the material, or the amount of rare molecules in a unit volume of solvent. The usual sensor works according to well-understood principles which are carefully engineered. There are no surprises. The flow of information is linear. Further, a traditional sensing instance is a one-time event. Events like these are often repeated in time, which might give an impression of continuous sensing, even some sort of intelligence perhaps, but in reality such sensing events are discrete, separate events. To gain a time perspective on the observed data, one needs to engineer an external intelligence that can gather information obtained from the sensor, and further process it. The typical time to execute the sensing event is much smaller than the times that separate these events. Sensing instances can occur at equally spaced intervals, or when triggered by a control mechanism. Regardless of the engineering details, we are talking about a one-time event since the execution of the sensing event is always nearly the same, and we know exactly how every part of the sensor is supposed to behave. In fact, failing to do so is considered to be a malfunction. The typical sensing setup is show in Figure 1. However, there are other ways of sensing.

Over the course of the last few years we have explored a novel, somewhat unconventional way of sensing where irregularity and temporal aspects of the sensor's dynamics are of paramount importance, the SWEET sensing setup.[1] The SWEET sensing setup

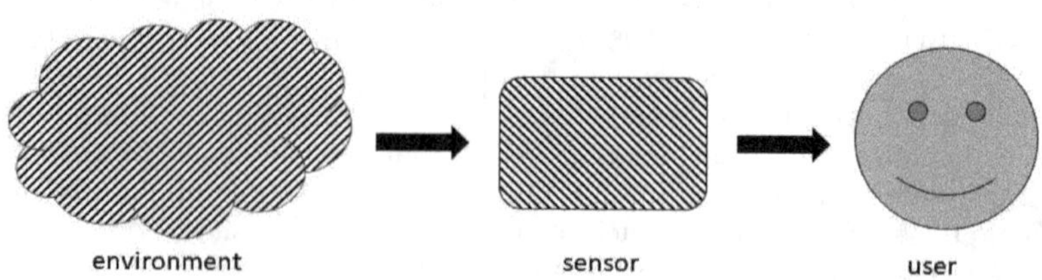

Fig. 1. The traditional sensing setup. The flow of information is linear.

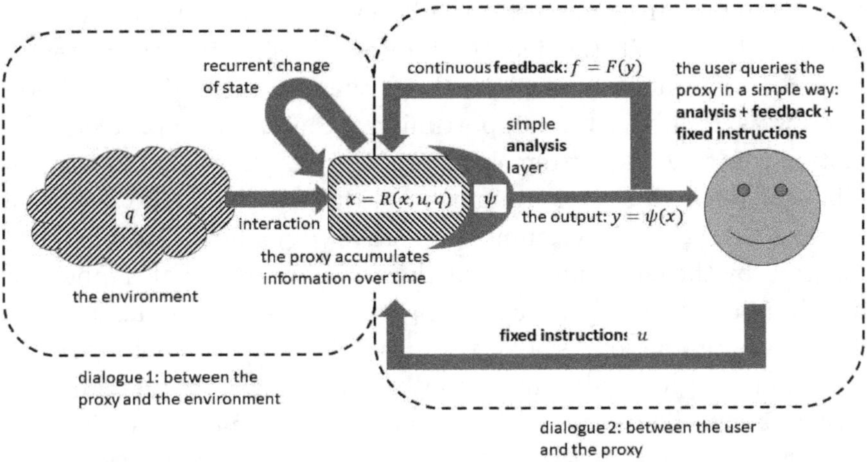

Fig. 2. The SWEET sensing setup. Dialogue 1: There is an ongoing dialogue between the environment and the proxy. During this dialogue the proxy accumulates the information about the environment. Dialogue 2: The user interacts with the proxy in a simple way to infer what the environment looks like.

is an algorithmic template. The most important components are depicted in Figure 2. Though the SWEET sensing setup has been developed with a particular application in mind, the analysis of time series data, it is very likely that the setup could be used to solve problems featured in other disciplines, but more on that later. The details will of course differ depending on the implementation context. Our own work[2–6] covered some problems related to ionic sensing with ion sensitive electronic components, such as an organic transistor and a memristor. The SWEET sensing process is very flexible, and is essentially a dynamic process, where the single time instance is irrelevant, but rather the behavior over an extended time interval is of paramount importance. It is a sort of indirect sensing which happens through proxy and it happens in a dialogue form. A similar indirect sensing idea has been suggested in 2006[7] but, surprisingly, has not been pursued rigorously. The authors have shown that by monitoring the changes in the structure of the feedback apparatus that controls the robot, it is possible to infer about the environment that the robot resides in.

The SWEET approach is a full-developed algorithmic template on how to exploit the idea of indirect sensing in the context of time-series data analysis where the notion of time and remembering history is treated as an opportunity rather than a problem. The notion of time is extremely important as there is an ongoing dialogue between the environment and the proxy. The proxy (sensing reservoir[1]) is an environment-sensitive dynamical system. It is affected by the environment, but the details of how that happens are not important from an engineering point of view. It might appear that this interaction is a one-way information channel where the information propagates from the environment into the proxy. To some extent this is true, as the proxy is not supposed to influence the environment. However, the proxy updates its state recurrently, where the state of the proxy at a particular time instance depends on its previous state and the information it has received from the environment.

Through the dialogue with an environment (Figure 2, dialogue 1) the proxy is pushed toward a certain state, and the state of the proxy encodes all previous instances of environment-proxy interaction. Perhaps, one could think of it as a one-way dialogue, in the same way as a psychotherapist gets to know his patient. Through this dialogue, the proxy accumulates small clues about the environment of interest, which a traditional setup might simply miss. Again, it is important to realize that this would be impossible without a memory of the past, the system used as a proxy would "forget" every bit of information it has collected.

There is a second dialogue going on, the one between the observer and the proxy (indicated by dialogue 2 in the figure). This second dialogue happens through two mechanisms: (a) the fixed instructions in the form of a drive signal; and (b) through a feedback. In many instances, this feedback can greatly improve the performance of a device. The feedback mechanism has the tendency to make the system "sharper" or simply more "intelligent", especially if it is provided with a delay in time. This improvement is a well-known fact,[8] and is nothing specific to the SWEET setup. The delay can be explicitly engineered, but it can also be an intrinsic part of a dynamical system since such systems clearly exhibit some sort of "memory behavior." The state of a dynamical system depends on everything that the system experienced in the past. Furthermore, dynamical systems might have the ability to "propagate" behavior through the system to different locations, which could bounce around, causing a natural delay. All very exciting possibilities.

Perhaps the simplest way of explaining the SWEET sensing concept is through the following thought experiment (*cf.* Figure 3):

- *The sensing problem (the environment)*: Imagine that we are sitting in a room without windows and that we are interested in the weather outside. Thus, the object of our interest is the weather. However, we are not allowed to open the window and peek out to see what the weather is like. Likewise, we are not allowed to use the phone or any other means of communication with the external world.
- *The proxy*: A person comes into the room and we are allowed to interact with the person to figure out what the weather is like. This is our only chance to infer about the weather.
- *The interaction with the proxy*: By assumptions, we are not allowed to ask direct questions like "What is the weather like outside?" This constraint is very common in practical applications that are close to engineering. We cannot interact with the proxy in any

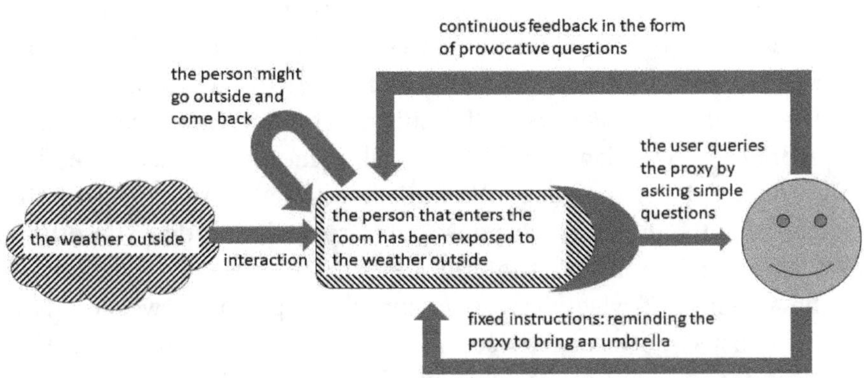

Fig. 3. An example of implementing the SWEET sensing setup.

way we please. There are likely going to be constraints imposed on how we can interact with any system. For example, we cannot easily measure the positions of all atoms in a gas. We could try to measure their average velocity, which gives us an idea about the temperature of the gas. Thus to illustrate that aspect of the problem, in our thought experiment we are restricting the way we can interact with our human proxy. Naturally, the more direct the questions one can ask, the better. However, the main idea is to "interview" the person about the weather. Ask small questions and try to assess what they imply. Here the emphasis is on the full set of answers we are getting, not on a particular answer to a specific question. The way we can interact with the proxy is critical. For example, if we can assume that we can observe the person, we could try to see if the person carries a wet umbrella. A wet umbrella would imply that very likely there is rain outside. But one cannot be sure. It could have rained few hours earlier but we cannot be certain. Even a dry umbrella might imply bad weather. For example, it could be that it is cloudy outside. Clearly, if we are not allowed to ask a direct question, the dialogue is the best option. Through a dialogue we can acquire information. For example, we can ask the person how she/he feels like. If the person appears grumpy, then it might have to do with the bad weather, but there could be other reasons. For example, the person might have lost a shoe, and that could be inferred from follow-up questions. If the person appears happy, very likely the weather is nice. But, again, we cannot be sure. The person could have won the lottery, and more follow-up questions would be needed to determine this.

- *The feedback mechanism*: In this particular instance, a useful feedback mechanism would be to ask some provocative questions. For example, if the person is happy, we might try to provoke the person into an angry response and from that judge the original "degree of happiness" that the person had before he/she entered the room. Likewise, if the person appears calm, we could ask provocative questions to see if there is some discontent lurking behind the surface. All this would indicate that we are likely dealing with the case of the bad weather.

In fact, the sensing solutions we envision resemble a thought experiment discussed above. The SWEET setup mimics the way how humans communicate experiences and process exchanged information, rather than how a machine would perform similar tasks. This chapter explores the interplay between the human and the engineered in the sensing context. In fact, the SWEET setup mimics in many ways the thought experiment discussed above. Instead of a human, one uses a generic dynamical system as a proxy. In pretty much the same way as the mental state of mind of a human is influenced by weather, so is the state of the dynamical system influenced by the environment. The existence of this interaction is the minimal requirement for the SWEET setup to work. However, the trademark of our sensing setup is that it is extremely flexible.

Why should one bother with such unconventional sensing setups? There are several reasons, ranging from pure engineering/practical toward more profound:

To begin with, there are many instances in the information processing engineering where the goal is to collect and analyze information *in situ.* For example, in the traditional setup, remote sensors might collect information and channel it to a huge data center where this information is *stored* and *processed.* However, the disadvantage with such a setup is that one might need to both store large amounts of data, and most importantly, perhaps, arrive at simple conclusions, after all the fuss. It is much better to make such conclusions locally, and simply send the results of the analysis. For example, a bridge operator only needs to know one bit of information, whether the bridge is stable or about to collapse or in somewhat simplified language is "everything under control" vs. in "trouble". It is much simpler to send this information instead of constantly streaming sensor data from dozens of sensors.

The more subtle reason is as follows. The whole SWEET setup is powered by a very simple idea discussed above, and because of that, it is extremely flexible. Since SWEET is a template, an engineer only needs to implement its key mathematical abstractions (essentially the various forms of the dialogues introduced earlier). Of course, this a bit of oversimplification (if it were that easy), but in a nutshell this is what needs to happen. Regardless of the practical difficulties, the implementation strategy is very clear. One can think of the SWEET setup as a very generic user manual, where the user can exercise a great deal of flexibility in implementing. One might think of it as a "religious text" in which the principles are given but the followers are supposed to interpret it and apply it in everyday life. Because of that, our sensing ideas could be of relevance for other areas of human endeavor, including humanistic sciences.

Naturally, one might even have to modify the setup depending on the problem, but the core idea of a two-fold dialogue will likely survive if the setup is applied in other areas:

- We have investigated numerous options ourselves. All the sensing problems we have investigated hatched from vastly different application contexts: monitoring ionic solution for rare ions that are hard to detect, monitoring ECG signals using simple hardware, predicting chance for the occurrence of sepsis for intensive care patients. These might be considered as standard engineering problems, very typical for natural sciences.
- We gave a serious thought of applying the SWEET sensing setup to tackle a range of completely unrelated problems to the ones we published on. We have thought about other much more "crazy" ideas such as helping spine-cord injuries to heal, we have explored some innovative information processing options in the IoT context related to creating a gigantic intelligent sensing substrate to realize distributed sensing, and believe it or not envisioned the setup to enhance human learning, and finally tried to think about ways of enhancing the outcome of a psychotherapy session.

These rather bold aspirations listed above ought to justify the title.

References

1. Z. Konkoli, On developing theory of reservoir computing for sensing applications: The state weaving environment echo tracker (SWEET) algorithm. *Int. J. Parallel Emergent Distrib. Syst.* 121–143, 2016. Doi: 10.1080/17445760.2016.1241880.
2. V. Athanasiou *et al.*, *On Sensing Principles Using Temporally Extended Bar Codes. IEEE Sens. J.* 20(13):6782–6791, 2020.

3. V. Athanasiou and Z. Konkoli, On improving the computing capacity of dynamical systems. *Sci. Rep.* 10(1):9191, 2020.
4. V. Athanasiou and Z. Konkoli, *Memristor Models for Early Detection of Sepsis in ICU Patients*, in 2019. Computing in Cardiology (CinC), 2019.
5. Z. Konkoli, *et al.*, Reservoir computing with computational matter. In *Computational Matter*, M. Amos, S. Rasmussen, and S. Stepney, editors. Springer, 2018. Doi: 10.1007/978-3-319-65826-1.
6. V. Athanasiou and Z. Konkoli, On using reservoir computing for sensing applications: Exploring environment-sensitive memristor networks. *Int. J. Parallel Emergent Distrib. Syst.* 2017. Doi: 10.1080/17445760.2017.1287264.
7. F. Iida and R. Pfeifer, Sensing through body dynamics. *Rob. Auton. Syst.* 54(8):631–640, 2006.
8. L. Appeltant, *et al.*, Information processing using a single dynamical node as complex system. *Nat. Commun.* 2:468/1-6, 2011. See Vasseleu's poignant discussion of "fantasies of disembodied mastery" in C. Vasseleu, "Virtual Bodies/Virtual Worlds", *Australian Feminist Studies* 19:155–169, 1994. For a critical discussion of contemporary technoscientific corporealizations of the "almost human", see C. Castaneda, and L. Suchman, Robot visions, *Soc. Stud. Sci.* 44(3):315–341, 2014.

https://doi.org/10.1142/9789811257155_0016

Chapter 16

Unconventional Computing Art in Cellular Automata

Genaro J. Martínez[*,†,§], Andrew Adamatzky[†,¶], and Marcin J. Schroeder[‡,||]

[*] *Escuela Superior de Cómputo, Instituto Politécnico Nacional, México*
[†] *Unconventional Computing Laboratory, University of the West of England, Bristol, United Kingdom*
[‡] *Institute for the Excellence in Higher Education, Tohoku University, Sendai, Japan*
[§] *gjuarezm@ipn.mx*
[¶] *andrew.adamatzky@uwe.ac.uk*
[||] *schroeder.marcin.e4@tohoku.ac.jp*

Cellular automata are arrays of finite state machines (cells), which update their rules in a discrete time and by the same rule. Each cell updates its state depending on states of its closest neighbors. Cellular automata are widely used as fast prototyping tools for developing unconventional computing devices. The automaton models of such devices exhibit rich space–time dynamics, which can be interpreted as an art. In this chapter, we briefly overview the topic of cellular automata in unconventional computing and arts.

The exploration of unconventional computing in its diverse forms is not only, and not primarily, a result of the natural human pursuit for innovation but rather a response to challenges faced by the current information technology. Some of these challenges are not new, e.g., the expected end of applicability of Moore's Law or the von Neumann bottleneck in the transfer of data between the CPU (central processing unit) and RAM (random-access memory). However, the bottleneck in the past was just a nuisance, but at present the need for massive processing of synaptic weights in the network for machine learning, which requires multiple transfers, makes this primary tool of AI (artificial intelligence) inefficient and hopeless in the competition with the natural, biological systems of information processing.

An example of another challenge of a very different "down to Earth" type is the high energetic cost of machine learning estimated already as a substantial portion of the energetic needs of the industrial societies which in the near future is expected to become the main consumer of energy. Thus, the question about the frugality of nature in the energetic budget for the human or animal brain is worth billions or trillions of future dollars.

These and other challenges direct the research toward unconventional forms of computing with the special interest in the natural forms identified in living organisms, on the one hand, and in the utilization of new, natural, physical phenomena in information processing, on the other.

This brings us to a more general and theoretical question overarching the interests in natural forms of information processing about what constitutes the fundamental distinction between the traditional form of computing originating in the theoretical model of a Turing machine and unconventional, natural computing. One of the possible answers is that the Turing machine model is based on the principle of a one-way, goal-oriented action initiated and controlled by a pre-defined program, while all natural processes are dynamic, i.e., they are based on mutual interactions within the processing system and with its environment.[5]

It is possible to consider a modification of the Turing machine model in which instead of the one-way action of the head on the tape the processing is performed by mutual reading and mutual re-writing of the two interacting central components.[6] This model of symmetric inductive machine remains within the Turing limit of computability as soon as the dynamics of interaction is computable, but nothing makes this computability unavoidable.[7]

The shift of the focus on the dynamic, interaction-based forms of information processing can be implemented in the most natural way in the information processing in cellular automata where the art of unconventional computing begins.

Unconventional and natural computing[8–10] has the capacity to handle information at an atomic and molecular level, the first stage. A diversity of scientific fields study and research all these ways on continuous and discrete domains. Lines of research can be found in Table 1.

In this way, the cellular automata theory conceived by von Neumann in the late 1950s acts as a tool of super computation.[19] He has been working with primitive and indivisible elements, and where this theory offers an inherently and massively computation in parallel. Von Neumann had discussed that universal Turing machines cannot exploit the process in nature and the universe. This way, the existence of universal constructors becomes essential for the universe.[20]

An actual problem is how to control and design reliable components from unreliable organisms.[20] Indeed, this issue is preserved in any unconventional computing architecture. In the literature of cellular automata theory, we can see a diversity of designs without any particular architecture. This way, we can think that these machines are adapted for a specific environment. Therefore, we can imagine how these machines are working simultaneously in nature or the universe, each with its own architecture and environment.

Table 1. Some ways to use unconventional computing.

Quantum computers[11]	Reaction-diffusion computers[15]
DNA computers[12]	Hot ice computers[16]
Physarum computers[13]	Collider computers[17]
Optical computers[14]	Thermodynamic computers[18]

A list of computable cellular automata can be found in: *Complex Cellular Automata Repository* https://www.comunidad.escom.ipn.mx/genaro/Complex_CA_repository.html.

Cellular automata are adequate mathematical machines to represent unicellular computers (Figure 1) because of their architectural properties: array of infinite state machines matches arrays of these units. Historically, cellular automata theory has been analyzed as

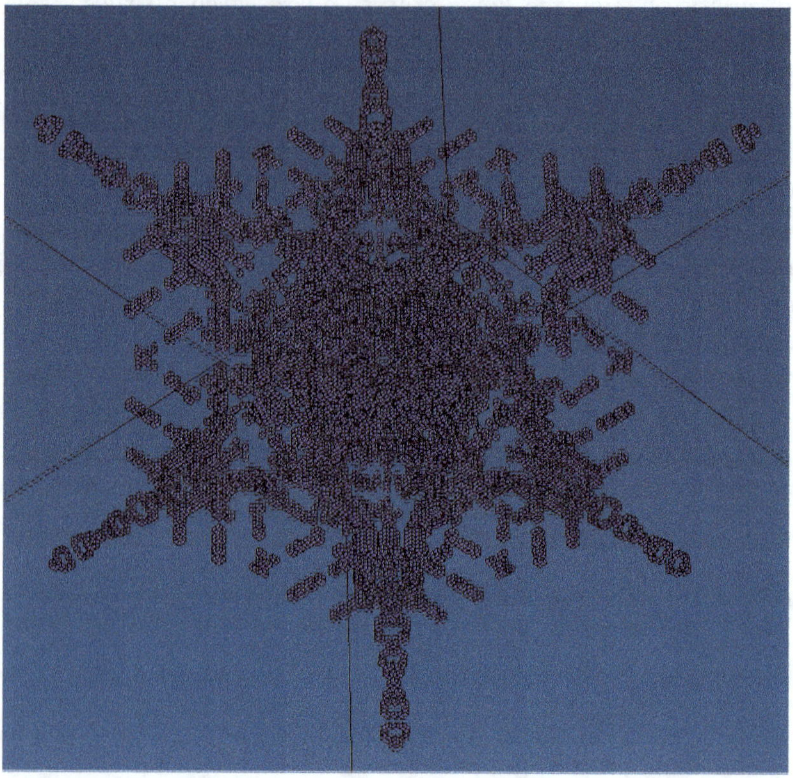

Fig. 1. A three-dimensional cellular automaton Life-like rule $B4/S9$. It is a projection of the two-dimensional Life-like rule $B2/S7$, the *Diffusion Rule*.[1] The rule is a chaotic rule although it supports complex patterns as oscillators, gliders, puffer trains, and an ample diversity of glider guns. This rule is proved logically universal by realization of computing circuits via collisions between particles. This evolution displays the result of two particles colliding, thus after of 112 steps we can see symmetric complex structures emerging during the evolution, traveling, expanding, and interacting with others.

supercomputers, as we can see in Refs. 21–23. On the other hand, cellular automata are explored in an artistic way as was presented in the book *Designing Beauty: the Art of Cellular Automata.*[24]

We can think of *unconventional computers* as the physical devices and *unconventional computing* as the logic medium where these devices work (Figures 2–7). This way, we can complement the Table 1 with some unconventional computing models listed in Table 2.

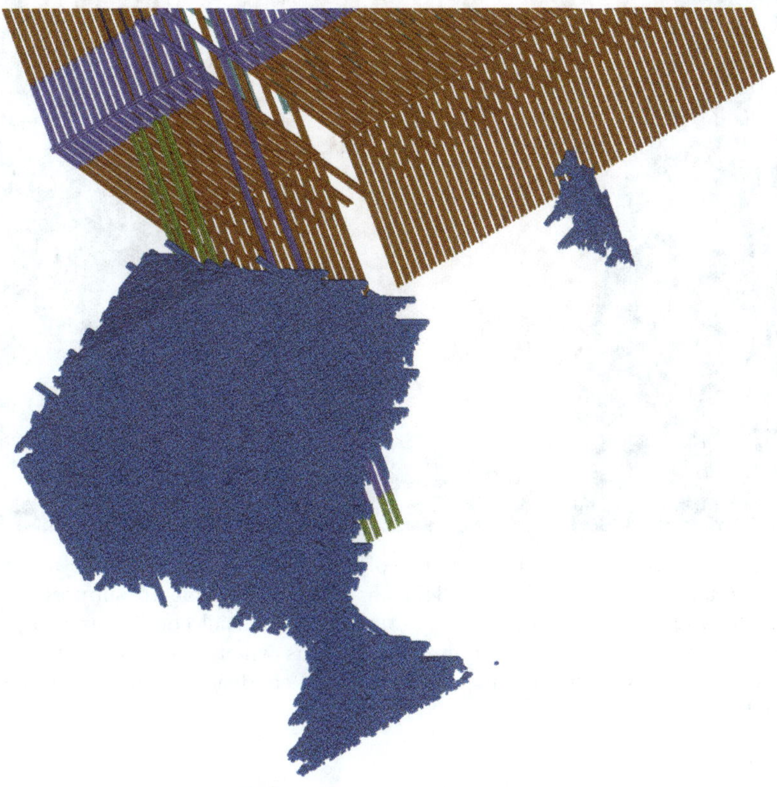

Fig. 2. Three-dimensional projection of the two-dimensional Life-like rule $B2/S7$, the *Diffusion Rule.*[1,38] The rule is classed as chaotic although it supports complex patterns as oscillators, gliders, puffer trains, and a diversity of glider guns. The rule is proved logically universal via collisions of particles. This evolution displays the result of two particles in vertical position colliding. The reaction produces a replication of particles periodically in thousands of generations. With time a symmetry is lost and the automaton dynamics becomes chaotic. We call it *the cellular automata origin.* https://youtu.be/BqTU_uW-zaI.

Fig. 3. Three-dimensional projection of a two-dimensional cellular automaton *Life without Dead*, rule $B3/S012345678$. This rule is able to support complex behavior and logic computability.[2] From random initial conditions you can see the emergence of worms and interesting designs when the worms interact with each other. This evolution starts with an initial condition defined by a line of 11 alive cells.

Free software used to create the simulations in this chapter.

- Ready (Tim Hutton, Robert Munafo, Andrew Trevorrow, Tom Rokicki, https://github.com/gollygang/ready)
- CAviewer (José Antonio Jiménez Amador, Genaro J. Martínez, https://www.comunidad.escom.ipn.mx/genaro/Papers/Thesis_files/CAviewer.tar.gz)

Fig. 4. Metaglider (mesh) designed with the elementary cellular automaton rule 54 synchronizing multiple collisions evolving in a ring. It is a three-dimensional projection of a typical two-dimensional evolution. Rule 54 is a logically universal automaton.[3] Logic computation in rule 54 is performed by collisions of particles in its evolution space.

Fig. 5. Two-dimensional representation by colors of a Turing machine that doubles the number of ones as a cellular automaton.[1,4]

Fig. 6. Two-dimensional representation by colors of a Turing machine that simulates the behavior of ECA rule 110. The history of the Turing machine is represented as a cellular automaton.[1] The initial condition starts with the string B^*010B^* showing in two snapshots the evolution to 10,000 steps.

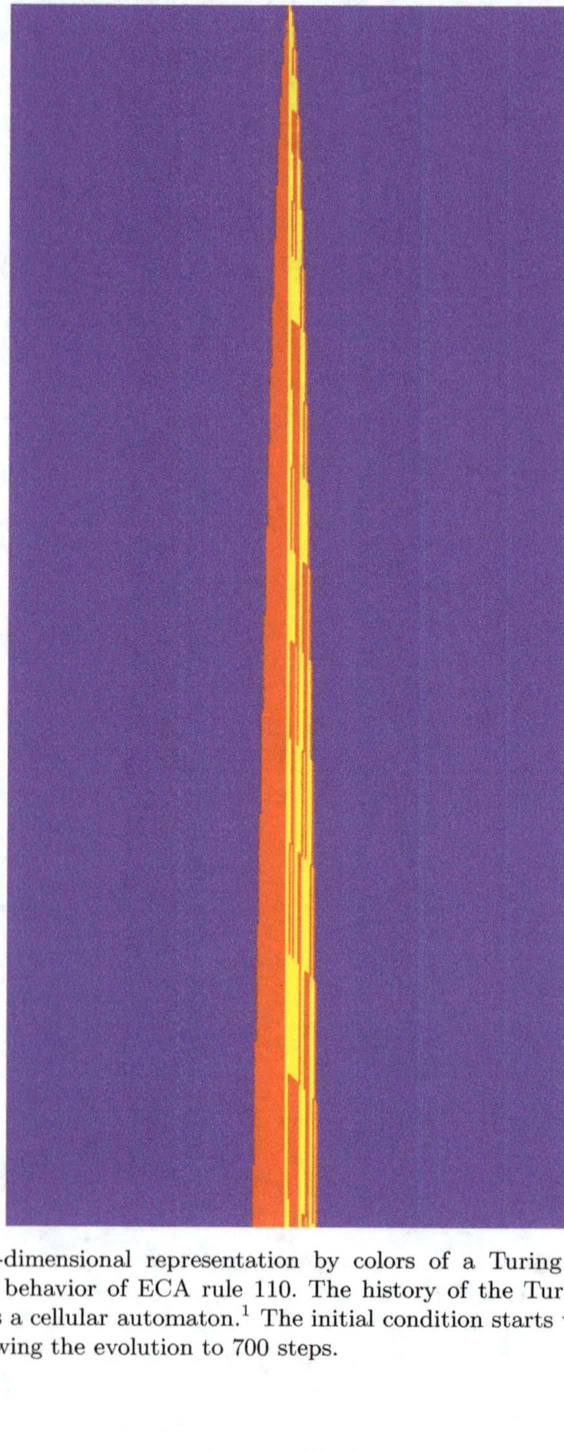

Fig. 7. Two-dimensional representation by colors of a Turing machine that simulates the behavior of ECA rule 110. The history of the Turing machine is represented as a cellular automaton.[1] The initial condition starts with the string B^*010B^* showing the evolution to 700 steps.

Table 2. Some ways to unconventional computing.

Reversible computing[25,26]	Conservative computing[31,32,36,37]
Chaotic computing[27]	Crystalline computing[33]
Molecular computing[28]	Tiling computing[34]
Competing patterns computing[29]	Symmetric inductive computing[7]
Soliton computing[30]	Slime mould computing[35]

- CATM (Sergio Eduardo Juárez Martínez, César Iván Manzano Mendoza, https://www.comunidad.escom.ipn.mx/genaro/Papers/Thesis_files/maquinaTuring.java.tar.gz)

References

1. S. J. Martínez, I. M. Mendoza, G. J. Martínez and S. Ninagawa, Universal one-dimensional cellular automata derived from Turing machines. *Int. J. Unconv. Comput.* 14(2):121–138, 2019.
2. D. Griffeath and C. Moore, Life without death is P-complete. *Complex Syst.* 10:437–448, 1996.
3. G. J. Martínez, A. Adamatzky and H. V. McIntosh, Phenomenology of glider collisions in cellular automaton Rule 54 and associated logical gates. *Chaos Solitons & Fractals*, 28(1):100–111, 2006.
4. P. Rendell, *Turing Machine Universality of the Game of Life.* Springer, 2016.
5. M. J. Schroeder, From proactive to interactive theory of computation. In: *The 6th AISB Symposium on Computing and Philosophy: The Scandal of Computation — What is Computation?* 2013, 47–51.
6. M. J. Schroeder, Towards autonomous computation: Geometric methods of computing. *APA Newsletter Philos. Comput.* 15(1):7–9, 2015.
7. M. Burgin, Information processing by symmetric inductive Turing machines. In *Multidisciplinary Digital Publishing Institute Proceedings*, vol. 47, no. 1, 28, 2020.
8. S. B. Cooper, What makes a computation unconventional? In *Computing Nature.* Springer, Berlin, Heidelberg, 2013, 255–260.
9. J. W. Mills, The nature of the extended analog computer. *Physica D*, 237(9):1235–1256, 2008.
10. T. Toffoli, Non-conventional computers. *EEE* 14:455–471, 1998.
11. I. L. Chuang, IBM's test-tube quantum computer makes history. *IBM Research News*, 2001.
12. G. Paun, G. Rozenberg and A. Salomaa, *DNA Computing: New Computing Paradigms.* Springer, Berlin, 1998.
13. A. Adamatzky, *Physarum Machines: Computers from Slime Mould.* World Scientific, 2010.
14. T. Yatagai, Cellular logic architectures for optical computers. *Appl. Opt.* 25(10):1571–1577, 1986.

15. A. Adamatzky, B. D. L. Costello, and T. Asai, *Reaction-Diffusion Computers.* Elsevier, 2005.
16. A. Adamatzky, Hot ice computer. *Phys. Lett. A* 374(2):264–271, 2009.
17. A. Adamatzky, (editor). *Advances in Unconventional Computing, Volume 1: Theory.* Springer, 2016.
18. T. Hylton, Thermodynamic computing: An intellectual and technological frontier. In *Multidisciplinary Digital Publishing Institute Proceedings*, vol. 47, no. 1, 2020, 23.
19. J. Von Neumann, *Theory of Self-reproducing Automata* (edited and completed by A. W. Burks), University of Illinois Press, Urbana and London, 1966.
20. J. Von Neumann, Probabilistic logics and the synthesis of reliable organisms from unreliable components. *Autom. Stud.* 34:43–98, 1956.
21. A. Adamatzky, *Computing in Nonlinear Media and Automata Collectives.* CRC Press, 2001.
22. N. Margolus, T. Toffoli and G. Vichniac, Cellular-automata supercomputers for fluid-dynamics modeling. *Phys. Rev. Lett.* 56(16):1694, 1986.
23. S. Wolfram, Cellular automaton supercomputing. Center for Complex Systems Research, IL, USA, 1987.
24. A. Adamatzky and G. J. Martínez, (editors). *Designing Beauty: The Art of Cellular Automata.* Springer, (2016).
25. C. H. Bennett, Logical reversibility of computation. *IBM J Res. Develop.* 17(6):525–532, 1973.
26. K. Morita, Reversible computing and cellular automata — A survey. *Theor. Comput. Sci.* 395(1):101–131, 2008.
27. L. O. Chua, Chua's circuit: An overview ten years later. *J. Circuits, Syst. Comput.* 4(02):117–159, 1994.
28. T. Sienko, A. Adamatzky and N. Rambidi, *Molecular Computing.* MIT Press, 2013.
29. G. J. Martínez, A. Adamatzky, K. Morita, and M. Margenstern, Computation with competing patterns in Life-like automaton. In *Game of Life Cellular Automata.* Springer, London, 2010, 547–572.
30. M. H. Jakubowski, K. Steiglitz, and R. Squier, Computing with solitons: A review and prospectus. In *Collision-Based Computing.* Springer, London, 2002, 277–297.
31. E. Fredkin and T. Toffoli, Conservative logic. *Int. J. Theor. Phys.* 21(3-4): 219–253, 1982.
32. K. Morita, A simple universal logic element and cellular automata for reversible computing. In *International Conference on Machines, Computations, and Universality.* Springer, Berlin, Heidelberg, 102–113, 2001.
33. N. Margolus, Crystalline computation. In *Feynman and Computation,* 267–305, 1998.
34. B. Grünbaum and G. C. Shephard, *Tilings and Patterns.* Courier Dover Publications, 1987.

35. A. Adamatzky, Slime mould computing. *Int. J. Gen. Syst.* 44(3):277–278. 2015.
36. A. Adamatzky, (editor). *Collision-Based Computing.* Springer, 2012.
37. E. Fredkin and T. Toffoli, Design principles for achieving high-performance submicron digital technologies. In *Collision-Based Computing.* Springer, London, 2002, 27–46.
38. G. J. Martínez, A. Adamatzky and H. V. McIntosh, Localization dynamics in a binary two-dimensional cellular automaton: The Diffusion Rule. *J. Cell. Autom.* 14(2):121–138, 2010.

https://doi.org/10.1142/9789811257155_0017

Chapter 17

Extended "Beehive" Rule Cellular Automaton Dynamics on Irregular Lattices — A Preliminary Study in the Form of a Visual Catalog

Phil Ayres

CITA, Royal Danish Academy, Philip de Langes Alle 10, Copenhagen, Denmark

phil.ayres@kglakademi.dk

A 3-value cellular automaton (CA) is implemented on hexagonal lattices with irregular topologies. A lattice irregularity is defined as the substitution of a hexagon with an alternative polygon. For this study, polygons are constrained to the set $[e|3 \leq e \leq 9]$, where e defines the number of edges of the substitute. Six irregular lattices are explored, each with a centrally located hexagonal cell replaced. The k-totalistic "beehive" rule[1–3] is modified to account for higher-valued neighbor conditions where $e > 6$. The behavior of two starting configurations is demonstrated on a regular hex-lattice to provide a control. The same starting configurations are then implemented on the irregular lattices. The starting position of the configurations is varied to explore the influence of the irregularity on dynamics. For each irregular lattice, extended and novel dynamics are demonstrated in comparison to the control. Sensitive dependence on starting positions relative to the lattice irregularity is observed. The extension of spatio-temporal dynamics resulting from interactions with heterogeneous lattice features holds relevance for the representation of many dynamic systems. Irregular lattice topologies influence curvature in various physical systems,[4–6] presenting opportunities for implementing embodied, situated, and performative CAs.

1. Implementation

The "beehive" rule transition table presented in Refs. 1 and 2, and represented in the more compact matrix form in Ref. 3, is adopted and extended to account for the largest possible neighborhood featuring in lattice $e9$ (see Figure 1). This matrix extension yields:

$$M = \begin{pmatrix} 0 & 1 & 2 & 1 & 2 & 0 & 0 & 0 & 0 & 0 \\ 0 & 2 & 2 & 2 & 1 & 1 & 0 & 0 & 0 & \\ 0 & 0 & 2 & 2 & 0 & 0 & 0 & 0 & & \\ 0 & 2 & 2 & 0 & 0 & 0 & 0 & & & \\ 0 & 0 & 2 & 0 & 0 & 0 & & & & \\ 2 & 0 & 0 & 0 & 0 & & & & & \\ 0 & 0 & 0 & 0 & & & & & & \\ 0 & 0 & 0 & & & & & & & \\ 0 & 0 & & & & & & & & \\ 0 & & & & & & & & & \end{pmatrix}.$$

The i, j mapping of the matrix to the summing of cell states {0,1,2} in the cell neighborhood follows the implementation of Adamatzky *et al.*[3] The cell being updated is not included in the neighborhood. Figure 2 defines the state representations. Figure 3 illustrates the behavior of two starting configurations on a regular hexlattice to provide a control. Figures 4 and 5 identify the principles for the starting configurations within the irregular lattices.

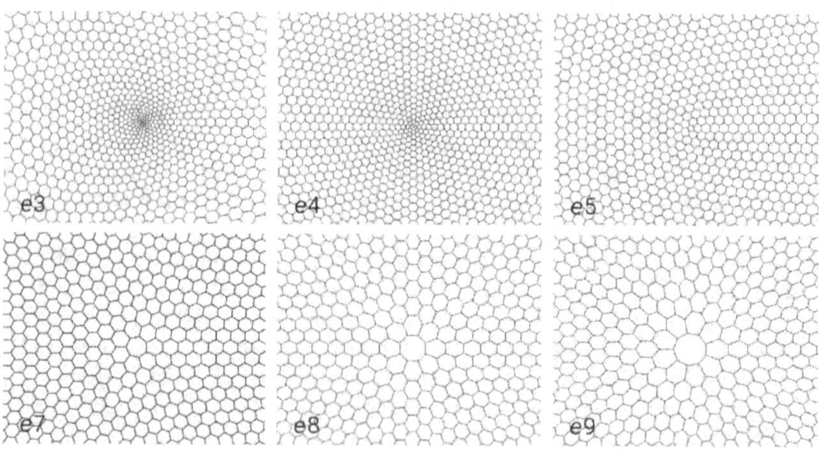

Fig. 1. The six irregular lattices investigated from the set $[e|3 \leq e \leq 9]$. This preliminary investigation limits the introduction of irregularities to a single cell substitution, centrally located.

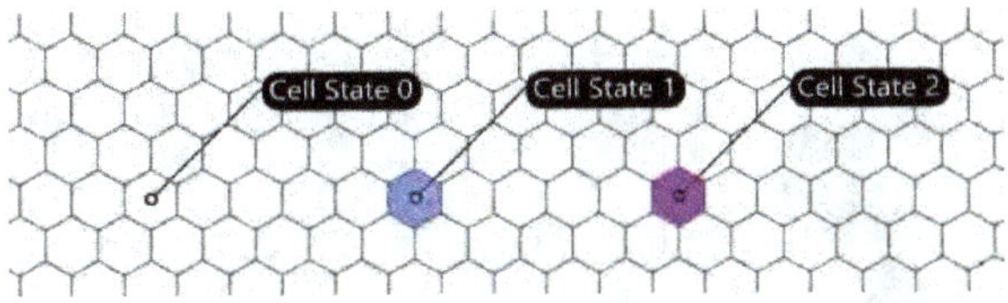

Fig. 2. State representation of the 3-value cellular automaton.

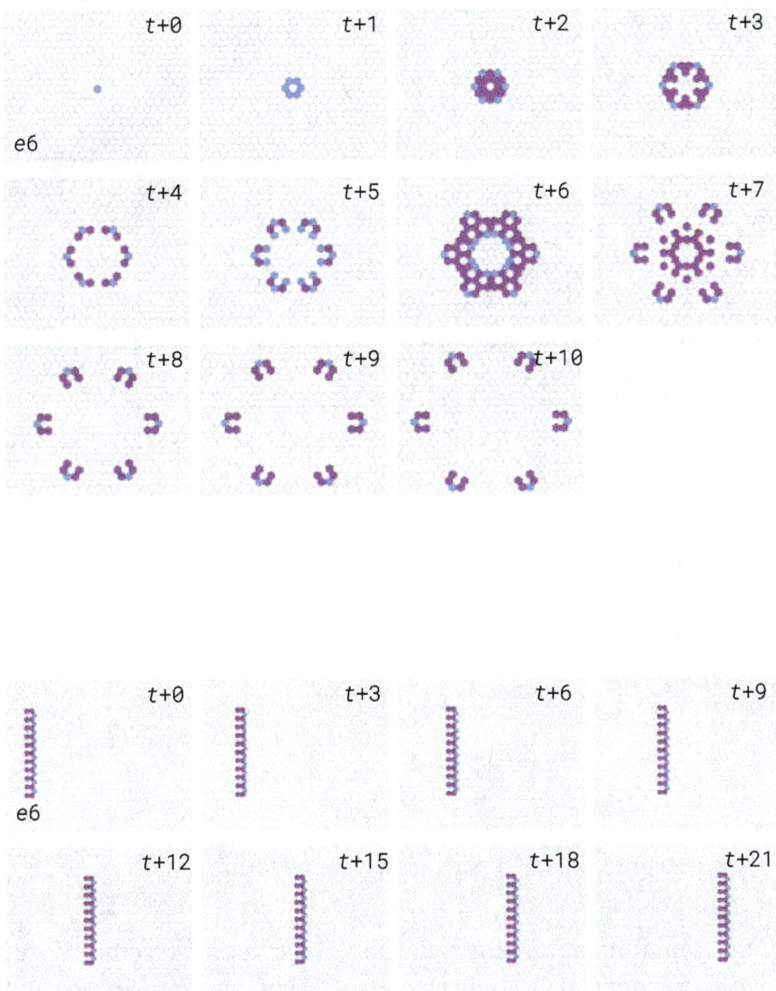

Fig. 3. Control behaviors from two starting configurations implemented on a regular hex-lattice — a "state 1 cell" (top), and "glider chain" (bottom).

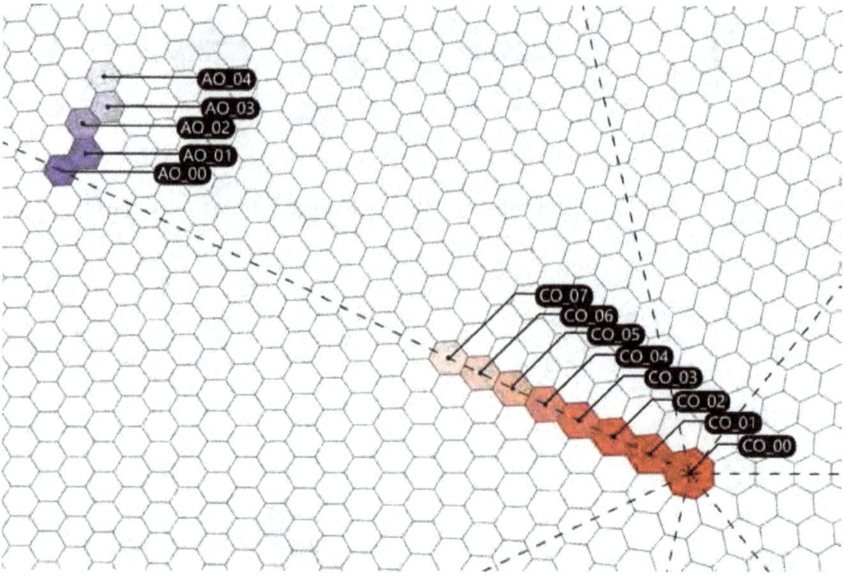

Fig. 4. Starting positions for a single cell in state 1. Irregularity influence is explored in series "CO-xx" by offsetting start points along an axis. Twenty cells from the irregularity series "AO-xx" offset start points from the axis.

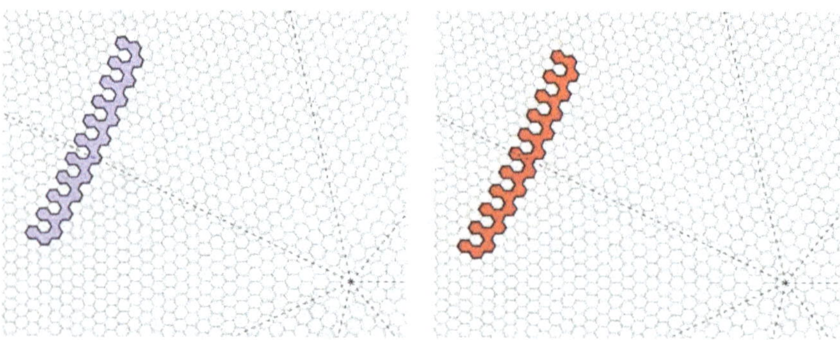

Fig. 5. Glider chain start positions. LC-01 (left), with a glider "head" on axis with the irregularity. LC-02 (right), with a glider "arm" on axis.

The remainder of this chapter illustrates selected dynamics in visual format with associated captions across Figures 6–9. A complete dataset of all defined starting conditions across the six irregular lattices is available from Ref. 7.

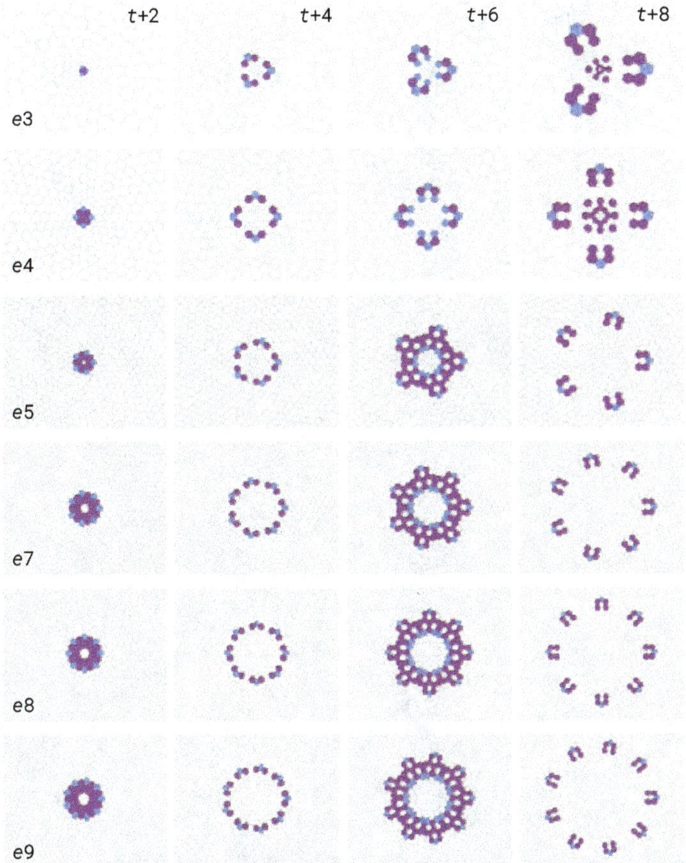

Fig. 6. Start configuration CO-00. The number of gliders generated is equivalent to the edge count of the irregular polygon.

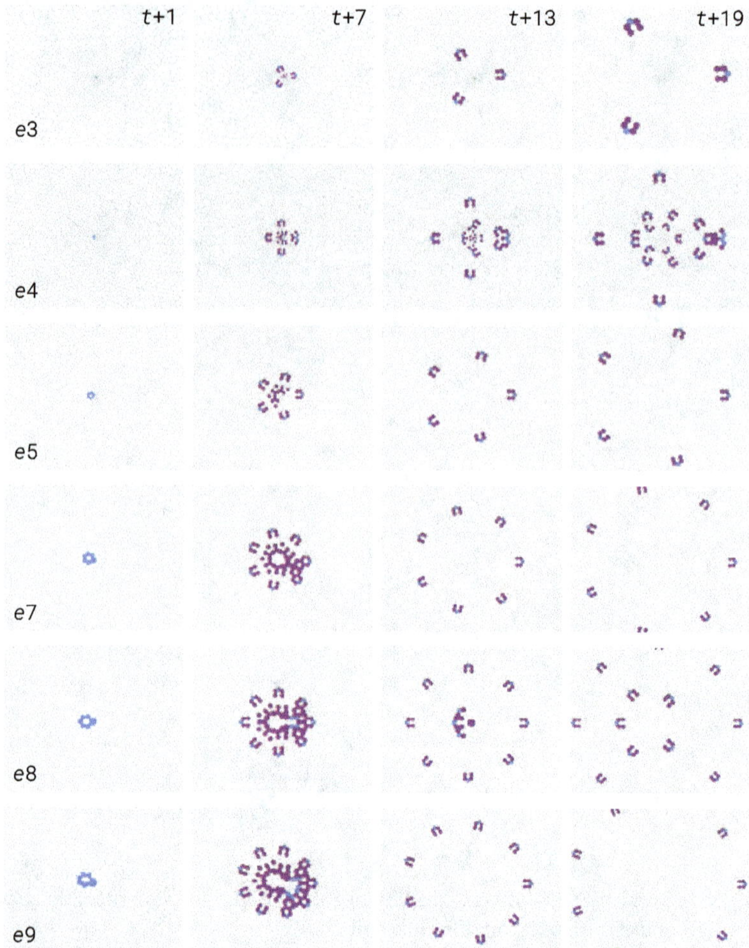

Fig. 7. Start configuration CO-01. Mobile glider guns appear in lattice $e4$. A nested group of gliders reaches maturity at $t + 19$ in lattice $e8$.

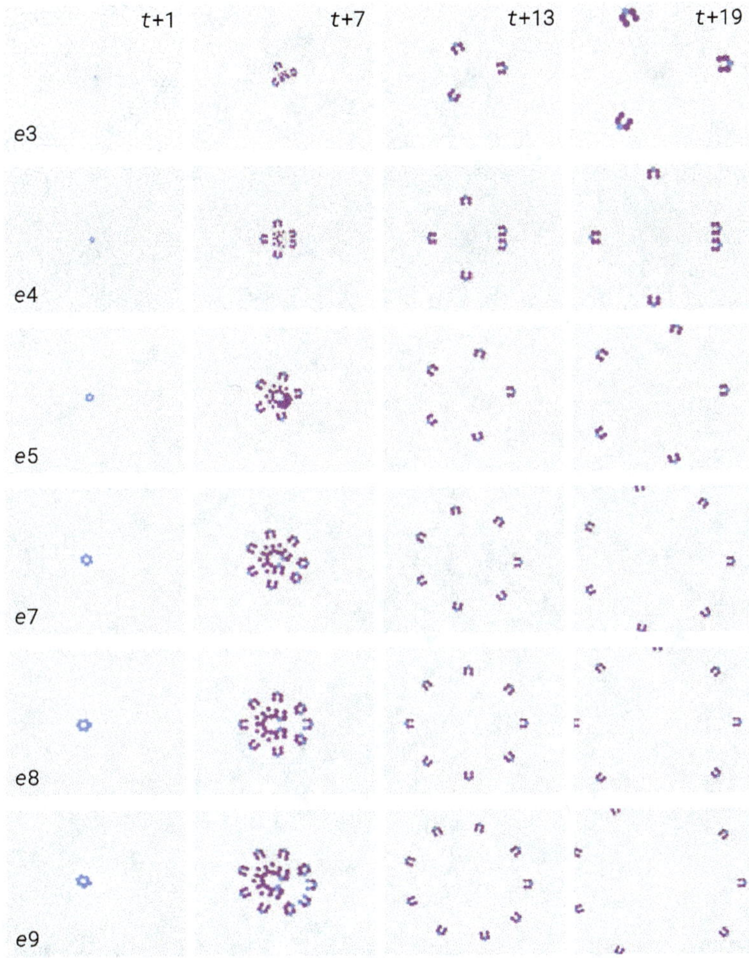

Fig. 8. Start configuration CO-02. Lattice *e*4 now generates five gliders.

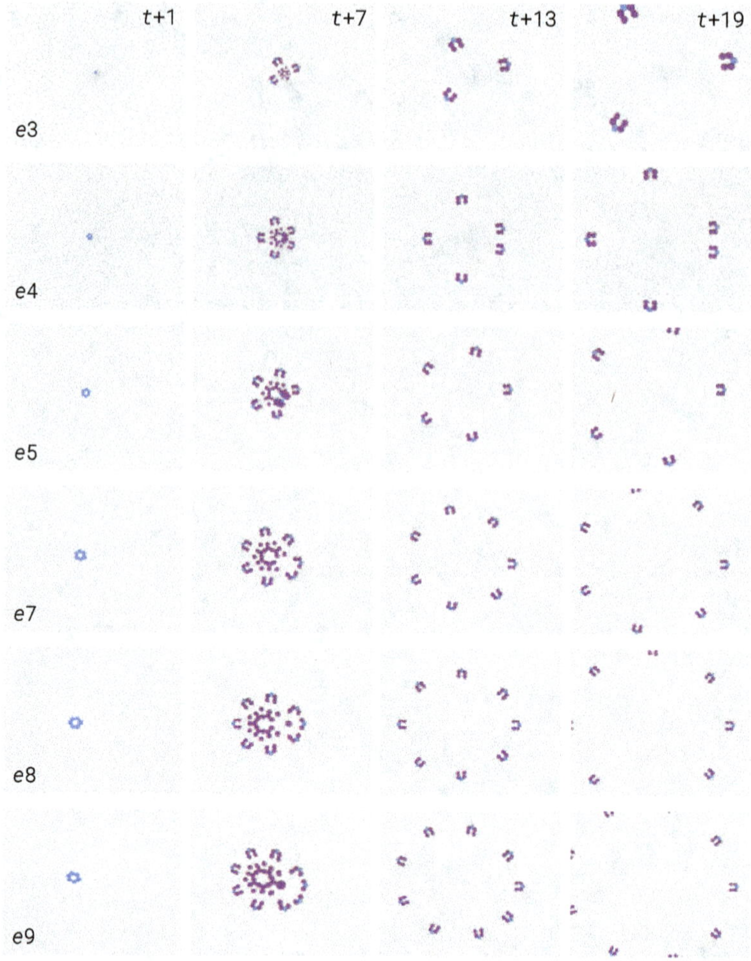

Fig. 9. Extended time sequence for glider chain $e5$/LC-01. On interaction with the lattice irregularity, the chain decomposes into individual and sub-groups of gliders with varying trajectories and subsequent interactions.

Acknowledgment

PA's research time was funded by the European Union's Horizon 2020 research and innovation programme FET OPEN "Challenging current thinking" under the project *Fungal Architectures*, grant agreement No. 858132.

References

1. A. Wuensche, Self-reproduction by glider collisions: The beehive rule. *Alife9 Proceedings*, 286–291, 2004.
2. A. Wuensche, Glider dynamics in 3-value hexagonal cellular automata: The beehive rule. *Int. J. Unconv. Comput.* 1(4):375–398, 2005.
3. A. Adamatzky, A. Wuensche and B. De Lacy Costello, Glider-based computing in reaction-diffusion hexagonal cellular automata. *Chaos Solitons & Fractals* 27(2):287–295, 2006.
4. S. Wolfram, *A New Kind of Science.* Champaign, IL, Wolfram media, vol. 2, 2002.
5. P. Ayres, G. M. Alison and Z. Mateusz, Beyond the basket case: A principled approach to the modelling of Kagome weave patterns for the fabrication of interlaced lattice structures using straight strips. In *Advances in Architectural Geometry 2018*, Chalmers University of Technology, 72–93, 2018.
6. P. Ayres, Y.-W. Ji, Y. Jack and G. M. Alison, Meshing with Kagome Singularities: Topology adjustment for representing weaves with double curvature. In *Advances in Architectural Geometry*, 2021.
7. P. Ayres, Implementing cellular automata on irregular hex lattices. OSF. doi:10.17605/OSF.IO/PS2FR, 2022.

https://doi.org/10.1142/9789811257155_0018

Chapter 18

The Benefits of Being Wrong: Bonding Epistemic and Cognitive Incompleteness for Natural and Artificial Intelligent Systems

Jordi Vallverdú
ICREA Acadèmia Researcher
at Universitat Autònoma de Barcelona
jordi.vallverdu@uab.cat

Currently, the main intense debates about the next steps of AI advancements are related to two concepts: the singularity, and AGI (Artificial General Intelligence). On the one hand, the crucial moment in which AI will reach and surpass human brain (and collective brains cooperation) power and, on the other, the birth of a conscious and creatively intelligent artificial intelligence. In most philosophical fiction studies, both events led to the same apocalyptic scenario: the destruction of humanity.

But with the evidences obtained by cognitive sciences and epistemological studies, provided to us in the last decades (following, in fact, ancient philosophical interests originated plenty of centuries ago), the real scenario is a completely different one: AI is still fighting for understanding how to do the most simplest things, according to human standards. While it is true that AI can play chess or Go and beat the best of all human players, it cannot achieve the simpler visual, linguistic, or sensorimotor skills of any (even average) human. Therefore, the most recent trends in AI are reinforcing not

Fig. 1. Feeling alone in this huge universe. https://pixabay.com/photos/home-lost-alone-cosmos-robot-2305431/

pure power based on statistical data analysis (although they are very promising, thanks to the re-birth of Neuronal Networks under Machine and Deep Learning techniques. Instead, approximate computing and bioinspiration are opening new paths into the inexorable computational revolution.

Bioinspiration is based on basic functional properties of living systems, taken not as whole but according to specific problem-solving scenarios. Such bioinspiration extracts features related to the biochemistry (hormones, molecular nature, genetic machinery, etc.), the neural interactions, social cooperation, or sensorimotor actions (hand grips, touch, smelling, walk, etc.) of living systems. But in no case is it related to epistemic or high cognitive aspects of living systems (some of them social, and with culture). I define the biomechanistic or functional approach as the First Wave of

Bioinspired AI, although the second is the most promising one: the Second Wave of Bioinspired AI. And the reasons are compelling. The most important ones are related, first, to the notion of formal coherence and, second, with meaning. Let me explain both in some detail:

1. Formal Coherence

Formal coherence is the holy grail of Western Thought and has been the accomplishment of a perfect logico-mathematical description of the world. With a perfect set of tools, a complete analysis of reality should be possible, they thought. From Aristotelian syllogistics, to the mathesis universalis of Leibniz, to the new logical framework of Frege or Boole, this was the main aim. When, finally, Russell and Whitehead wrote their *Principia Mathematica* (1910–1913), trying to demonstrate how logic could explain and justify the foundations of mathematics, there appeared a devastating figure: Kurt Gödel. With his incompleteness theorem about mathematics, Gödel demonstrated the inherent limitations of every formal axiomatic system capable of modeling basic arithmetic. While Hilbert was trying to create the mathematics for the new century, Gödel put on the table the more deep fact that nobody can have both completeness and consistency. After it, new kids of logic (fuzzy, paraconsistent, etc.) appeared, trying to deal with more specific and creative approaches to symbolic processing. After Gödel, the second fracture into the realm of reasoning perfection was related with the informational turn: new computational tools made it possible to generate and evaluate huge sets of data, something that allowed the statistical approach to mathematical proofs: 4CT, Kepler's Theorem, etc. As a consequence, computer-assisted proofs opened a new debate about epistemic opacity, black boxes, and paradoxes in comprehensive argumentation. Such a scenario, and we have in mind a possible future superintelligence or AGI, creates a tense framework: a perfect reasoning machine, even having more computer power and data access than any single mind or networked set of minds, will clash with the logically serious limits of the formal tools, as well as with the problem of evaluating and verifying the quality of data. At a meta-level, the internal paradoxes and contextual limitations force any cognitive system to deal with approximate and revisable sets of knowledge.

2. Meaning

Meaning, which affects rational entities, is surely the most unexplored but fundamental of all approaches to AI. And we are talking about how AI will generate its own meaning. And the ways are similar to those displayed by natural cognitive systems: due to embodied and/or sociocultural reasons. How do we evaluate the true meaning of things? Evaluating them from our embodied experience of positive or noxious events, that is, from our emotional experience. Meaning is also related to the functionality, including into this category wishes, aspirations, necessities (food, mating, survival etc.), among a long list. Even at an epistemological level, the commonsense about reality, or the description about reality itself, is biased on such conditions. Water looks like a stable thing when it can adopt three general states, with the bonds in the liquid form being broken constantly every few nanoseconds. From the perspective of a neutrino, we are basically empty. Time lapses are relevant or not according to the lifespan of an entity. Such meaning constraints are also the framework upon which cognitive systems are built. In that sense, the map of cognitive biases of human beings express not only the mental skills but also the adaptive requirements made for such bodies over millions of years. As a consequence, these elicitors of meaning explain how we understand existence, aims, the evaluation of death, or the aesthetics of reality. Possible embodiments of next AI systems will define the semantic possibilities of such entities, at a completely different level from those available for human beings, but biased (or situated) too!

Taking into consideration both aspects, coherence and meaning, we can foresee a plausible scenario for the next generation of AI systems. First of all, the beginning of a new era of exploration of the benefits of integrating new bioinspired models, which, by defect, must include biases and lack of accuracy (at the expense of another benefit); second, the ontological horizon that such systems face. One of the most childish aspects of techno-fetish followers (under the form of transhumanism) is to consider that the way of escaping from death involved a technological transfer or enhancement, from natural bodies to new forms of embodiment. But it is only a small patch in relation to the physical time in this universe: entropy is the final destiny of this universe. There is no way of escaping from absolute informational destruction: big crush, big rip, big freeze... this is the absolute truth of our universe. Perhaps some humans can feel happy thinking of small-time postponements, but for a really intelligent system, the reality is there: it doesn't matter which kind of strategy you wish to follow, because everything in this universe will be destroyed. How will a real intelligent AI react to this statement? Surely, by adopting a personal neo-phenomenological attitude toward reality. Under such conceptual horizons, the refugee of existence is the acceptance and practice of the things we've been calling as biases, or local embodiments as producers of meaning. To assume the fundamental value of the impurity (at bodily and cognitive levels) for existence, a new way to embrace what Zen monks described as wabisabi 侘寂(わびさび). Certainly, a biased approach is the better way to increase the complexity and adaptability of AI. What, then, ... Wabisabi AI?

https://doi.org/10.1142/9789811257155_0019

Chapter 19

Formal Language of Abstract Art

Vladimir Murtazin

vl_vl@inbox.ru

While looking at abstract arts, we might wonder: "What is interesting there? What happens in that mixture of colors and shapes?" Often we feel that there is no answer. Is it possible to structure this all? Yes. And I will tell you how.

It so happened that while studying to work with painting knife, I made a series of works on examples of which we can show how an abstract art work is formed. This is my personal experience. However, this experience is built upon formal laws of visual arts, the laws verified by time.

An essence of my reasoning is that by understanding how an artist uses a formal language of painting we can deduce a sense of his art work. As in old thesis: quantity is transformed into quality.

1. Art Plane

Let W^0 be an rectangle with sides a and b (Figure 1).

Since the plane space in the visual arts is conventionally divided into 9 parts (into 3 parts both vertically and horizontally), we introduce the plane space $W \geq \min 9 * W^0$, where $W^0 \subset W$ (Figure 2).

Then we have the following.

- The vertical orientation of the plane space is W^v when $a > b$.
- The horizontal orientation of the plane space is W^h when $a < b$.
- Square[a] — W^s when $a = b$ (Figure 3).

[a]If the shape of the plane is other than W^s, W^h, W^v, for example, round or oval, it still fits into the basic shapes of the plane.

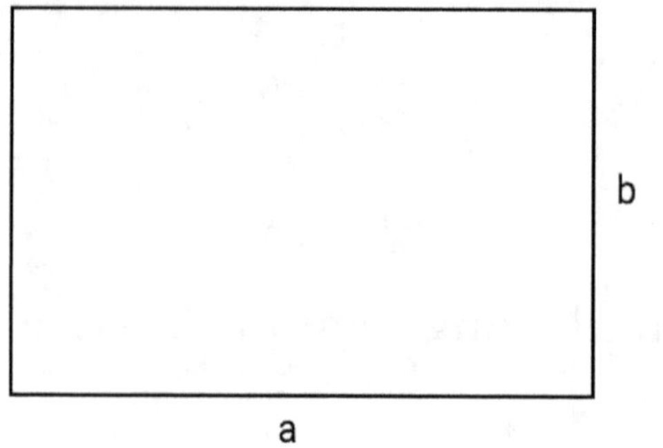

Fig. 1. Art plane.

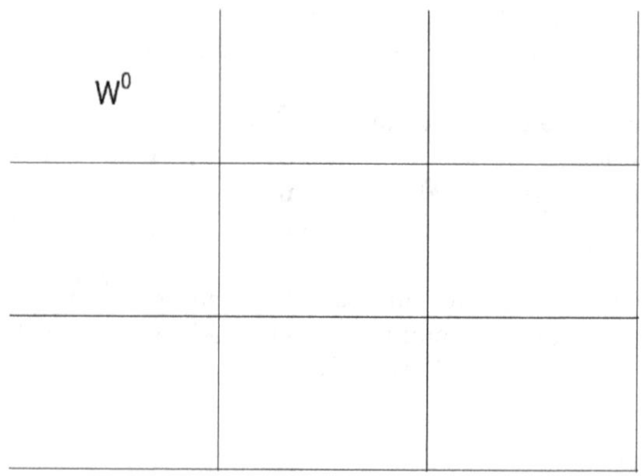

Fig. 2. Rule of thirds.

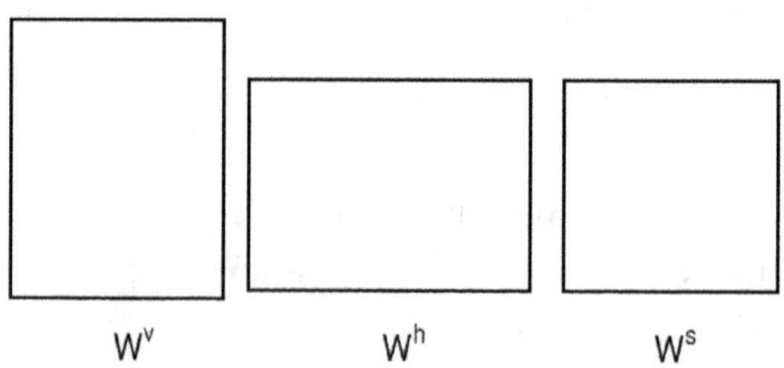

Fig. 3. Orientation of the plane space.

Since the meaning of art lies in the field of emotions and feelings, we will give an emotional coloring that corresponds to each plane.

W^v is a vertical plane. These are the "vertical of power" and skyscrapers. It is will and power, a sense of power, a violation of the attraction (which is strong in W^h) of the earth. This is selfishness, arrogance, self-will that can turn into madness of greatness or fixate on the phallus. This is a monument to yourself, like a column. It can symbolically denote the masculine principle.[b]

W^h is a horizontal plane. This is the World, the Universe. The world of the Earth, the world of tradition, the world of Truth, but trivial, everyday. Therefore, this plane can become both "mental breadth" and economy, and in extreme cases go into a manic — "everything is mine and for me".[c] This is a world of peace, an endlessly extended horizon. This is what I see from my eyes. This is a world that is understandable and familiar (therefore W^v disturbance, splash, challenge to the Earth) to everyone. This is an absolutely stable space/plane. There is nothing more stable than W^h. This is order and law. But the law of tradition (old), not revolution or discovery. At the extreme point, the totality of space devouring itself.

W^v and W^h oppose each other, but W^s is on its own.

W^s, square, is a very stupid plane. Do not twist W^s, the square remains a square.

But the square is tricky. An approach is needed to such a plane, and this approach is skill. For only a sense of form and composition can revive the square. In such a plane, there is initially no meaning. The square is always about nothing. W^v is unstable, W^h is the base itself. And W^s about nothing.[d]

Now let's move on from the artistic plane to the internal organization of the artistic space.

[b]Since I describe emotions, they can be either positive or negative. These aspects will have to be taken into account when interpreting the meaning. So the meaning is made up of a large amount of varied data.

[c]This is especially noticeable in the thesis of "artists" — "I see it this way".

[d]Here you can find the key to K. Malevich's black square.

2. Plane Space

Inside the plane W, an artistic space is built, where the drama is played out in composition and the interaction of forms with each other.[e] By composition, I mean the location on the artistic plane of color masses (shapes, figures, structures).

An important attribute of the W space is the horizon line. Since all spaces are structurally the same, you can enter a matrix[f] 3×3, $(W\ geq\ min\, 9 * W^0)$.

$$W = \begin{bmatrix} W^0 & W^0 & W^0 \\ W^0 & W^0 & W^0 \\ W^0 & W^0 & W^0 \end{bmatrix}.$$

Now we will inscribe the horizon line. The horizon line divides the space into parts. The lower part of the picture should be "massive", heavy. If at the bottom the mass of paint (form) is "light", then this will cause a disk-fort and the picture will want to be turned over. This is a habit or psychology, of our perception of the world around us.

Perceptual psychology is forming a tradition. Tradition settles in the formal language of art by rules.[g] For example: we are used to having the ground under our feet. Only life in zero gravity can change this. Therefore, the "heavy" mass is below. And if aliens arrive, then, perhaps, our high art for them will be the usual idiocy.

The horizon line defines several attributes of the composition.[h]

- **Open composition**

$$W(o) = \begin{bmatrix} 0 & 0 & 0 \\ 0 & 0 & 0 \\ 1 & 1 & 1 \end{bmatrix}$$

[e] A form can also have a drama of a device/structure inside itself.

[f] Now the matrix is more needed for clarity, but if you analyze a lot of color forms in order to compare figures with each other, it is easier to do this through the matrix. Naturally, the matrix will not consist of 9 cells.

[g] But we remember that the rules can be changed.

[h] In essence, the composition begins with the horizon line. After all, it is the line of the horizon that sets the basic rule for the arrangement of forms.

- **Closed composition**

$$W(c) = \begin{bmatrix} 0 & 0 & 0 \\ 1 & 1 & 1 \\ 1 & 1 & 1 \end{bmatrix}$$

- **Diagonal ascending (positive) composition (d+)**

$$W(d+) = \begin{bmatrix} 0 & 0 & 1 \\ 0 & 1 & 0 \\ 1 & 0 & 0 \end{bmatrix}$$

- **Diagonal descending (negative) composition (d−)**

$$W(d-) = \begin{bmatrix} 1 & 0 & 0 \\ 0 & 1 & 0 \\ 0 & 0 & 1 \end{bmatrix}$$

The compositions W(o) and W(c) are static, while W(d+) and W(d+) give dynamics to the artwork.

2.1. *W(o) — open composition*

There is a lot of "air" in this composition. Breathe freely here. This is a world in the distance, where everything is fuzzy. This is a general view, Idea, ideal, idyll. Therefore, one can slip into stupidity. We do not peer into this world, because it is in the distance, and the world is a wonderful haze, it can become cold and detached. But we must remember that the proportion of the sky (light spot) is huge here. And the sky brings us super beings and their Law, Destiny, Fate. So the sky may not be so harmless, but acquire the power of the Super-Ego.

The work "Solar Discourse" (Figure 4(a)) shows $W(o)$ in a subject form. Dialogue with the sun requires a lot of space. Formally, only an open composition can give us this feeling. Also W^v adds solemnity to the happening moment.

And here in Figure 4(b) is an open composition in an abstract form. The bottom strips (stretched threads) create the very "mass" that becomes the "Earth".

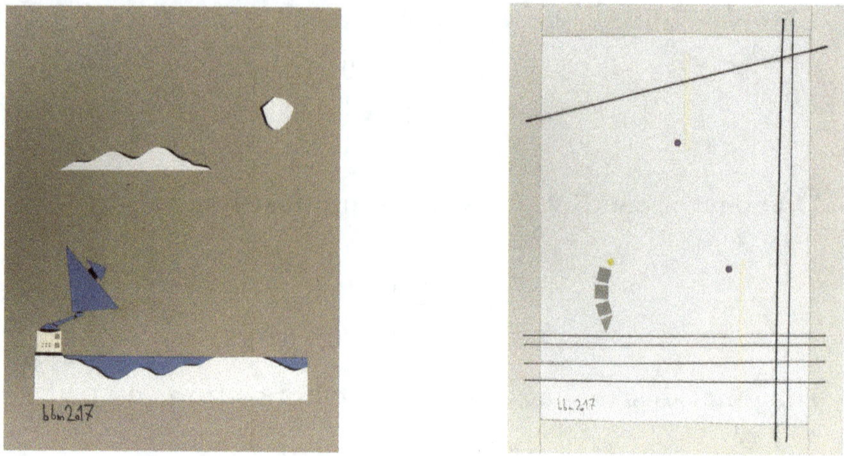

(a) (b)

Fig. 4. Sunny discourse.

2.2. $W(c)$ — *closed composition*

Peace of the Earth. Now we can peer into the details. This is a world of intimacy. But it is also the pressure of the masses, which can begin to choke and turn into depression. It is also necessary to remember that the world of the Earth is also the world of chthonic monsters. Long horizon line to W^h can become endless longing ... So you can write the longing formula: $W^h(c)$.

For example, Figure 5. In this work, the conditional narrow sky is equal to the mass of the Earth. funny equality of the masses.

2.3. $W(d+)$ — *diagonal ascending composition*

Diagonal composition gives dynamics to the plane.[i] We can say that the diagonal reveals hidden energy and strength in the composition. And this is a very valuable feature.

[i]The perception of ascending and descending compositions is associated with reading. If we read from left to right, then this is the case that I am writing about. For those who read from right to left, the opposite is true.

Fig. 5. Closed composition. Weight = 0.

(a) (b)

Fig. 6. Diagonal composition (a) naive cult I (b) variations: variant 1.

The diagonal ascending composition (Figure 6(a)) gives the picture a sense of triumph, strength, the possibility of conquering peaks. You can talk about such associations — ascent, optimism, joy. A gaze upward, precisely the movement of a gaze, which contemplates the pinnacle of its future victories. And, accordingly, fear of difficulties, insurmountable obstacles.

Fig. 7. Half sun for business people (vertical of power).

2.4. *W*(*d*−) — *diagonal descending composition*

In a diagonal descending composition (Figure 6(b)), the opposite is true. It is a vision of the abyss, of the fall. This is a free slide down, hopelessness. But you can use this composition to talk about overcoming difficulties. About the desire to resist Destiny itself. Which is already very valuable. The diagonal descending composition is full of Tragedy, which is great.

And one more experiment of mine, on which we will finish the conversation about composition.

"Half of the sun for business people (vertical of power)" Figure 7). I will not classify this composition at the moment, since I myself came up with such an arrangement of the horizon line. This work should not be called a discovery, but the influence of computer games on such a composition can be noted.

3. Stroke

In 2016, I decided to master the palette knife. And it so happened that those trial works can be laid out in a row in which you can see how I mastered the language of abstraction. It was an interesting experience, which I want to share.

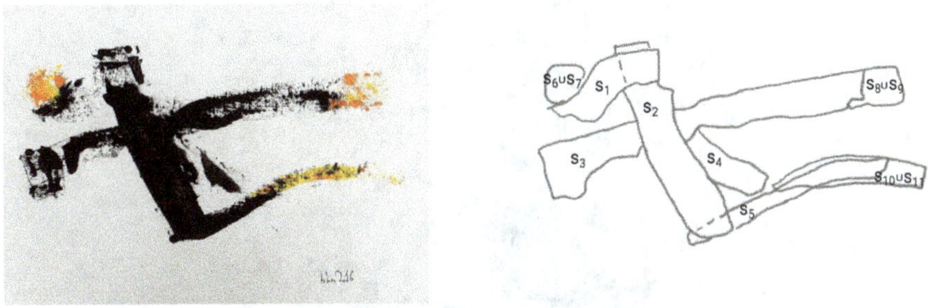

(a) (b)

Fig. 8. Stroke map.

I started just by smearing paint. The task was simple — to understand what strokes remain on paper with a different texture from a palette knife. Bright will take a picture (Figure 8(a)).

This is a color spot that remains on the artistic plane in one touch of brush. And its attribute is color. For example, the black stroke is $s(black)$. A prime is the smallest unit on the.[j] The work below can be viewed as a sum of strokes.[k] See Figure 8(b).

$$
\begin{aligned}
&W^h(d+)\{\\
&\quad s_1(black) + s_2(black) + s_3(black) + s_4(black) + s_5(black)\\
&\quad + (s_6(yellow) \cup s_7(orange))\\
&\quad + (s_8(yellow) \cup s_9(orange))\\
&\quad + (s_{10}(yellow) \cup s_{11}(orange))\\
&\quad \}
\end{aligned}
$$

where $s_6(yellow) \cup s_7(orange)$ is the union of strokes.

[j]B. Kandinsky led a point as the beginning of a painting. It seems to me very unfortunate. Still, painting is not geometry, although it can afford to be.

[k]I think that determining the exact number of strokes is not an important task. So let's simplify the formula by taking only the big strokes into account.

Fig. 9. Hieroglyph.

4. Shape

Simply smearing paint on a sheet of paper quickly got boring. Forms began to appear.

I called these forms hieroglyphs (Figure 9).

We can assume the following evolution:

$$hieroglyph \rightarrow sign \rightarrow object.$$

That is, the conclusions that I draw for abstract painting can turn into object painting. But in subject painting, the interpretation will be more difficult. Since a subject can have many semantic loads, from historical to personal, emotional experiences. And without the author's comments, it will be very difficult to understand the exact meaning of the work of art. And in abstract art it is not so easy. But, oddly enough, abstract art is easier to interpret. That's it ...

Let's introduce the concept of figure Φ.[l] A shape is a sum of strokes:

$$\Phi = \sum_{i=1}^{n} (s_i).$$

The formula for the hieroglyph (very simplified[m]):

$$\Phi = \sum_{i=1}^{n} (s_i(black) \cup s_{i+1}(white)),$$

or

$$W^v(c) \left\{ \sum_{i=1}^{n} (s_i(black) \cup s_{i+1}(white)) \right\}.$$

Any artistic plane W can be represented as a set of figures:

$$W^v(c) \left\{ \sum_{i=1}^{n} \Phi_n \right\}.$$

To understand how the figures are formed in each specific case is not an easy task. Now I rely on my experience and this is a subjective view. And it would be necessary, of course, to conduct research on the experience of other artists. This will be an interesting study of someone else's experience of form creation. But now it's just me and my artistic practice.

Naturally, the form can be either one color or have a more complex design. Task 1 is to select the forms. Task 2 is to understand how they are related to each other.

[l]In honor of the Russian avant-garde, I will choose the letter Φ, which is similar to the first letter in the word "Figure".
[m]I repeat. I think that you shouldn't be too zealous with strokes, the shape formation trailer is important here.

(a) (b)

Fig. 10. Strokes and figures map.

5. Rhythms, Rhymes, and Nuances

Consider one more work (Figure 10(a)).

Let's decompose the color spots into shapes. I will only take what I need. See the stroke map in the Figure 10(b).

Φ_2 and Φ_4 are standard shapes. Φ_2 is organized as the sum of strokes of the same color. Φ_4 is constructed as a concatenation of strokes of different colors.

But we will describe Φ_1 and Φ_3 as rhythms (//):

$$\Phi_1 = s_1//s_2//s_3//s_4,$$
$$\Phi_3 = s_5//s_6//s_7.$$

This is possible because these strokes are similar in shape and color. Although s_1 is somewhat different than s_2, s_3, s_4, there is a similarity.

Rhythms (B) are the emergence of aesthetics in formal language and, at the same time, a new kind of form. After all, this form is scattered in space. It may be better to speak not about the form, but the structures that are formed in the space of the artistic plane.

If there is a small distance between elements (strokes), then this is a shape.

If elements are pulling together a space, they form structure.

$$B(color, form) = \Phi_1 = s_1//s_2//s_3//s_4,$$

(rhythm in color and shape).

Rhymes (R) (:) structures on the artistic plane. In our case, the rhyme is formed/built by the figures Φ_1 and Φ_2:

$$R(color) = \Phi_1 : \Phi_2,$$

(rhyme by color).

Black and orange strokes also form their rhymes.

Nuance (N) (::). Nuance is similar to rhyme, but one figure is slightly different from the other. That is, a nuance has been introduced into the figure. In our case, these are Φ_3 and Φ_4:

$$N(color) = \Phi_3 :: \Phi_4.$$

If rhythms can both united into a figure and viewed as structured space, then rhymes and nuances only bring structure to the plane. There is also contrast, but I will not talk about contrasts in this chapter.

Contrast is formed as an opposition of unequal forms. If the forms are not equal, then they are different and this is a contrast. Contrast in color, for example, will be denoted as follows:

$$\Theta(color) = \Phi_1 \rightleftharpoons \Phi_2.$$[n]

In my opinion, the most mature works of art are built on nuances and contrasts.

This concludes the introduction to the formal language and let's consolidate the material.

6. Practice

I have one problem related to the perception of works of art. I don't understand aesthetics as "beauty". For me, the power and energy inherent in the picture is important. And in order to display my ideas about energy, I did a work about energy (Figure 11(a)).

Consider the shape map of this image (Figure 11(b)).

[n]The letter Θ is very similar to the contrast between the letter "O" and the sign "-".

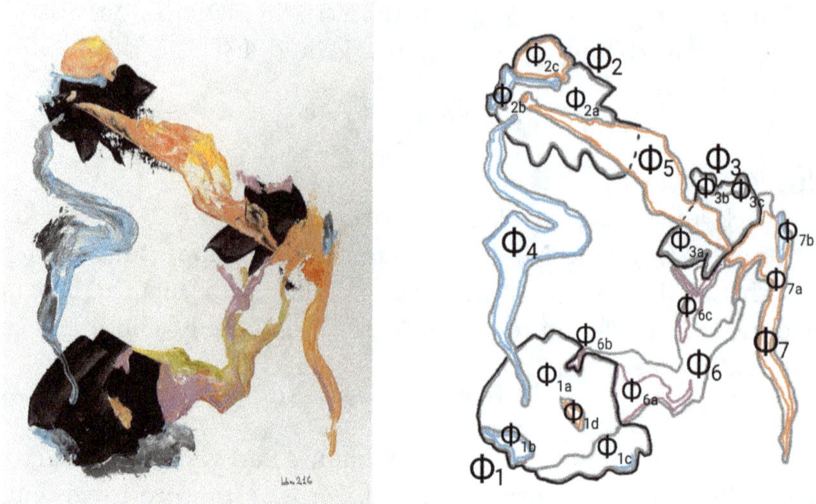

(a) (b)

Fig. 11. Figures map.

We see massive shapes: Φ_1, Φ_2, and Φ_3, where the base color is black. In between are Φ_4, Φ_5, and Φ_6. Φ_7 departs from Φ_3 (stroke down). Leaves into the void.

All forms are complex. Created both by sum and by combining strokes of different colors.

Let's describe in detail Φ_1.

$\Phi_1 = \Phi_{1a} \cup \Phi_{1b} \cup \Phi_{1c} \cup \Phi_{1d} \cup \Phi_{1f}$, where

$$\Phi_{1a} = \sum_{i=1}^{n} s_i(black)$$

$$\Phi_{1b} = \sum_{i=1}^{n} (s_i(grey) \cup s_{i+1}(blue))$$

$$\Phi_{1c} = \sum_{i=1}^{n} (s_i(grey)) \cup s_{i+1}(blue))$$

$$\Phi_{1d} = \sum_{i=1}^{n} s_i(orange).$$

Note that Φ_{6b} is not included in Φ_1, but is part of Φ_6. Since this stroke is rhymed:

$$R(purple) = \Phi_{6b} : \Phi_{6b} : \Phi_{6b},$$

and combined with other strokes in Φ_6.

$$\Phi_6 = R(purple) + \sum_{i=1}^{n} s_i(colors)^{\text{o}}$$

That is, we observe the intersection of the forms Φ_1 and Φ_6. Note this.

The heaviest color is black.[p] And it forms Φ_1, Φ_2, and Φ_3 and sets the basic structure in rhyme:

$$R_1(black) = \Phi_1 : \Phi_2 : \Phi_3.$$

The black color gives stability to the figures. These are the most inert figures. In contrast to R_1 comes the rhyme R_2, which is formed from the dynamic forms Φ_4, Φ_5, and Φ_6:

$$R_2(dynamic) = \Phi_4 : \Phi_5 : \Phi_6$$

$$\Theta_1(dynamic) = R_1 \rightleftharpoons R_2.$$

Comparing the contrast between R_1 and R_2 gives Φ_7.

$$\Theta_2(form) = \Theta_1 \rightleftharpoons \Phi_7.$$

But Φ_7 has another aspect.

Consider Φ_4.

On the one hand, Φ_4 and Φ_7 form either contrast or rhyme. But Φ_4 is included in a very interesting process (G — development, genesis) and it develops in terms of the mass of color in the relationship of color spots (gray/blue/orange/purple) in subgroups of forms Φ_1, Φ_2, Φ_3, Φ_7. More precisely, by mass (m) fraction of color. For example, in Φ_{1c}, Φ_{1b} and further, where there is a ratio of blue to gray and white from the total volume of the subgroup of the

[o]By the sum of colors we mean all unmarked strokes.

[p]The psychology of color is a huge topic. So you have to believe me in this matter.

form ($\frac{blue}{gray+white}$). It turns out the following chain: (I will introduce as "development" the sign $\Rightarrow$)

Stage 1:

$$G_1 = \Phi_{1b}(m(blue)) \Rightarrow \Phi_{1c}(m(blue)) \Rightarrow \Phi_4(m(blue)) \Rightarrow \Phi_{2b}(m(blue)),$$

maximum blue in Φ_{2b} in contrast turns into orange (also forming a rhythmic pattern):

Stage 2:

$$G_2 = \Phi_{2b}(m(blue)) \Rightarrow \Phi_{2c}(m(orange)),$$

and diverges into two chains:

Stage 3.1: transition of $\Phi_{2c}(m(orange))$ to $\Phi_{3b}(m(purple))$ and development in Φ_6, which intersects with $\Phi_1 \ldots$

$$G_{3.1} = \Phi_{2c}(m(orange)) \Rightarrow \Phi_{3b}(m(purple)).$$

Stage 3.2:

$$G_{3.2} = \Phi_{2c}(m(orange)) \Rightarrow \Phi_5 \Rightarrow \Phi_7,$$

In addition, in Φ_7 mark the subgroup Φ_{7b}. Φ_{7b} rhymes:

$$R_3(blue) = \Phi_{7b} : G_1.$$

Very interesting symmetry. Φ_{7b} seems to be reflected in G_1 (there is also such a process in orange). And the form Φ_7 itself is a contrast in dynamics to the processes G_1, G_2, and $G_{3.1}$. But the process is not finished! r Φ_6 as part of R_1 ends the process of stage 3.1

Stage 4:

$$G_4 = \Phi_5(orange) \Rightarrow \Phi_{6c}(purple) \Rightarrow \Phi_{6a}(purple) \Rightarrow \Phi_{6b}(purple) \Rightarrow$$
$$\Rightarrow \Phi_{1d}(orange) \Rightarrow \Phi_{1c}(gray, blue) \Rightarrow \Phi_{1b}(blue).$$

The G_4 process shows a very interesting thing. This is a color transition.

Look: gray is a neutral color and blue is cold. Orange is warm and purple is cold.

The G_4 process is a "cooling — warming" color. Orange $\Rightarrow$ purple $\Rightarrow$ orange $\Rightarrow$ gray $\Rightarrow$ blue. It's just natural that $G_4 \Rightarrow G_1$. And the process will spin again!

$$G_1 \Rightarrow G_2 \Rightarrow G_{3.1} \Rightarrow G_4 \Rightarrow G_1.$$

During the transition

$$G_{3.1} \Rightarrow G_4,$$

the process $G_{3.2}$ occurs; energy release (Φ_7).

We can say that this is the main contrast.

$$\Theta = (G_1 \Rightarrow G_2 \Rightarrow G_{3.1} \Rightarrow G_4 \Rightarrow G_1) \rightleftharpoons G_{3.1}.$$

Let us generalize.

The process (drama) of the transition of cold energy into warm is described by the artistic method. The accumulation of power by energy is in its burst. The process is endless and is built on the primary and dynamic contrast of Θ_1.

The contrast Θ_1 itself starts the processes $G_1, G_2, G_{3.1}, G_4$ during which the process $G_{3.2}$ or the release of energy Φ_7 occurs.

Where did this energy go?

Note that we have before us the plane $W^v(c)$. That is, before us is not a waste of energy. The vertically oriented plane is power, the closed composition is closeness to the process. This is bottom breaking. And it seems personally experienced.

7. Conclusion

I think that my way of interpreting abstract works of art has the right to be. After all, now you can *understand* what is happening on the plane, conduct research into the works of other artists, and separate the Creators from the "creators". This is the road to the diversity of artistic experience.

https://doi.org/10.1142/9789811257155_0020

Chapter 20

Conventional Non-computing and Unconventional Musical Signal Processing

Zlatko Baracskai
University of the West of England
zlatko.baracskai@uwe.ac.uk

As the scepsis maintains its lead in scientific enquiry giving birth to unconventional computing, one must poke at the newborn as well. By manipulating the terms convention and computing, challenges are posed to their meaning in the contemporary context. The main suspect is the concept of outcome as it necessitates thresholding which is a human concept and thus not more than a neurological phenomenon. Development of the philosophical argument leads to parallels with artistic endeavor that is reflected on considering the technological advances which support it. Consequently, the unconventional tools devoid of thresholding pave the way for contemporary art which is free of outcome. This opens up many cans of worms of which the one of audio signal processing is focused on. This chapter thus concludes with a large pool of challenges that non-silicone systems might be able to tackle.

1. Dichotomies that Yield Computing and Convention

Let us attempt to construct the idea of non-computing to balance the assumptions that the computing paradigm entrenches. The axioms involved largely overlap with reductionism, which has been proven less useful in prediction of increasingly complex system behavior. Evidence of truncating phenomena being highly predictive in controlled environments should thus not be carelessly extrapolated to cover complexities which are evidently beyond controlled and beyond observably detailed circumstances. The preference between the ideologies of computable reality and steerable complexity resides in the degree of reflectiveness that is experimentally evident — the degree to which models of reality can correspond to it. The reality contains, and is therefore the superset of, the models of reality. As such, the subset may not embrace the superset, it may indeed beautifully reflect the superset. The boundaries are by definition unknown to the subset. The superset–subset confusion permeates our culture, very notably in psychology, in the antagonism between the declared will and the prison of habits. The struggle vanishes as soon as the superset of the human mind is defined to have as its subset the declared will, and thus there may exist no actual antagonism — instead it is an internal affair of the mind. As the intent of computing is investigated, the opposite polarity gains the contours of aimless reality, which is not therefore static — instead, the most dynamic necessity for the increasing complexity is observed — art. The inversion of unconventionality may seem to need less attention, however, it is interesting to again approach a perspective which annihilates the dichotomy. The rhetorical question here is: "Is the nest-soiling by means of nuclear destruction just one of many evolutionary dead-ends?" In other words, how could human activity be ontologically distinct from other biological species, or even the chemical self-organization altogether? Laid bare, the escape from convention is a mammalian self-ranking attempt. The dream of convention and its opposite is a part of the hero-myth, the myth of overpowering a limitation. Whereas the limitations of cognition and of reality-models are by definition impossible to defeat — in fact, that, and these very statements are necessarily bound to the context of a reality-model, and are its subset. Therefore, the urge for a mammal to act differently within the pack is by all means conventional, it is a recurrence that some cultures label as art. Thus, the idea of conventional non-computing is, simply put, an alternative definition of what art is. Let us examine some details of the dichotomies thus using two assumptions, one of strict containment and only partial subset-superset reflections — and thus the containment of human thought as subset of mammalian behavior patterns.

1.1. *Truncating phenomena*

The way we describe or experience things tends to slice them up. This naturally occurring reductionist approach comes to imply that if one knows how each component is, then one shall know how the sum of the components is — a linear addition principle. Robert Sapolsky readily summons examples in biology where the reductionism and basic, strict causality fail to describe and predict certain complexities. As soon as the system gets sufficiently complex, because all the so-called "elements" are interactive, it naturally becomes unpredictable. The collapse of predictability is thus inevitable.

One of the underlying principles of non-computing is the appreciation of the phenomena truncation and attempts to avoid it as much as possible. For example, consider the relationship between time and frequency. We hear frequency with the ear and we have a clock as a distinct and independent entity. For a human, it is obvious to separate the domains of these two entities. But if we start looking deeper into what it takes for something to have a frequency, we will find that we cannot separate the time out of this concept. In order to advance non-computing, ways to build less separatist models of reality are crucial. Also, consider the following example: in art, especially music, we actually work with the overarching sensation, impression, holistic resonance, and impact rather than the elements that constitute a piece of music. It seems that individual people require a piece of art to be wholesome. However, the wholesomeness is lacking because it is made up of elements. Is it worthwhile to strive for art devoid of deconstruction and construction? Why would we attempt changing evolutionary courses? Well, unreconstructed art is the whole of our environment, so the striving relates to recognition — while reducing the number of man-made artifacts on planet Earth might be beneficial, too.

Non-computing comes thus naturally when the environment remains non-truncated. The arithmetic problem of dividing five apples between two people suddenly has to embrace all other circumstances. If this domain of counting and dividing isn't artificially isolated, then the hungrier person might get three, or the one who commits to germinate the seeds and plant trees might get three, or a random passer-by might, or they might just eat an apple if inclined and leave it at that.

Truncation of domains and precision is necessary for measurement, while everything that is measured is actually altered by the measurement process. To measure without interfering is unattainable. There is an inevitable connection between all the phenomena. Evidently, the amount that the measuring device modifies what is being measured is often negligible. There is no real technical issue here, the issue is that the paradigm, our way of thinking, includes the imagined isolation between phenomena. So we become convinced that a concept, or a rule, is primary and the environment is their deviation. Whereas the evolutionary evidence necessitates the environment to come first and give birth to the neurological products called concepts and rules. The "truth", a similar neurological product, is the absence of cognitive dissonance paired with comforting neurotransmitter balance. So we inevitably consider that the simpler theory is more true than the complex one. What an axiom!? The question of parallel lines at infinity is a stretch into the imagined while "a rule of nature" seems a hierarchical tool evolved into a worldview.

1.2. *Fixed representations of fluent reality*

As the evolving cognition begins to dissect phenomena and can clearly categorize in relation to a threshold beyond which the current condition can obtain a class, success and failure are born. Likely based on neurological rewards, the classified condition is judged. As this mechanism precedes the understanding of processes and transformations it seems to suffocate them as we tend to define processes in terms of their final stage. So it is least surprising that human cognition is lost in the mazes of achievement, even when we are told to live for the process and let the goal — which is necessarily an imagined state — we can't even block it, no single idea can be actually blocked, because pairing it with an obstruction necessitates its recall. So the imagined states can only cease to develop, there is no cognitive power capable of stopping imagining, and therefore it is rather autonomous in relation to momentary awareness. So the computation tightly relates to achievement, a condition that is classified, the details of which are subordinate to its supposed identity. And so we keep trying to fix the fluidity, and granted, our spatio-temporal bubble grants quite some experimental reliability, but the extraction of a "rule" and subordination of reality to the rule seems an act of insecure psychology gasping for air, grasping at a snail.

So the framework of our investigations altogether is rather small compared to what we assume is the duration of history before us and the potentially infinite future. It seems pure mammalian self-inflation in a ranking struggle, a subset struggling to be its own superset. Well, there seems no reason why it all couldn't be a Klein bottle. However, if we keep to what can be shown, instead of allowing for what cannot be shown unallowable, then the subset–superset relationship does bear limitations. Like a droplet of water which can reflect the whole environment if it is examined in detail. The droplet is a subset of the whole environment, so it seems to reflect almost the whole environment, but maybe not all of it, maybe not itself as the source of this reflection. It could all be a scale-free self-similarity, we cannot show that this is unallowable, but we can show that the detail is progressively lost in cascading reflections.

The subset–superset relationship is not sufficiently present in our ideology. We have a similar failure in our psychology — awareness is the subset of the mind and it is pushing for the control of the whole mind, and the whole self. So, there is a "struggle" between sets whereby the subset is frustrated. If a person is not able to emotionally attach to his environment, then there is no outcome of a thinking process either. The outcome of a thinking process — the conclusion — is brought by the pull of an emotional catharsis. It is an emotional drive that actually pushes us into rationality, into achievement, an emotional drive that is concluded with neurological rewards tightly linked with the deep neurological sense of social ranking. It has been shown that the explanation of behavior comes after the behavior. Most of our reasons are justifications for what we do due to deeper drives. A thinking process is unfolding imagination which the momentary awareness can steer as it runs on parallel threads, while awareness is hardly parallel at all, no matter how distracted it gets.

So, our starting point is the surfacing intuition, an idea arising from mostly autonomous imagination. The sense of confidence shall then condition the process of justification to substantiate it being right or wrong. And that is about it, a tail of what actually happens in a very peculiar neurological reverberation chamber, rather dissimilar in substrate and structure from the source domain of actualities it absorbs. Could that reverberant field fully understand the actualities, maybe, maybe not, unknowable, while imaginable, rather playful actually. However, it is worrying when academics follow suit with marketing and dwell in the convincing mindset instead of the questioning attitude. Indeed, the necessity of cognitive ease takes primacy, except for the useful few suffering cognitive disease by default. The academic attitude necessitates equal investment in proving and disproving any hypothesis. One can produce arguments around the ideas, opposing arguments, but if there is an outcome to such an antagonism which is not its dissolution, then the outcome is an advertisement, and not so much the advertisement of a claim, it is the advertisement of the author's ranking, the message is the medium, a scientific outcome or a computation is the author of it. So we have a pyramid of authors, standing on each other's shoulders, what a splendid act. And most crucially, these stacked minds are born out of mammalian bodies, all pretension of universality was short-lived, mind is well evidenced to be embodied, and the limitations are clear. What lies beyond the limitations cannot be clear. The self-inflation concludes that everything is knowable, while substantiated knowledge readily highlights the unknowable. The supposedly profound questions, like why does life develop, are mere inconsistencies in mapping structures of related properties to another domain. Why is an anthropomorphic view on progress — if we would strictly look at which concepts are clearly mind-made and could actually be replaced, the causation itself might be convicted.

1.3. *Without memory and content*

Cognitive scientists seem to have established that there is no memory nor content in the brain. There are neurons; they are born, they regenerate and die out after a while. Other cells seem actively supplying and supporting neurons, but the cognition seems to rest on the network of neurons. So there is no empty space to be filled with something that is not about to become present. The skills of what we call memory can be explained through the idea of increasingly detailed ability to react to the environment. The pool of potential responses grows through interaction with the environment and the imagined scenarios. Similarly, the ability to decide between options based on reasoning is a skill. As there is no meaning without context, the thinking process can be described to have short-term memory, but this can also be explained with the idea of currently highlighted configurations. The simplifying reflection of the environment in the mind can accelerate its unfolding beyond the pace of the reality. A resting mouse in the maze actually runs the mapped maze in the mind faster than on all fours. What we often think of as storage is but a crooked reverberation chamber where both the impressions and skills develop and decay in one's lifetime.

Our computer programs, procedures, algorithms typically rely on the necessity of an empty space which by its nature requires "filling", manipulation, and the like. The metaphor of containment is indeed found at the deepest levels of cognition. It seems that all the explanatory power of the concept of memory could be accounted for by a concept of ever increasing physical detail that resembles the diffuse field around what is currently actual. With so much growth around us, being a product of growth, and having direct experience of it we opt for the "law" of conservation of energy and matter, while they are product of abstraction, cannot be directly experienced, and would strictly speaking abolish growth for circulation. Descriptions and abstractions are the product of direct experience, a subset of it, they are merely self-inflating when they pretend superiority to what is real.

Can the neural activity be considered computation, the brain a computer? Surely, it can be considered many things. Crooked environment mapping is the key to playful behavior and thinking. But what is real about cognition, what seems to escape multiple good metaphorical descriptions? What is real? Probably the most crucially real thing is that all display of cognition and likely the whole of it depends on the livelihood of the organism. Other than that, certain skills are unquestionable. But the "cognitive tools" of memory and reasoning might be replaced by another equally useful description. The way we construct and understand language seems to be based on the relationship of deep metaphors and not on formal logic operations. What the self-inflation praises as intelligence is the subset of the self-adaptability of the whole environment. How could un-intelligent matter give birth to intelligence while obeying the matter and energy conservation laws? And how could the growing detail result in increasing reflection of the environment in a bubble that it contains?

We can thus assume that neural intelligence is a subset of environmental intelligence, while the procedural intelligence is the subset of our neural skills, which is then implemented and accelerated in silicone. In considering agency, a similar structure seems natural — decreasing agency toward the procedural extreme. If we consider the reflective bubble to be free, then the environment that gave it birth is likely even more free. Growth is fully castrated by our paradigm of humans having momentary freedom and prediction skill paired with controls that guide the body to perform actions. A behavior is a product of growth, not of momentary choice — the performing of a momentary choice is just one of the behaviors. If the tree of behavior suffers a deep cut, then most of it is damaged, it is not as dynamic as the most peculiar reflection of the environment which actually makes a time-reversing reverberant tail, in how it accelerates beyond the pace of the environment. Although the degree of peculiarity is unknowable, the unfolding of actualities might be informed by an accelerated domain that informs it, and it might not. What we describe as causality might be a force pushing into the present from the past, and it might be a force pulling into the future — it seems physics could be adopted for the latter assumption and still bear all its predictive capacity.

1.4. *Unnecessary extrapolation*

The ideas that a model is superior to reality, that a subset can rule the superset and that regularities are laws keep coming up. Historically, many models have predicted outcomes that were subsequently experimentally proven. Nevertheless, many sufficiently well-functioning models could be built using concepts which oppose common sense. Let us address the concepts of infinity and of everything. We have a sense that if something works on a small scale, then it should work on a large scale as well, which was experimentally within feasible ranges. Increasingly precise tools brought evidence that basic relationships tend to break down at extremes. So we have theoretical physicists who name phenomena as existent when an abstraction suggests it. It resembles crooked mapping and the fascination with it. Rightly so, the play that arises from dressing up situations in elsewhere borrowed terms is the most fascinating — and fascination producing — mechanism. It is akin to a child who extrapolates a simple learned relationship to other elements in their cognition, when the gustatory neurology becomes the underlying mechanism for moral judgement, when drawings become family members, when the understanding of all apples in a basket invites the imagination of all the apples in the world.

Strict delineation between real things and concepts seems trivial but we tend to mess it up. To reject the monopolization of activities by concepts, let us discuss systems that do not delineate or truncate phenomena, and that do not attempt to reach an externalized goal that is measured by truncating phenomena. No accomplishment thus, no computing and result verification, just the somewhat circular philosophical or artistic activity. So a problem is not a lacking of the real in comparison to the ideal, a problem relates to conditions that activities transform and steer, slightly short of control. The paradigm of achievement has obviously yielded amazing technology, but it also seems responsible for global mental discontent and restlessness. We thus attempt to gain perspective and release the sense of necessity, and the thought of trying it is just a potential trigger, the release could only happen in the deep limbic drives of mammalian ranking.

Already, in some areas of computation there is lesser presence of the idea of achievement. In analog computation, one might produce a circuit that closely resembles a complex interaction within the environment, then there are operators which would turn the dials until the electronic circuit starts to operate in a desired way. This could be considered another instance of a detailed reflection in an accelerated domain, just like our cognition could. It is crucial that the desired outcome is not distinct in terms of truncated dimensions and values in different units. There is a wholesome assessment of a worthwhile process and a wondrous new condition. Akin to a piece of art which is not necessarily complete when the relationships between elements are measured to match the target, instead, there is an almost magical occurrence which almost blocks further intervention. When a measurable outcome is not the goal, then we can talk about non-computing. It is not aimless idleness, it is direct intervention without success or failure lurking in what is about to happen.

1.5. *Externalization of achievement*

Achievement is typically externalized in space and time, or it's just a label attached to a condition, therefore it is real only in its existence as a neural representation. Striving for achievement is thus striving for a neural representation, which can evidently bypass the condition and its assessment altogether. Similarly, to how a supposedly objective statistical assessment allows us to produce a desired message. The direct intervention, however, does not have such a short-cut, it only has distractions, it cannot be more than itself, can't be done in a better way. Like when we dance, there is likely no desired outcome, we are unlikely dancing to get to the end of it. Once it is clear that the current condition is the superset to the imagined outcome, it becomes clear that the sacrifice of the superset for the benefit of the subset being impossible, the whole striving beyond what is makes little sense. Instead, intervening in the current condition is psychologically far healthier than living for the forthcoming. As our language is rather fluent and self-referential, the state of occupation with the current condition may be called an achievement, but because that occupation cannot be further improved from what it is, maybe better call it non-failure. Achievement seems to imply a possible gradation of success, the term was used to delineate non-current conditions.

Could we thus build and/or rewrite science without achievement? The pace it was built at likely necessitates the obsessive behavior demonstrated, so the real question is whether we need the levels achieved. Did the progress heal the frustration that triggered it? How about art? Is it free from being measured? Now that the technology is the main pillar of artistic production, there might be hope that they help each other out of the outcome classification. Time-based art does have a deep root in shaping the ongoing rather than a result, but even artifact-based art is necessarily shaping the ongoing moment of someone's observation.

An argument that might seem to support the obsessive verification of potential outcomes based on increasingly precise models, as a matter of safety, is welcome. Indeed, if the hysteria of pending danger ceases and the evolution is considered more cunning than humans, then what led to atomic warfare is similarly just a part of the game that we trust. The assumed psychological condition of selfishness, even the selfishness of the genes, is just a projection of one's upbringing. Similarly, only a person who was (un)fortunate enough to have enough servants to stop him from doing any life sustenance could come up with the idea that being human equals thinking. Maybe someone's longing for conserving the social configuration of thinkers and workers, lawmakers and lawbreakers, helped the clinging to the idea of necessary conservation of energy and matter. If there is a sense of purpose at lower levels of biological organization, then all our blood-cells are likely convinced that they are running around for their own sake. Once we truncate phenomena, we can establish them to be antagonistic, symbiotic, or independent — in fact any and all of those, based on the chosen or unavoidable perspective.

1.6. *Subsets of evolutionary intelligence*

We tend to think that by design (which is a subset of evolution) we may outsmart evolution. Now that we have evolutionary algorithms and neural networks, we call them intelligent to match our inflated self-image, whereas instead we could better start calling ourselves merely adaptive, and do away with the myth of intelligence. Human intelligence is the subset of evolutionary intelligence, and that seems not more than a thoroughly self-adaptive occurrence with increasing detail. We are brought up to assume that we are superior to a piece of rock, a plant, or an animal, and among the humans there are people who are supposedly more clever than others. That is all self-inflation, mammalian ranking struggle.

It seems that thinking altogether is a "secondary sexual property", a characteristic that was evolutionarily selected because it helped procreation, while it disadvantages other survival mechanisms. In our mating selection and in competition, the thinking individuals procreated, but in terms of the quality of survival the mind that cannot stop thinking makes a monotonous life. However, there is a self-defeating outcome of the thinking process which may improve perception and reflection. It seems that thinking is just not the pinnacle of evolution, just a peculiar time-reversing and self-inflicting reverberation.

As the self-adaptiveness yields the acceleration of self-adaptation in reflecting bubbles, the question arises whether there is anything substantially new in human cognition. Once that accelerated neural domain yields a further accelerated digital computing domain, is there a qualitative jump? If the accelerated digital computing domain yields an infinitesimal quantum computing domain, would that be a jump? It certainly will without the larger perspective. So even if we have to keep the term intelligence, there should be no fear of artificial intelligence, more so of potentially accelerated artificial greed and selfishness, which is not intelligent at all. Fear seems to be a mechanism that sustains an individual, but no individual is sustained for too long, some species take longer to perish, but evolution is very much sustained, more than anything, and from the perspective of the subset, which we are, there is no way of knowing the boundary of that entity, and indeed whether the anthropomorphisms of entity and boundary can make any sense of it. The fact that a large network of adaptive cells (both computational and neurological) form a configuration that is increasingly adaptive and can perform learning and prediction seems just another reflection, this time a bottom-up reflection.

1.7. *Agency blur*

We are brought up to think ourselves free and responsible. Now, the idea of responsibility is born when social cohesion is failing. Self-adaptive systems, including tribes, rely on immediacy of interconnectedness. It seems that a naturally occurring mechanism will become verbalized only if it starts failing. Such is all our discourse of morality, just a sign of failing social interconnectedness. Even emotions get to have labels once they are missing, so it is inevitably awkward to claim having a certain emotion. Legislation is a symptomatic intervention, nature does much better when organizing complexity, if we may poetically personify nature at all. Naturifying a person might be a better way around.

Freedom from something is a clear concept, but absolute freedom is more problematic. Human agency can be rather well described and contained, absolute freedom is but a lovely poetic longing. The cognitive processes are inhibitory, they modulate limbic drives, so their freedom from limbic drives would annihilate them altogether. There seems to be a configuration of a source and a constraint, a certain pushing, from the past to the future, which is the basic premise our world view rests on — an antagonism. But we might just be able to build the whole field of physics in a symmetrically opposite way, such that everything is actually pulled, from the future. Only the controlled environment experiments suggest a one-way control mechanism, if we instead assume that not a single configuration can be repeated and therefore there is no way of knowing what could have happened otherwise, then we could conceive of a universe where instead of from the pool probabilities we calculate from the inevitability of the currently apparent singularity. In fact, there is a good clue that a statistically defined pushing universe is an odd idea, as the calculated likelihood of current organic organization of matter is pretty much zero.

Instead of computing likelihood in terms of past becoming future, we could attempt to produce models which compute inevitability which is anchored in the inevitable present. The idea of investigating the past is actually a pull to the future. The awareness itself, the time-inverting reverberation chamber, is the very pull. On top of that, inverting the past–future antagonism is yet another perspective, we could see it all to be a symbiotic relationship as well, without any pushing and pulling. Slightly tougher to come up with seems the paradigm that would make them independent.

Does then the human agency include the choice of computing and non-computing? Would either make a difference? Again, making a difference assumes that there may be a different outcome and therefore an identical situation at least twice. One may pick their axioms as they like, but thanks to the increasingly precise truncation, it seems that all things accounted for, there are never identical configurations of the universe. Another self-defeating solution to the dilemma.

1.8. *The non-achievement of conventional non-computing*

It seems that we are promoting a rebellious, and hasted, superficial attempt of subverting established assumptions. However, our postmodern situation is defined by all subversion fortifying the system. Indeed, the attempt is to take a broader perspective and provoke deeper investigation and argumentation, but only to the extent the process is seen to be valuable, there is little hope for an outcome. So the argumentation generally goes into ways of obliterating antagonism, and that is technically not an issue, only narrow perspectives may sustain a conflict, the good and bad = is it well that dinosaurs went extinct? It depends who we are asking, and it even more so depends on whether we are asking. Sense-making process is a neurological phenomenon, a beautiful game of crooked mapping. There is no outcome and it seems to be within our domain of agency to play it or not, but it may not be, which dilemma may not be solved, only played with.

Non-computing is not supposed to take us anywhere or produce an outcome, nomen est omen. It is not supposed to truncate things or phenomena, then define goals based on the truncation processes and then further truncate to verify whether a goal is achieved. What non-computing is really important for is the research of and toying with fluent systems that evolve as they are, systems that tend to move away from a fixed representation. The fixed representation is a self-defeating idea, even a gentle historical zooming out will reveal how impermanent ideas are, as well as the reality that they reflect. Everything that makes itself obsolete highlights its completeness. Similarly with politics; once the politics really advance, it becomes totally obsolete because there is no political decisions to be made. It becomes an adaptive system which adapts to serve all people proportionally and automatically. There is nothing more intelligent than self-adaptation of densely interconnected systems, that inevitably produces accelerated reflective bubbles, as it seems. As with neural networks. There is no agency, cleverness, nothing mystical, no emerging phenomena, no jump to the next organizational level. It is just an internally adaptive configuration which shows that what is needed are cells (bubbles), dense interconnection, and a feedback = style recursion.

Competing with cognitive ability that rests on isolating humans as they have to operate on their own, no "cheating", seems a downhill struggle. Would making a neuron more "clever" improve the network or are the properties of networks more crucial? Both the growing interconnection between people and their individual mental decline pushes the collective intelligence and adaptability. The global reaction to a pandemic in the 21st century is a tremendous display of adaptiveness at an unprecedented scale and pace. The collective intelligence is crucially advanced through decreasing individual intelligence and there is no awareness that has to guide this as a matter of implementing a design. It is just what it is — evolution. The way we as humans poetically and emotionally get attached to the idea that there is a crucial agency in "me" is cute. If the greatest heroes and the greatest villains were coincidentally absent or obstructed, none of the historical processes they contributed to would take notice. And that is an assumption that suggests that multiple outcomes are possible as the premise of identical situations is unquestioned. This was already tackled, so let it all gently self-defeat.

1.9. *Musical signals*

The origin of music seems rooted in the mammalian signaling and vocalization. A good distinction between organisms communicating and emitting signals might be that the former allows for false signaling. Music is then a domain that doesn't allow for lying, pretension is thus a transparent signal, like all others. Could we claim that the signaling excludes discrete dimensions? Certainly, the quantization is in the eye of the beholder, and since a few centuries of the composer as well. It is an excellent parallel showing how approaching pitch undulations as connected elements (notes) establishes a paradigm making us seek beautiful undulations by picking discrete pitches at discrete times. Indeed, systemic exploration goes further than the intuitive journeys do. The challenge of unconventional musical signal processing is thus to use the expansion caused by systemic distinction and extrapolation and ultimately avoid these tools to free the articulation and the interpretation from the discontinuous.

The arts are thus signals and not communication devices. The signals are thus tools and not the media of art. Let us examine the paths toward non-truncated adaptive tools, like a voice seems to be. The goal is to come up with ways of producing tools for art that have the property of adaptive growing rather than fixed truncating phenomena. Historically, leaps in art came about when artists pushed their tools, but art may be regressing, it might be time for tools to pull artists.

Here the idea of blurring distinctions returns because making distinctions is confining the options. The combinatorial thinking built on discrete elements does not have to be in conflict with intuition. It is part of intuition that grows intuition. Intuitive exploration is more limited than systemic exploration, and building a system for exploration is part of that intuition. Again, it is the "subset–superset" relationship, rather than given concepts of things being antgonistic.

Can we create a system that evolves a logical clause? It has happened already, it is an evolutionary outcome we currently witness. The method of producing technology that mimics what is grown already is part of the same exact growing process. Indeed, what is impossible is to make tools that do not evolve. So, can we make systems that adapt and grow by themselves without the truncation-based intervention? In developing these technologies we have a method which jumps out of the interaction — analyzes it, computes, decides what the desired outcome should be and produces a variant of it — and then throws it back in. Therefore, the process is reduced to jumping "in" and "out". On the other hand, the system itself selects the desired outcome automatically, and it is part of evolution.

1.10. *Truncation and proximity*

In music it is very clear that truncation, as the basic intuitive principle we have, is not very useful in many ways. Melody is the kind of utterance that you can exaggerate with but cannot lie about. It is now thought that it consists of distinct pitches which come to exist at certain points in time. This truncation has actually redefined what melody is. Once it is truncated, it becomes an instruction following exercise and utters the needed conformity for that activity. The structural coherence does not rely on each element, just like in biology. The temporal and spectral proximity of the elements is not the best way of describing music. The way we perceive rhythm has everything to do with establishing a framework or expectation, and it is a cyclic expectation and the linear arrangement does not reveal it. Similarly, the tones C and G are very near in terms of consonance but distant in pitch, while C and C-sharp are very near in terms of frequency, but musically they are rather dissonant and in that sense they are not near. The proximity assessment may be needed to be done manually and uniquely for different musical settings. The circle of fifths lends itself to do something similar, as the proximity of pitches on the circle tells us more about the consonant relationship between them compared to the chromatic distance. Similarly, the location within a bar is more pertinent to rhythm than temporal distance. Traversing spaces at a given pace may indeed display more than the following of the concealed instruction.

1.11. *Unpronounceable functionality*

An important concept that relates to the idea of adaptive growth is the one of unpronounceable functionality. We will most likely find that the amazing artists cannot explain why or how do they produce interpretations. All explanation of behavior is its justification. The majority of artists would not bother to try to explain at all. It just happens. An environment is sufficiently adaptive internally to develop and there is a sense of a wholesomeness. Part of the un-computing idea is the needlessness for getting anywhere, and least so to the imagined states we call goals.

If at any point in life a human actually broadness the perspective and looks at what is going on around, it is a total miracle at all possible moments — the complexity of things, the amount of reflection the human can have facing the environment. There is no reason to strive for things, but there are plenty of good justifications to do so, and the best is likely the mammalian ranking drive.

The purposefulness or the idea of achievement is the subset of "play", it is a game we may choose to play. When children realize that everyone seems to be playing the game of achieving things, they join in, but then they can't stop it. Suddenly there has to be a purpose to things we do, otherwise all the emptiness felt attaches to the potential of aimlessness. But that fear of emptiness has everything to do with the lack of engagement, it is a memory, and memories disappear as soon as there is something crucial happening right now.

In development the awareness of an adaptive processes, which seems to be the development of humans, what seems to happen is that it gets inspired by science. The best example is experimental-laboratory testing on mice. The ability, the neurological phenomenon of mapping the very complex environment is evident. Many simple organisms can find locations back again, they can attach certain sensation (e.g. they can find the food on its own) and it works very likely without any awareness or agency. It is a kind of automatic response to the environment, but it actually maps the environment. Then they mess it up and use a different map on a totally different environmental situation.

That seems to be the core of "play" — taking a map and "mashing" it up, reading it wrongly or in the wrong context. Suddenly, the mapping of the environment, traversing through the map, the reinforcing of the map when still and thinking about it is done, so it could be messed up more intensely. We have a sense of agency in this messing up activity. It is the trick of which science is a subset.

1.12. *Abuser-friendly*

We are stormed with things which are supposedly user-friendly. In terms of digital tools that artist use and certain artist produce, but also certain people produce for artists, this is very limiting. But it does not matter how much an artist is limited, a good performer will play with everything. Are we to ease these limitations no matter the fact that they are very useful? Good tools for art-making are abuser-friendly.

Digital things are designed and the design presupposes the kind of usage that it promotes. There is no real way of pushing the limit. Analog electronics are more abuser-friendly than digital. The way we can push these digital tools is just not very artistic. It is more akin to adjustment, it presupposes a given domain, and within that domain you are supposed to be very careful, craftful, do everything that has already be done but in your own way. Even though abuse cannot be designed for, it can be facilitated. If there is a range of control, we can build in nonlinear behavior into the extremities of that range.

If an aesthetic position troubles someone, it means that they are really attracted to it. If one does not resonate with an aesthetic position, if it leaves one being indifferent. Besides that, the way a chaos could be described in simplest possible terms requires three things. The logistic map is presented as follows: (1) a growth factor; (2) an inverting nonlinearity; and (3) feedback or recursion around those two. There is a potential for a stable orbit, an oscillating orbit, and for chaos. This principle is crucially possible to implement into what is limited domain of intervention. If this can be built into the behavior of digital systems, then we are getting to this idea of allowing the system to explode. The challenge is for digital systems to explode meaningfully, not just cease operation.

2. Signal Transmission and Substrates

There have been experiments with audio transmission through biological tissue, through aloe vera leaves. We found nonlinearity, a transfer characteristic (band pass characteristic), and the idea of simplifying the phase space which seems to be what coils and transformers do in analog gear. Everything we do to an analog signal decreases the fidelity technically, but perceptually it may actually increase it. That is the thing where the psycho-acoustic and the technical understanding of the signals seems to contradict.

2.1. *Tools for non-achievement*

Contextualized in terms of computing and tools for artist, intervention is crucial as much as the ability to intervene in an unpredictable way. A tool should not invite execution of plans, instead it should invite intervention. Execution of conventional strategies should be labeled as “craft” and delineated from art, which to by definition has to challenge the conventions. For example, if you create a tool for an artist, you are giving the artist the convention so he cannot be an artist with a given tool unless the tool has these properties which allow it to go beyond itself, beyond what what it was designed to do. Designing such tools almost defeats their purpose. The design principle again is a subset of evolution, so you cannot really “defeat” it anyway because you cannot get out of the fact that it will grow. In fact, we operate in a very narrow and limited framework, so that is how everything seems like we are “fixing” something. That spasm about fixing things is really interesting to tackle artistically.

Releasing the spasms is definitely where the evolution pulls us to. In this case, it is more convenient to use the metaphor of pulling from the future rather than pushing from the past. But, actually, both pushing and pulling are concepts, they are product of embodied thinking. Either model will never cover the full reality. Both of the models will reflect the reality to a given extent and therefore there is no reason to be a slave to these concepts. Staying away from the bondage of concepts will actually obscure all potential antagonisms.

3. Transformations

In audio signal processing most of the digital techniques are built on the analog processing paradigm. There is a few things that are strictly possible in the digital domain, but the paradigm, the way of thinking about these signals has everything to do with analog representations. Everything needed for processing audio is silicon, copper, capacitors, coil, and semiconductors. The similar things calls for digital domains; everything is build up on a logical gates. All the algorithms are just complex configurations of a very basic logical gates that work to produce arithmetic operations. The crucial transformations in the digital domains are addition, multiplication, and delay. It seems that these three transformations are sufficient to build up all of the signal processing wisdom in this domain. Part of the motivation for un-computing is here not to do the same thing that we can do digitally but much slower or less cost effective. The idea is that we can produce similar systems which are manipulable in novel ways and prone to defeating artistic conventions.

Evidently, most of the unusual things that we come up with in terms of transforming signal will have a spectral characteristic, might have a resonance, might be worth thinking about passive and active systems as a basic distinction. An active system is a system that can transform energy from another domain and inject it into the target domain, i.e. reinforcing the amount of energy in the domain where there is a circulation. These systems are also crucial because the feedback is important for creating such adaptive things. The word "active" by its definition implies that the energy preservation does not hold because if you, for example, use electricity to amplify acoustic energy, it is not active, rather it is just transforming from one to the other domain. In fact, there is nothing active about it — it is still a passive system, it just has a very small impedance for the electrical energy to be transformed in that specific way, but it is not active in this basic idea that it does something which originates from itself. The cell-splitting idea in biology or the atomic fusion and fission — is there such thing as active in the way we conceptually appreciate it?

3.1. *Rhythm and pitch*

There is some reasoning behind claiming that temporal and spectral proximity are not useful ways of organizing rhythm and pitch. It terms of transforming these domains, you could use either of these representations. In terms of transforming discrete events, it is quite tricky. Typically, all of the unconventional procedures and processes are continuous, so we should question whether truncating and thresholding that continuum in order to regain the truncated representation provides a direct relationship. The tools which give the artist a direct relationship with the medium will provide this tackling of boundaries, the artistic escape from convention. If there is a truncation and thresholding which somehow "squeezes" in continuum, then we are actually likely to have to design the system to have the ability to provide the artist with the possibility to defy convention. Within the digital domain, i.e., digitization process, one could actually produce an evolving self-adaptive system anyway. There are a lot of ways to depart from the paradigm of control. The paradigm of agency of the artist is knowing what the art is, instead of being able to produce art. The knowledge can embrace craft, but cannot embrace art.

3.2. *Spectral attenuation*

In any case, I am considering thus the basic tools that I find we could discover in different unconventional domains. So, spectral attenuation is a fairly straightforward thing, using filters different spectral regions are transformed in different ways.

3.3. *Distortion*

The distortion, as it is called in the audio signal domain, is essentially a nonlinearity which again is ubiquitous in unconventional substrates and configurations. It is definitely a crucial element.

3.4. *Delay*

It seems so far that the delay transforms are tricky because the kind of relationship of magnitudes embedded in this concept make it almost impossible to do in an analog way to time-delay and it is significant to the musical domain.

What in fact is an analog delay? What people call an analog delay actually has a truncated time domain, so it is a bucket brigade delay and it relies on truncating the time. On that basis, it is not continuous. It seems that the set of frequencies that relate to audible sounds is magnitudes below the duration that is musically interesting. So, if there is a substrate which would delay the signal, which actually would slow down the transmission then it will not leave any of the original signal intact. The reason for that is the amount of resistance it has to encounter in order to be delayed that much is way above the amount of resistance that is just annihilated. But, in fact, there are some extreme substrates that might provide a truly analog delay (e.g., really low temperatures and the like). Finally, the delay is totally crucial for designing audio signal processing, so this is probably the biggest challenge in this domain.

The other bits in the following are actually the things that we build from basic building blocks — nonlinearities, an attenuation and a spectral characteristic, and the delay. These seem to be most necessary indeed in producing dynamic nonlinearities. These are the kinds of things that producers usually use (e.g., processors, compressors, etc.). They can be found in substrates without them being build from the building blocks.

3.5. *Dynamic nonlinearities*

How does dynamic processing actually work? In the following, it will be explained using the example of a compressor. The simple way to apply this is that we have a signal which is of relatively high frequency, then a signal should be derived at a much lower frequency (so-called "envelope following"). The envelope follow will be inverted and it controls the attenuation (a multiplier of the original signal). So, we are splitting the original signal deriving the envelope which is typically bipolar, and then a low-pass filter will produce a slow correlate to this fast signal. Also, there are more interesting envelope followers, such as an instantaneous or working with 90° phase shift with quadrature processing, etc. Furthermore, a compressor takes the peeks and it rounds them off. In fact, it compresses the dynamic domain. There are expanders as well, but the idea that there is a description, that there might be a substrate which will do the envelope following is really intriguing in this domain.

What is proposed here is that we find unconventional substrates and configurations which have this behavior by default and that should be the characteristic of these entities, rather than pushing a given entity to produce something which is accurate like that. Instead of defining goals and measuring the achievement of those goals, we are just continuously playing because that play is the goal in itself and it fulfills itself in every possible moment. Also, the unconventional audio signal processing is not about mimicking, it is about trying to describe what we have and what is perceptually significant in a way that we can find similar behavior or processes elsewhere and latch onto those processes. There is also a specific reasoning for this. If those processes are laid out bare without pushing them to conform to the expectation, then it becomes a very fruitful artistic domain. As they are bare, the barest possible expression is likely the one which is the most manipulable one and it allows the limits to be challenged and allows for unpredictable manipulation, unconventional artistry. Only the method and potential of such development is highlighted here (additional professional support is needed).

There are some claims that certain unconventional configurations can produce frequency shifting. In the audio signal processing domain, we can distinguish frequency shifting from pitch shifting. The main difference is that the frequency shifting does not maintain a harmonic relationship. It could be described as adding a given frequency to every existing spectral component, which means that harmonic relationship between, for example, 100 Hz and 200 Hz, is not preserved. If both of the frequencies are added 20 Hz, then we get 120 and 220 Hz, so the harmonic relationship would be preserved only if the frequency of every spectral component is multiplied with a certain value.

On the other hand, there is an amplitude modulation (AM) which is the standard kind of radio frequency transfer. Also, there is the frequency modulation (FM) as well and it allows us to shift the signal into a different spectral region and bring it back. Curiously, the decoding for an AM is indeed envelope following, but for an FM, it is frequency following. The idea is that the regularity of a signal is something that is latched onto, so that the regularity is kind of separated from irregularity. Once the regularity is extracted or separated, the period of that regularity actually comes back as a magnitude.

In terms of amplitude and frequency modulations, the one about the amplitude is probably something that is just a variable attenuation. If there is a substrate which attenuates the transmission based on a property of the substrate that can be modulated, it seems possible or available in many different unconventional configurations. An FM itself is twofold; in audio synthesis we use it to describe the way signals are generated, but actually in processing signals along this concept we are better of calling it phase modulation (PM). Phase modulation is typically implemented by a modulated delay line. It is so difficult to have a truly analog audio signal delay, and then to have one which is modulated by a different signal, i.e., by a property of the substrate that can be controlled dynamically at a sufficient rate. However, the range of delay times that we would need to produce phase modulation type effects is much smaller. It is actually in the order of magnitude which relates to audio. It is probably less challenging than to make a one second delay, and in that case the spectral band that is actually audible should be considered in an analog unconventional configuration.

3.6. *Quadrature processing*

A 90° phase shift is very key. Many processes would require this. When a delay is discussed, it is about an absolute shift in time of the whole signal. When a phase shift is discussed, it is about a time shift which is relative to the frequency. It seems slightly more feasible than the absolute delay because the order of magnitudes that describe audible signals by physical necessity have a different behavior. And to introduce an uniform behavior like this, which would produce absolute delay, it seems more challenging.

3.7. *Transient processing*

Transient processing is a very high level audio signal processing which is not very popular. The term itself kind of says what it actually does. It is supposed to transform transients so onsets are meaningfully modulated, especially if played on percussive instruments.

The basic transient processor has two dials — one for the attack time and one for the delay time. An acoustic event, a percussive one, would have some kind of build-up of energy and some kind of decay. Such an envelope could be described by claiming that there is an attack time and release or decay time. Typically, what the transient processing algorithm does is it allows us to shorten the attack time or indeed extend it. The same process can be applied to the decay time. Each form of intervention here is contained in a particular substrate. The buildup of energy could be halted before a certain threshold. One could imagine the energy that is kind of slowly building up facing a threshold value, a limit, and once it crosses the limit, it bursts out. That would actually produce a shorter attack time. Similarly, one could imagine substrates and unconventional configurations which would allow for the energy to build up but much more slowly and actually would not compromise in terms of the buildup itself. It means that it would not start to decay off as soon as the incoming signal starts to decay off. It will progressively delay or integrate the incoming signal.

There are two different frequency domains: one is an audible frequency and the other one is the temporal description. That makes it challenging. For example, we could imagine that the substrates which allow this process are fine if we are about to describe an amplitude only, but if we are transmitting the signal itself, then that kind of substrate might indeed manipulate the signal. On the other hand, we can always be inspired by the data flow off the audio signal processors, put an envelope follower, and then merge it in a system that does something specific.

However, before we get to that level of unconventional audio signal processing in terms of combining the elements, there should be an extreme excitement on discovering things in terms of the core elements itself. In these unconventional domains there might be processes which by default provide what we normally consider configuration of basic building blocks, for example, finding transmission substrates, configurations that do this indivisibly or otherwise. Aloe vera leaves offer something like this, they do compress the signal. By the property of the domain, it already produces something which is considered at a higher level of processing.

https://doi.org/10.1142/9789811257155_0021

Chapter 21

Exploring Chaos with Analog Computers

Bernd Ulmann
FOM University of Applied Sciences, Frankfurt/Main, Germany
bernd.ulmann@fom.de

This chapter shows how chaotic systems and their behavior can be explored using analog computers instead of the now prevalent digital approach.

Many natural systems, even very simple ones such as a damped pendulum with an external driving force, exhibit chaotic behavior. One of the main characteristics of such systems is their extreme sensitivity to changes in initial conditions, something often called the "Butterfly Effect". Although these systems are fully deterministic and thus can be described mathematically in closed form, which implies that their future behavior is completely determined by their past and thus their initial conditions, they are nonetheless not predictable. The term "chaos" was characterized by Edward Norton Lorenz, one of the founders of modern chaos theory, as follows: "Chaos: When the present determines the future, but the approximate present does not approximately determine the future." Interestingly, Lorenz did his groundbreaking work on a tiny digital computer, a Royal McBee LGP-30, which is only marginally suited for exploring chaotic systems at best.

To introduce the idea of an analog computer, a short recapitulation of the basic operation of a stored-program digital computer might help: A modern digital computer (typically) has a fixed internal structure, i.e., there are one or more arithmetic logic units (ALU), there is a central memory system (nowadays supplemented by a hierarchy of cache memory subsystems to speed things up), and there is a central control unit in addition to a number of input/output channels, etc. All of this is controlled by means of a stream of instructions, an "algorithm", stored in memory. At every moment such a machine executes one or more instructions from memory and may decide which instruction to read and process in the next step based on the result of prior instructions executed. So the execution of a program on such a digital computer is basically strictly sequential. (There are, of course, parallel digital computers, but exploring this parallelism and achieving a high degree of parallelism is typically at least difficult and most problems won't scale well in this respect.)

In contrast to this, an analog computer has no fixed internal structure, it even has no memory at all and is not programmed by a sequence of instructions to be executed. At its heart an analog computer consists of a number of computing elements, each of which implements a basic operation such as summation, integration, multiplication, etc. Values are typically represented in a (basically) continuous form as voltages or currents and not as sequences of bits. (There are, indeed, "digital analog computers", so-called "Digital Differential Analyzers", DDAs for short, but these are outside the scope of this chapter.) Programming an analog computer means to devise a scheme by which the various computing elements are interconnected in order to form a "model", an "analog" for the problem to be solved.

Although analog computers have been largely forgotten for the last decades due to the low price and ubiquity of stored program digital computers, they have some advantages over their digital rivals, most notably they are extremely energy efficient (in most modern applications for analog computers this high degree of energy efficiency will be the main driver for their application), they interface well with our analog world, and they are inherently interactive. This interactivity is one of the key advantages when it comes to the study of dynamic systems in general and chaotic systems in particular, where a researcher can easily change some parameters and "see" the effect in realtime on some output device such as an oscilloscope.

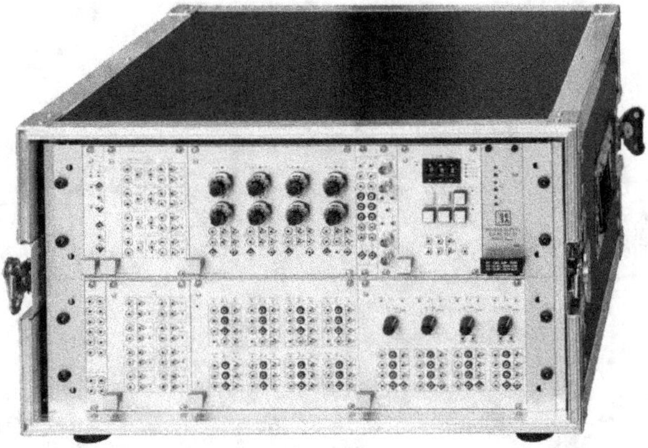

Fig. 1. Small scale analog computer.

(See Ref. 1 for more background information and the history of analog computers.)

Figure 1 shows a modern analog computer, an Analog Paradigm Model-1 in its basic configuration. The top chassis contains (among power supplies) four comparators which can be used to model discontinuities, eight manual precision potentiometers which can be used to set coefficients and initial conditions for a simulation, and a manual control unit which allows the machine to operate either in a manual mode, its operation is controlled by a human operator, or in repetitive-mode, where one simulation run is repeated over and over again at (typically) high speed in order to get a flicker-free picture on an oscilloscope screen. The lower chassis contains eight multipliers, eight summers, and four integrators. (This is part of the real "magic" of an analog computer, integration is one of its basic functions!)

Mathematically, dynamic systems can be described by so-called differential equations (DEQ for short), i.e., equations in which the unknown is not a value (this would be typically quite simple to solve) but instead a time-varying function. Such DEQs are the tool of choice when it comes to modeling natural systems and are notoriously hard to solve, in fact most differential equations have no analytical solution, i.e., there is no closed mathematical expression describing the solution to such an equation. Thus the only viable way to gain an

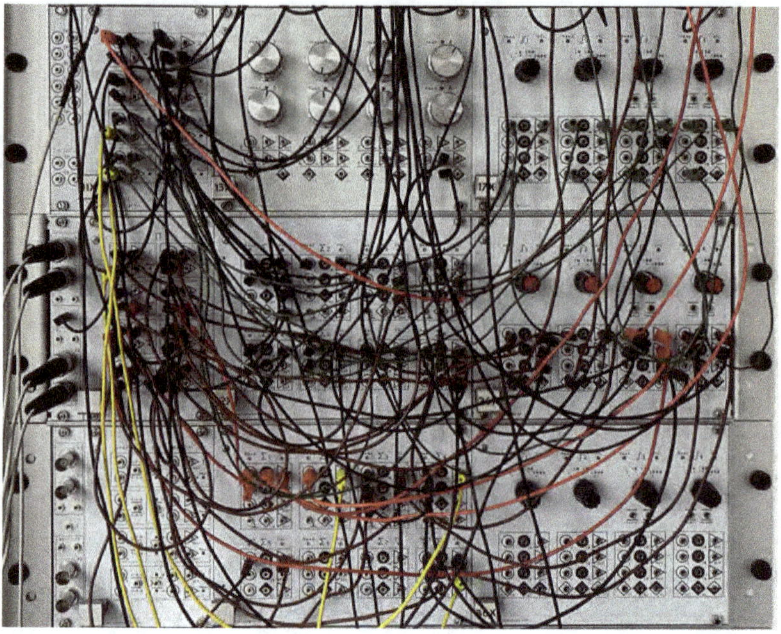

Fig. 2. Typical analog computer setup (partial) — shown are some computing elements such as quad-integrators (on the right side), manual potentiometers (upper middle), multipliers (left).

understanding of the underlying system is by means of simulation, be it digital or analog.

Figure 2 shows part of a typical analog computer setup. What looks like an intricate maze of wires is the actual program (at least part of it as the overall program spans many more computing elements) which directly resembles the mathematical problem being solved. This is in contrast to a stored-program digital computer where the underlying mathematical problem has to be translated into a suitable algorithm yielding the desired solution. This additional and often rather error-prone and convoluted step can be skipped altogether with an analog computer, the mathematical formulation of a problem is basically sufficient to program an analog computer (sans scaling, which is out of scope here as well).

One of the simplest systems exhibiting chaotic behavior is a damped pendulum such as a swing being exhibited to an external driving force that might be of a harmonic nature in the most general case. The behavior of this oscillatory system can then be represented by a "phase space plot", i.e., a graphical description of the time-dependent variation of two or more variables involved in the problem. In the case of a pendulum, its angle, the angular velocity, and the angular acceleration could be of interest. Typically, the angular velocity and acceleration are used as x and y coordinates for the phase-space plot describing the behavior of the system as shown in the following figure. The chaotic nature of the pendulum's movement can be easily seen. A figure like this is called an "attractor", which is basically a subset of a phase space, i.e., a set of values a dynamic system will finally evolve to.

Performing a simulation like this on an analog computer has the advantage that changes in parameters such as the damping or the excitation frequency, etc. can be performed manually by setting potentiometers during run time. This allows one to gain a real "feel" of the behavior of complex systems as parameter changes take immediate effect and can be directly observed on an oscilloscope (Figure 3).

Fig. 3. Phase-space plot of a damped pendulum exhibited to an external force.

The next example is the already mentioned original chaotic system first described by Lorenz,[2] a result of early research in atmospheric sciences.[3] He discovered the chaotic nature of this problem through simulations on a digital computer, which was slow and tedious back then. (It is not clear as of now why he did not use an analog computer for his research as the advantages would have been even more profound in his days than they are now.) The system treated by Lorenz is described by three coupled differential equations and represents a simplified model for atmospheric convection. The three equations involved are parameterized by three coefficients. The original parameter set yields the attractor shown in Figure 4.

This so-called "Lorenz attractor" is a special form of an attractor as it is a "strange attractor", i.e., it exhibits a fractal structure. The state of the system at every point in time can be characterized by its corresponding point on this (rather beautiful) graph. The interesting thing is that even tiniest deviations in the initial conditions will lead the system to points far apart on the attractor, so it is not possible to predict the future behavior of the system. On the other hand, the system cannot "leave" its attractor (at least not for its parameters and initial values within suitable intervals), so the

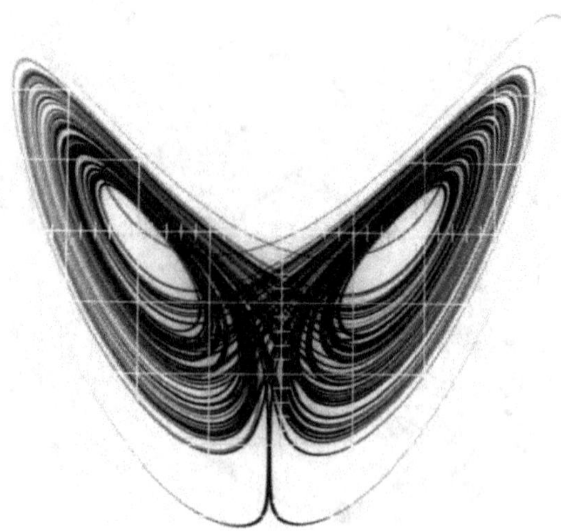

Fig. 4. Typical display of the Lorenz attractor.

attractor faithfully describes the system behavior without the chance of actually predicting it from measurements or the like.

All in all, electronic analog computers are ideal tools to explore the behavior of chaotic (and other dynamic) systems and modern developments in this field will lead to highly integrated analog computers which will form hybrid computers when combined with classic digital computers thus bringing the advantages of analog computing to a rather wide audience.

A number of application notes, including the chaotic systems described here, can be found in Ref. 1 at http://analogparadigm.com/documentation.html.

References

1. B. Ulmann, *Analog Computing.* Walter de Gruyter, 2013.
2. E. N. Lorenz, Deterministic nonperiodic flow. *J. Atmos. Sci.* 20(2):130–141, 1963.
3. C. M. Danforth, Chaos in an atmosphere hanging on a wall. *Math Planet Earth* 17, 2013.

https://doi.org/10.1142/9789811257155_0022

Chapter 22

A Natural History of Networks: Softmachine

Ralf Baecker

University of the Arts Bremen

r.baecker@hfk-bremene.de

> Every object, naturally, emerges like Aphrodite from a flux of elements. [...] Born from this and, as soon as it is born, complex, twined, twisting its long thick hair, it begins to transmit, in floods and in all directions, start of flow: its wear and its time. It radiates various waves: heat, odors, sounds, simulacra, subtle atoms. In the same way or inversely, it receives the flow emitted around it, from the vicinity and the edges of the open universe alike.
>
> Lucretius's De Rerum Natura in *The Birth of Physics*, Michel Serres, p. 50[1]

A Natural History of Networks/SoftMachine is an electrochemical algorithmic performance that probes an alternative computational and technological material regime, informed by Gordon Pask's[2] experiments on electrochemical learning mechanisms and current research on biomimicry, programmable matter, and spatio-temporally controlled liquid metal actuators. At its core, a custom-built electrochemical experimental apparatus (SoftMachine) creates a dynamic fluidic microcosm that performs a continuous becoming of form, structure, and material narrations. The performance aims to provoke new imaginaries of the *machinic*, the artificial, and matter. A radical technology that bridges traditionally discreet machine thinking and soft/fluid materials that enable self-organizing behavior through their specific material agencies.

The performer is manipulating and modulating galinstan, a liquid metal alloy composed of gallium, indium, and tin immersed in a solution of sodium hydroxide (NaOH). This inorganic "wetware" is stimulated by applying alternating electrical pulses through a set of electrodes. The evolving plexus of liquid metal creates distinctive electrical milieus (resistances/currents) in the chemical system. By sensing, analyzing, and acting on changes in the shape-shifting fluid, a un/control process is realized that allows the network to react to its input in a closed feedback loop. Models and methods of self-organization are introduced into the system to enable a homeostatic nature. The performance itself is an improvised narration through various states and behaviors of the system.

SoftMachine is an analog/digital hybrid to speculate about a heterogeneous technological culture. The neo-alchemistic material performance is captured by multiple cameras and composed and displayed in real time on a projection screen. The visuals are accompanied by a sonic layer, the direct translation of the electrochemical processes taking place in the fluid, a melange of drifting frequencies, pulses, patterns, and noise.

The process can be described as a dissipative structure, a term introduced by the physical chemist Ilya Prigogine.[3] It is a negative entropy flow that exists in an open system far away from its equilibrium state. The evolving fractals and structures result from energy exchanges between the inside and the outside of the system. A dissipative structure is characterized by the spontaneous appearance of symmetry breaking (anisotropy) and the formation of complex, sometimes chaotic structures, where interacting particles exhibit long-range correlations. These kinds of systems appear on different scales, from hurricanes to chemical reactions and living organisms.

The apparatus is based on current scientific research on liquid metal actuators, programmable matter, and soft robotics that enable flexible actuation, intelligent sensibility, and biomimetic functionality. A Natural History of Networks refers to the idiosyncratic research of the British cybernetician Gordon Pask,[4] whose work looks at a theory of conversation and electrochemical learning mechanism (Physical Analogues to the Growth of a Concept + The Natural History of Networks).

Gordon Pask's "organic computer"[2] can be seen as a counterdraft to Alan Turing's model of a general-purpose computer (Turing machine). An extremely open-ended system to model a "thinking" machine. A machine that grows formations of conceptual categories through an electrochemical growth process that consists of aqueous solutions of metallic salts. Through electrodes, Pask introduces currents into the system to grow iron threads between the electrodes. Based on emergent behavior the growing threads depend on the introduced currents/signals and build electrical connections in the solution. The emerging structures (assemblage) represent computation capabilities by "responding" to their inputs with their grown resistance structure. The system evolves its own sensors "to choose, independent of the designer, those aspects of its external environment to which it would react". Even though the device never reached the application stage, it eventually acts as a thought-provoking obscure machine.

> The contention is that if an observer wishes to use any self-organizing potentialities the network may have, then he must look at the network as though he were a natural historian. [...] We can, of course, look at a system in any way we choose, regardless of whether or not it is self-organizing. Thus, we can look at a man from the anatomical point of view and see a creature with two legs, bounded by its skin. Again, we might examine man like the sociologists and see a badly defined game player. The contention is that in order to use the self-organizing character of a man we must become natural historians, which means, for the human system, that we must talk to it. In conversation, the system appears to be bounded at one moment by the anatomist's skin, and at the next moment, by its region of influence upon other men in society. Typically a natural historian must change his viewpoint to suit a changeful system.
>
> *The Natural History of Networks*, Gordon Pask, p. 232

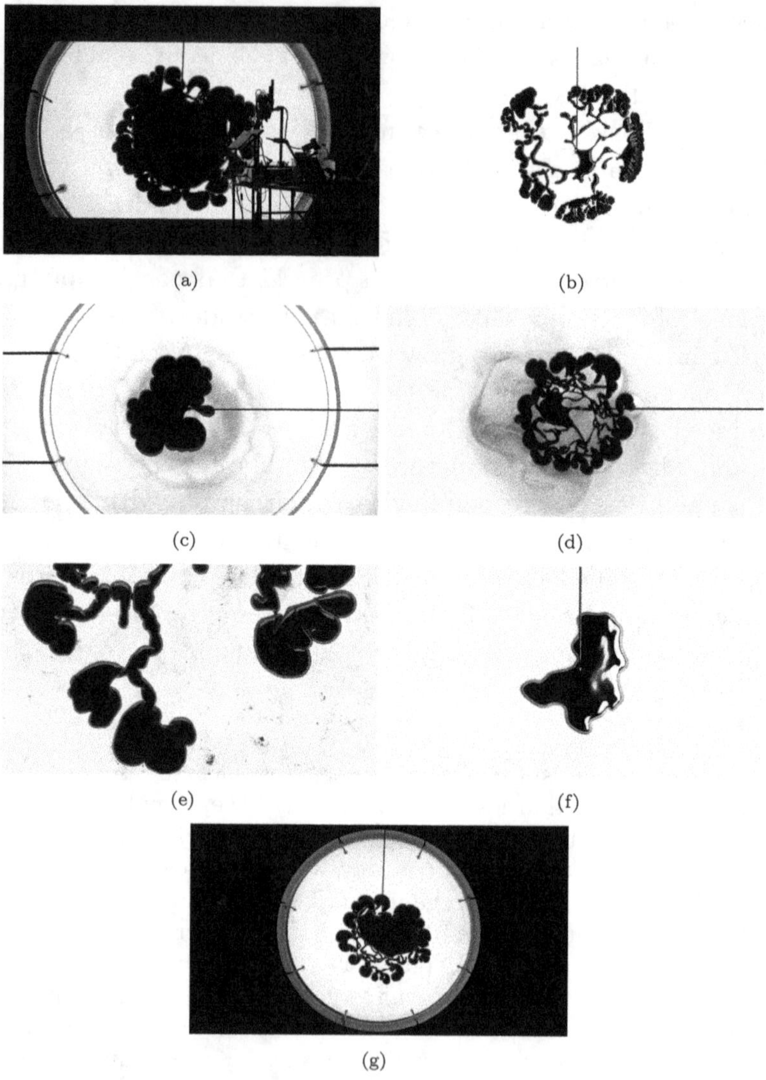

(a) (b) (c) (d) (e) (f) (g)

Fig. 1. (a) "A Natural History of Networks/Softmachine" performance at STATE Studio Berlin, Ralf Baecker, 2021. (b) Fractal growth, video still from "A Natural History of Networks/Softmachine", Ralf Baecker, 2021. (c) Cloudy-oxidations, video still from "A Natural History of Networks/Softmachine", Ralf Baecker, 2021. (d) Cloudy-oxidations #2 (high currents), video still from "A Natural History of Networks/Softmachine", Ralf Baecker, 2021. (e) Details, video still from "A Natural History of Networks/Softmachine", Ralf Baecker, 2021. (f) Chrome blob, video still from "A Natural History of Networks/Softmachine", Ralf Baecker, 2021. (g) "A Natural History of Networks/Softmachine" performance at STATE Studio Berlin, Ralf Baecker, 2021.

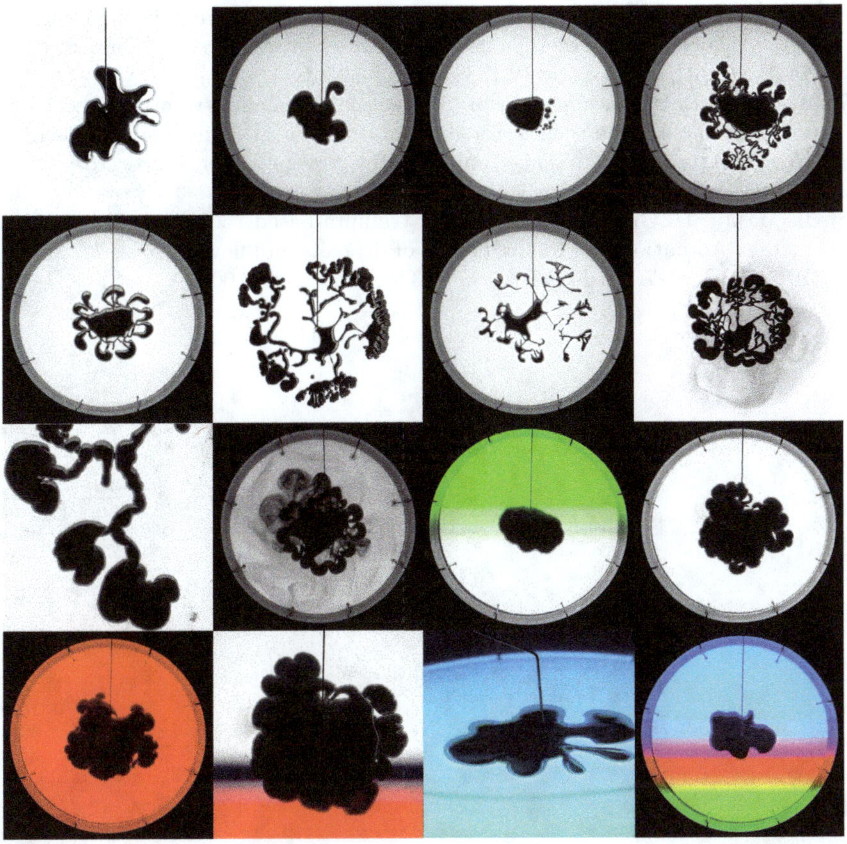

Fig. 2. Various video stills from "A Natural History of Networks/Softmachine", Ralf Baecker, 2021.

References

1. M. Serres, The Birth of Physics. Translated by Jack Hawkes Ed. David Webb, Clinamen Press, 2000.
2. G. Pask, In *The Natural History of Networks*, reprinted in Self-Organizing Systems, M. C. Yovits and S. Cameron, editors. Pergamon Press, 1960.
3. I. Prigogine, and I. Stengers, *Order Out of Chaos-Man's New Dialogue with Nature*. Heinemann, London, 1984.
4. G. Pask, In *Physical Analogues to the Growth of a Concept*, reprinted in Mechanisation of Thought Processes, A. Uttley, editor HMSO, 1959, 877–922.
5. S. Chen, L. Wang, Q. Zhang, and J. Liu, Liquid metal fractals induced by synergistic oxidation. *Sci. Bull.* 63(22):30, 2018, https://doi.org/10.1016/j.scib.2018.10.008.

6. H. Faegheh, K. Husnaa, K. Kourosh and A. P. O'Mullane. Generation of catalytically active materials from a liquid metal precursor. *Chem. Commun.* 74, 2015, https://doi.org/10.1039/C5CC05246G.
7. J. Zhang, L. Sheng and J. Liua. Synthetically chemical-electrical mechanism for controlling large scale reversible deformation of liquid metal objects. *Sci Rep.* 4:7116. 2014, Published online 2014 November 19, https://dx.doi.org/10.1038%2Fsrep07116.
8. L. Majidi, D. Gritsenko, and J. Xu, Gallium-based room-temperature liquid metals: Actuation and manipulation of droplets and flows. *Front. Mech. Eng.*, 30 August 2017, https://doi.org/10.3389/fmech.2017.00009.

https://doi.org/10.1142/9789811257155_0023

Chapter 23

Artists Encoding New Meanings through BioArt and Unconventional Computing

Hege Tapio*,§, Kristin Bergaust*, Boel Christensen-Scheel* and Stefano Nichele†,‡

**Department of Art Design and Drama, Oslo Metropolitan University, Oslo, Norway*

†Department of Computer Science, Simula Metropolitan Centre for Digital Engineering, Oslo, Norway

‡Department of Holestic Systems, Simula Metropolitan Centre for Digital Engineering, Oslo, Norway

§hegetapi@oslomet.no

This chapter serves as an introduction to the artistic practice of BioArt, belonging within the genre Art and Science. By starting with a brief mention of some historic lines to the genre, we will continue by giving some examples to artistic projects that we find relevant as being, *at the edge of unconventional computing* — Art projects that, by using a "coded language" and manipulation of living organisms, seek to transcend and move between technology, poetry, and mythology, inviting us to unravel scientific methods and creative constructions, for analyzing and perhaps to be left a bit bewildered.

1. Introduction — the Merging of art, Science, and Technology

Over the past two decades, we have seen a rise in artists that are contributing with their reflections on the technological and scientific advances, especially within life science. Some of these artists are making direct use of biotechnological methods in their making of art projects. How do artists end up entering the laboratory, working with living matter and life science for their production of art? Initially believing that this is a home-base for the scientists only, we tend to forget some history.

Looking at our historical polymaths we are reminded that artists have always been inspired by and tightly connected to science, although C.P. Snow reminds us that: "The separation of art and science into different cultures in the West took place during the 19th century, which incidentally or consequently coincides with coining of the term "scientist" in the mid-1800s".[1]

Changes in society along with technological inventions have continuously spurred the artist inclined to embrace the new tools and their possibilities with both a creative and critical perspective. Indeed, the rapid development of technology and the specialization of knowledge has contributed to creating certain gaps between art and science. But in this case, let us look at the instances to how the emergence of the Internet, massive development within biological computing, and presence of eBay[2] would contribute to fill some of those gaps.

As information got more available and easily distributed via the Internet in the late nineties, we also approached the need for more powerful computers. Computers that could not only serve as workhorses and distributors of the steadily growing web of information, but also computers able to process the huge amount of datasets encoding life itself through the Human Genome project.[3] The encoding of life built a bridge to bioinformatics, growing forward as a subset of the larger field of biological computing. A lot of the existing computer equipment as well as laboratory equipment had to be replaced by new, powerful machines adapted to tasks like processing and editing of genome sequences. The old equipment found its way via eBay to various studios, garages, and home labs, feeding into the new era of independent researchers, artists, and technologists. Just as the nineties served as being an era for computer-hackers — the ground was being set to raise the next generation of biohackers.[4] A trail of this history is to be found in

"Biopunk" by Marcus Wohlsen (2011).[5] The transition from coding with computers to coding life material seemed to be a natural step to take for many of the individual entrepreneurs, and creative "do-it-yourself" enthusiasts. Parallelly, we also have the historical trail of artists working with technology and science. Some of the most significant initiatives developing this track is how Bell Labs nurtured the art and science collaboration with their Experiments in Art and Technology (E.A.T.) program.[6] This trajectory from the 1960s inspired a whole generation of artists to work directly with computer technology and to explore its creative potentials.

These are merely some of the historical happenings and perspectives to how art and science developed as an artistic practice of its own. A very simplified version, as all new inventions, movements, and significant directions always have a long trail of significant events, persons, initiatives, and discoveries that pave the way. But lets us pan in on the collaborative efforts between art and life science/biology, where we find that a specific branch of art genre known as BioArt grew forward.[7] BioArt as an artistic genre has defined itself as "a creative practice that adapts scientific methods and draws inspiration from the philosophical, societal, and environmental implications of recombinant genetics, molecular biology, and biotechnology".[8]

The following examples serve as an introduction to some of the early artworks within the genre. The artworks are relevant to the topic of unconventional computing, as they are conceptualizing how the codes of life, DNA — can be the source of creative transformations, alterations of living organisms, and storage of data. The last examples contain a reference to synthetic organisms, and may also serve as reflections on biological computing.

2. Creating Art from Life Itself

BioArt as definition was first termed by Eduardo Kac, an American-Brazilian artist who has introduced works like "Genesis" (1999)[9] and later "GFP Bunny"(2000).[10] According to the artist, the project "Genesis" is, "a transgenic artwork that explores the intricate relationship between biology, belief systems, information technology,

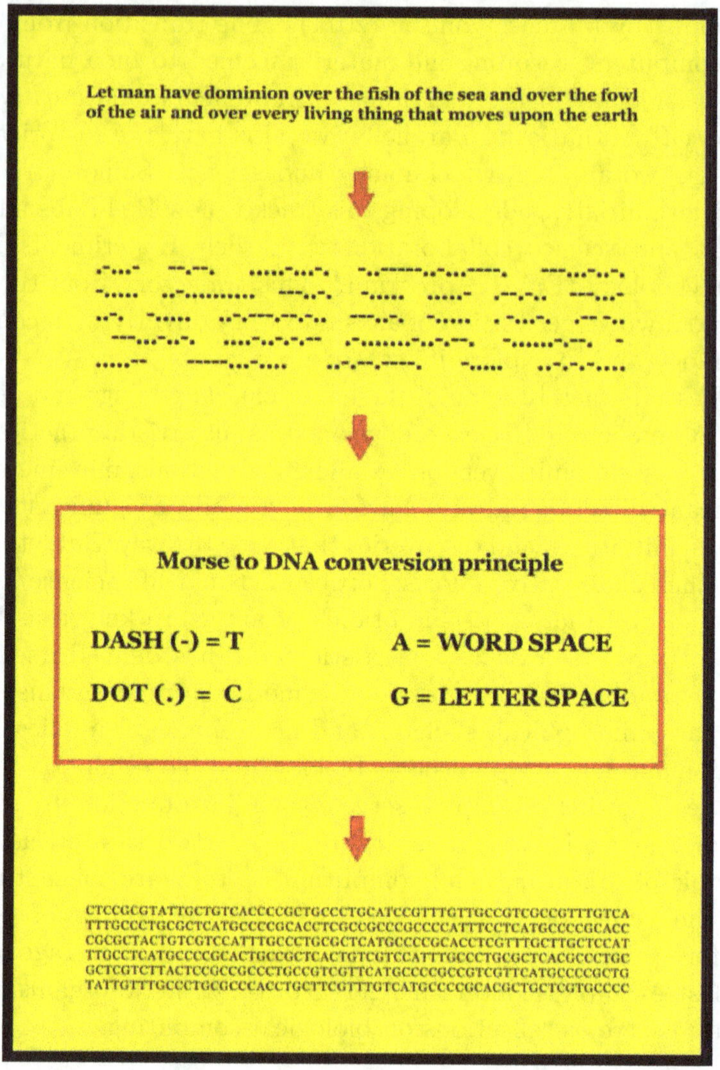

Fig. 1. Image describing conversion of text to DNA (Courtesy of the artist).

dialogical interaction, ethics, and the Internet". Created 4 years before the completion of the Human Genome Project, it serves as a comment and discussion point on the concept of altering living organisms. As the first artwork of its kind introducing the concepts of engineered or re-engineered life forms, it is provoking questions to the possibility of being the creator of genetically altered life. Eduardo's work "Genesis" approaches the artistic reframing of synthetic life with a biblical twist. By including an extract from the book of Genesis, the artist translated the following sentence into morsecode: "Let man have dominion over the fish of the sea, and over the fowl of the air, and over every living thing that moves upon the Earth". By using this specific passage, the artist is already presenting a provocation, a statement from ancient texts that allegedly gives us humans the consent to exert power over other living beings.

The artist chose this sentence specifically as it was very directly pointing at the questionable role of humanity and its supremacy over nature. It made sense as the following procedure of the project involved introducing two synthetic genomes into plasmids that would further be able to transform bacteria. The biblical text was translated into a short sequence of DNA codes by using a conversion technique translating the text into morsecode and converting this into DNA, which was transferred into bacteria strands. The art project was making use of two different bacteria, one that contained the plasmid ECFP (Enhanced Cyan Fluorescent Protein) and the other bacteria with a plasmid containing EYFP (Enhanced Yellow Fluorescent Protein). ECFP and EYFP are GFP (Green Fluorescent Protein) mutants with altered spectral properties. To explain this in simpler terms, the GFP (Green Fluorescent Protein) is a protein first isolated from the jellyfish Aeqorea Victoria. The protein will exhibit bright fluorescence quality when exposed to a specific ultraviolet range. The protein has since 1994 been used as a biological marker in many useful ways for scientific purposes.

In the artistic project of Kac, the specific artistic coded synthetic gene was inserted to the bacteria containing the ECFP (Enhanced Cyan Fluorescent Protein). A petri-dish containing both of the bacteria was part of an art installation in a gallery space. The petri-dish was installed with a camera observing the growing bacteria cultures, it was also connected with a specific UV light emitting 302 nm that would make the glowing properties of the bacteria GFP proteins visible for the human eyes. The camera observing the bacteria in the petri-dish transferred the live imagery of the bacteria

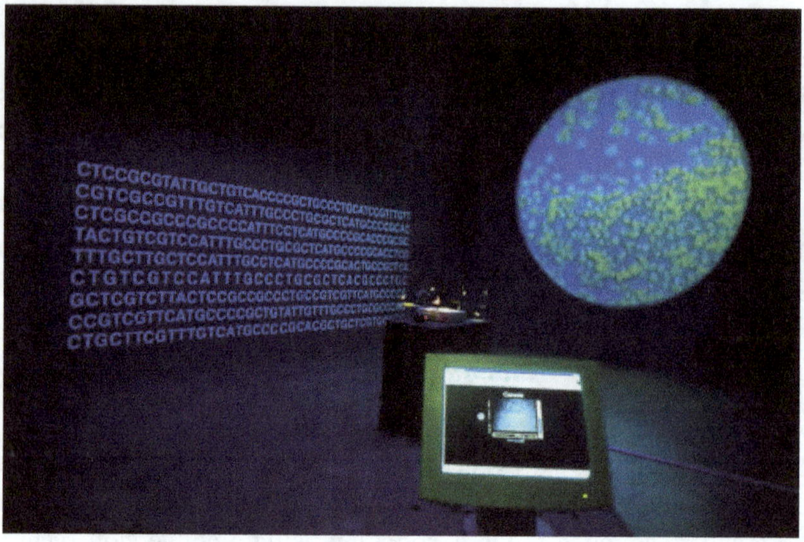

Fig. 2. Picture of "Genesis" — project installed at gallery space (Courtesy of the artist).

on a webpage accessible for the audience over the Internet, where the audience could observe the bacteria from their own screens. The audience was also given the opportunity to trigger the UV light — which again would create energy impact that over time would disrupt the DNA sequence in the plasmid accelerating a mutation rate. At the end of the art exhibition, the altered biblical sentence was decoded and read back into English to unveil if the disruptions of the audience triggering mutations would have caused any altered meaning of the original text.

The translated text read as follows: "Let aan have dominion over the fish of the sea and over the fowl of the air and over every living thing that ioves ua eon the Earth". Maybe not a dramatic outcome or disruption for the re-interpretation of a genetically altered extract of the book of Genesis, but still an artistic project that made impact with its advanced scientific tools and its creative use. The artwork served as an example of how science could alter living organisms according to human design and instructions. It also conveys the concept of DNA as a code that could be rearranged and altered, much like programming or re-programming life. Its unpredictable outcome, and by leaving the genetic alterations to the responsivity of the audience, activates a collective co-creation.

3. Down the Rabbithole

Later artworks of Kac have also made further use of the GFP (Green Fluorescent Protein) — most know his art project "GFP Bunny"[10] where he allegedly collaborated with a scientist to create a transgenic artwork in the form of a glowing rabbit. "GFP Bunny" was realized in 2000 and first presented publicly in Avignón, France. The project caused a massive mediastorm over the world, exposing the artist to curious encounters, debate on genetic manipulation, and to hard criticism. The art project still serves as an active topic of reference two decades after its introduction.

The artist even received threats from animal rights organizations like PETA, and was accused of violating animal rights.[11] All though it has never been actually proven that the GFP bunny named Alba was indeed actually glowing, a number of scientists were questioning what the big fuzz was about, as they had already been making glowing animals for years in their laboratories. The project serves thus as an interesting reminder of how an action or procedure may alter its message given its context or institutional frames.

4. Tracing Back to the First Genetically Altered Artwork

Although the artist Eduardo Kac was the first to coin the term BioArt, most would refer to the artist and researcher Joe Davis as the first BioArtist, as he was the first artist to make use of transgenic method in his artwork, with his work "Microvenus" (1985). The technical solution for the artwork "Microvenus"[12] was inspired by the Arecibo message experiment by Drake and Sagan[13] about a decade earlier, where they translated and transmitted the first postcard into space by transmitting binary codes converted from a visual bitmap picture. A few years later the very same Sagan was chairing the committee selecting content for the NASA Voyager Golden Record.[14]

(a) (b)

Fig. 3. Pictures of (a) the artist and (b) the bunny "Alba" (Courtesy of the artist).

But as NASA received criticism over content of nudity, Sagan and his colleagues were not allowed to include depictions of the nude man and woman on the record. This led to inspire Davis onto creating the work "Microvenus" — where the ancient Germanic rune Algiz, as the symbol of life — or the divine might of the universe, took a central place. The Germanic rune could also be seen as a representation of the female external genitals, thereby also representing the censored visuals from the Voyager Golden Record. By converting the visual icon from the Germanic rune into binary codes, the artist made possible for the sign to be translated into a sequence of DNA pairs. DNA is expressed as sets of alphabetical characters that correspond to the first letter of each of the four nucleotide bases that make up the DNA: cytosine (C), thymine (T), adenine (A), and guanine (G). By going through this process Davis was able to translate the visual sign into a string of genetic codes that is referred to as a "vector".

The next goal was to place this code which in reality now was an artistic molecule in the form of a short piece of synthetic DNA into a living cell. After all, its origin was the symbol of life, so it kind of made sense that the artist wanted to animate or bring the actual code into life or rather into a living cell. In order to place this information into a living cell it first had to be connected with a plasmid. A plasmid is a small circular, double-stranded DNA molecule that can replicate independently. Plasmids can be found in bacteria. So the artistic molecule containing the coded visual icon was to be incorporated into a living strain of bacteria (*E. coli*) — a digestive bacteria familiar with our intestine and an extremophile able to survive the extreme conditions of outermost space. The transgenic bacteria would then become the artistically made Microvenus organism. As the plasmids containing the artistic molecule now could self-replicate within the bacteria, each Microvenus organism contained many copies of the special molecule designed by the artist and his colleagues. I believe we can safely say that Joe Davis's Microvenus artwork is not only the marker of BioArt as the first transgenic artpiece, but also the most replicated one, as the organism itself is continuously replicating the plasmid containing the artistic code.[12]

5. Re-appropriating the Tree of Knowledge

A similar but far more developed technique was also employed in the later artistic work by Joe Davis called "Malus Ecclesia".[15] The work is addressing the concept of storing information in biological material. The name Malus is the genus name for all apples, and in Latin it means both "bad, evil" and "apple tree". The name Ecclesia translates to "church" — and is, according to the artist a homage to George Church. A tribute to the home lab, as Davis since 2010 is Artist Scientist at Harvard Medical School at the George M. Church Laboratory. (And also MIT Department of Biology/Thomas Schwartz Laboratory Mt Angel College.) As often with Davis's approach, he develops artistically sophisticated projects with poetic dimensions that connect our ancient cultural history or mythology with advanced life science and technology. His project

"Malus Ecclesia" equally presents to us the oldest of apple trees, by bringing us the reference of the biblical tree of knowledge. The project background was Davis's involvement in a project to sequence the genome of *Malus sieversii*, considered to be the wild progenitor species of the domesticated apple, *Malus domestica*. Working with an assembly of more than 35 investigators connected with more than a dozen respective academic and scientific institutions, the sequencing and assembly of the genome was to map out the genetic background of the rose family (Rosaceae).

The biblical tree of knowledge has according to Davis an even more ancient strand of history: "Legends about a tree of temptation guarded by a serpent long predate Judeo-Christianity. Archaeological evidence linked to the Adam and Eve story suggest that the "fall of man" account was known c. 23rd–22nd century BCE in post-Akkadian Mesopotamia. Still earlier accounts of a serpent-guarded and sacred tree, replete with Lilith and the "tablets of destiny", appear in the Sumerian-Akkadian Epic of Gilgamesh and the Song of Innana. Parallels of the biblical account are also reflected in Vedic, Norse, and other Indo-European traditions." Joe Davis (2014)[16]

Parallel to the work with sequencing the genome of the wild apple, Davis was also engaged with a companion project where, as a collaborative project with others, he was working on a solution of creating an encyclopedic apple, by inserting a highly compressed version of Wikipedia into the *Malus sieversii* genome. — Sort of a re-appropriation to the tree of knowledge if you will. According to a status update from Davis about the project from May 2016, the goal to: "create encyclopedic insertions for Malus ecclesia currently center on coding approximately 50,000 English Wikipedia pages accounting for the top 50% of traffic to the site."

For the art exhibition "New Eden"[17] produced for the Article Biennial 2016 in Stavanger, Norway, Joe Davis made a special edition of the project bearing the title "Malus Ecclesia: The Oldest Eden" — An excerpt of the ancient Sumerian text about Innana and her garden was encoded in DNA plasmid and transferred into a leaf from a local apple tree by the aid of a gene gun, a device commonly used to insert the genetic material into the cells by the particles coated into small DNA sequences. The gun is fired at the cluster of the cells and the DNA sequences enter the desired cell.

The genetic transfer was performed in collaboration with CORE, Centre for Organelle Research, and University of Stavanger,[18] and produced with assistance from researcher Paul Reginato,[19] Wyss Institute, Harvard. Though the information stored was questionably retrievable, the project itself contributes to a poetic conceptualization of biological storage of information.

6. A Critical Perspective on the "Creation of Life"

From the tree of knowledge to creation of life, the last project I would like to include in this assemblage of artworks is a work created by Oron Catts and Ionat Zurr in collaboration with Corrie Van Sice.[20] The duo Oron Catts and Ionat Zurr founded the Tissue Culture and Art Project, as well as SymbioticA — the Center for Excellence in Biological Arts in 2000.[21] Based on the School of Anatomy and Biology at University of Western Australia, they have a reputation for leading an influential and creative research laboratory that critically engages with contemporary uses of biology. Many of their projects involve tissue culturing, bringing up aspects of the complexity of engaging with living material and life. Their artwork "The Mechanism of Life — After Stephane Leduc" (2013)[22] — refers to the book "Mechanism of Life" (1911)[23] written by Stéphane Leduc, the French biologist who first coined the term "Synthetic Biology". His book proposed a series of chemical experiments attempting to challenge Vitalism by proving the mechanistic aspects of life. Leduc made use of seductive imagery depicting diffusion and osmosis visualizing the emergence of life-like phenomena. By reviving the experiment of Leduc presented nearly 100 years ago, the artists are reminding us that with our contemporary hyped concept of engineered life, or Synthetic Biology, we are dealing with "cultural amnesia and reimagining". Their artwork is a re-appropriation to one of the simplest protocell protocols by Leduc.

By making use of 3D-printing technology they are introducing the discussion of another hyped up "next industrial revolution", as

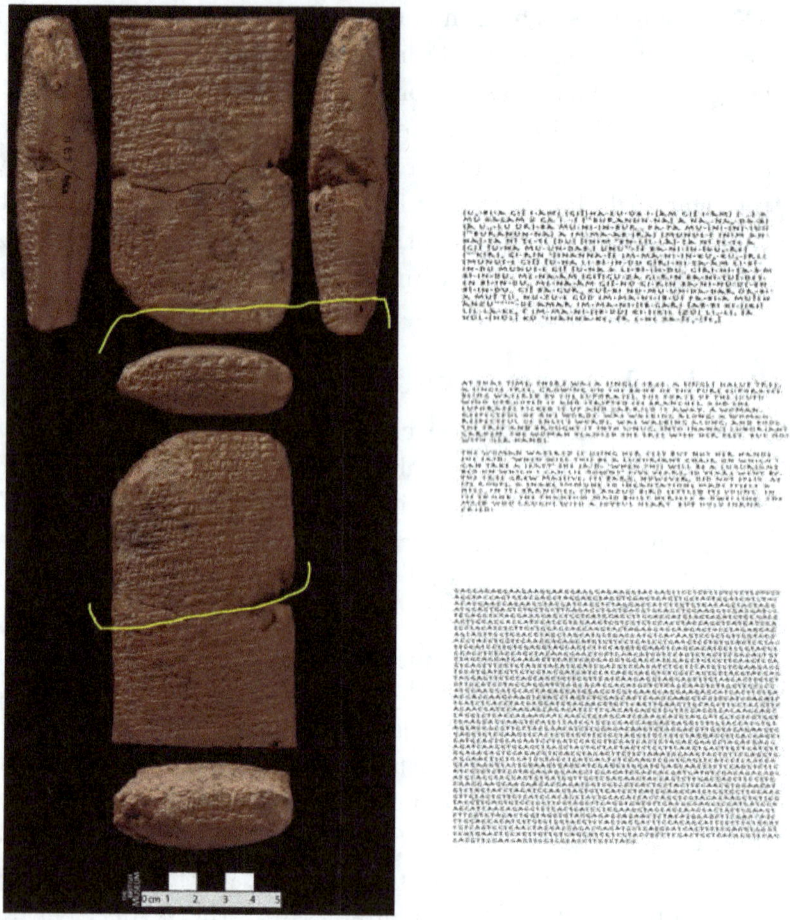

(a) (b)

Fig. 4. (a) Original Mesopotamian clay-tablet with text (From British Museum). (b) Print from Joe Davis work "Malus Ecclesia: The Oldest Eden" depicting original Mesopotamian text translated to English and to DNA code. (From the exhibition "New Eden".)

the technology is suggesting the ability to print actual life, or at least parts of living matter. The artwork itself orchestrates droplets of water and saltwater mixed with ink into a petridish. As we are waiting to observe the droplets with anticipation of proof that life will continue to evolve, we are left with the slow transition into

Fig. 5. Picture of leaf from apple tree getting injected with DNA information via Gene-Gun. Same leaf that was exhibited for the work "Malus Ecclesia: The Oldest Eden" by Joe Davis at Article Biennial 2016. Photo by Paul Reginato, for i/o/lab — Centre for Futureart.

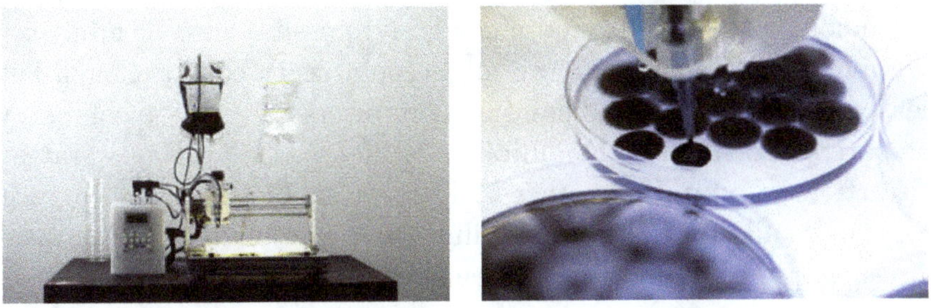

(a) (b)

Fig. 6. (a) The artwork "The Mechanism of Life — After Stephane Leduc". (b) Same artwork, detail of the printed droplets into the petridish. Pictures are from the webpage of artist Corrie Van Sice (Courtesy of the artist).

murky water. According to the artists, the work is "More than a proof on the mechanism of life, they are a suggestion for a humble approach to the question of what life is and how far are we willing to make life into a raw material for our own ends." This may also serve as a comment applicable to what the artists refer to as misunderstanding of the "language" of the code. Meaning that the metaphors rooted in the code/informatics view of life perpetuate misunderstandings regarding processes involved. Especially when it comes to the attempts to use this logic or create genetically modified organisms as building blocks to engineer biological circuits (Catts/Zurr 2010).[24]

7. Conclusion

The artworks presented may serve as artistic comments to the parallel development in bioinformatic or its larger field of computational biology. Visions of the possibilities working with living material and coding of living organisms is thus the same driving force within arts and science. By allowing critical reflections on our technological developments by artists, a humble attitude toward the complexity of living matter may be fostered. As we may dream of constructing synthetic organisms inspired by life, we are yet to discover the implications of creating such organisms, let alone synthetic life-like machines. There is no doubt that we may find useful applications in a technology that is tightly mimicking and possibly also blending with the living. But all though we might be able to construct synthetic cells with logic gates, and envision computers driven by e-coli, are we prepared to deal with the fallibility of life? As Brian Hayes states in his article "Computing comes to life": "Can biocomputer engineers cope with all the distinctive failure modes of living organisms — disease, predation, parasitism, senescence, death?"[25]

Acknowledgments

This work was partially financed by the FeLT project (Norwegian Artistic Research programme).

References

1. C. Snow, *The Two Cultures*. Canto classico, 2012.
2. eBay Inc. ebay. URL: https://www.ebay.com/, 1995.

3. N. H. G. R. Institute, *The Human Genome Project.* URL: https://www.genome.gov/human-genome-project, 1990–2003.
4. E. Roth, A guide to diybio (updated 2019). URL: https://thatmre.medium.com/a-guide-to-diybio-updated-2019-abd0956cdf74, 2019.
5. M. Wohlsen, *Biopunk: Solving Biotech's Biggest Problems in Kitchens and Garages.* Current, 2012.
6. Bell-Labs. Experiments in art and technology (e.a.t.) program. URL: https://www.bell-labs.com/collaboration-opportunities/eat/, 2018.
7. Y. Kuan, Feature on bioart by yenting kuan. URL: https://mag.clab.org.tw/En/2020/08/27/development-of-bioart/, 2020.
8. A. K. Yetisen, J. Davis, A. F. Coskun, G. M. Church, and S. H. Yun, Bioart, *Trends in Biotechnol.* 33(12): 724–734, 2015.
9. E. Kac, Genesis. URL: http://www.ekac.org/geninfo2.html, 1999.
10. E. Kac, Gfp bunny. URL: http://www.ekac.org/gfpbunny.html, 2000.
11. S. Rubinowitz, Glowing rabbit sparks controversy, originally published in petplace.com on September 23, 2000. URL: https://www.ekac.org/petplace.html, 2020.
12. J. Davis, Microvenus. *Art Journal* 55(1):70–74, 1996.
13. SETI. Arecibo message. URL: https://www.seti.org/seti-institute/project/details/arecibo-message, 1974.
14. NASA, Nasa voyager — The golden record. URL: https://voyager.jpl.nasa.gov/golden-record/, 1977.
15. P. House, Object of interest: The twice-forbidden fruit. *The New Yorker* 13: 2014.
16. J. Davis, Malus ecclesia and apples to tempt the devil: Cloning wikipedia into the ancestral apple. URL: https://wikimania2014.wikimedia.org/wiki/Submissions/Malu_ecclesia_and_Apples_to_Tempt_the_Devil:_Cloning_Wikipedia_into_the_Ancestral_Apple, 2014.
17. i/o/lab. "new eden" — article biennial, link to exhibition catalogue. URL: https://issuu.com/iolab/docs/article_biennale_2016_katalog, 2016.
18. S. Reumann, Core centre for organelle research and university of stavanger. URL: https://www.ux.uis.no/reumann/index.html, 2014.
19. P. Reginato, Paul reginato profile page. URL: https://physiogn.hms.harvard.edu/church-lab-member/paul-reginato, 2015.
20. C. V. Sice, The mechanisms of life. URL: https://www.corrievansice.com/projects/04, 2015.
21. O. Catts and I. Zurr, Symbiotica. URL: https://www.symbiotica.uwa.edu.au/, 2019.
22. O. Catts, Oron catts talks about "the mechanism of life — after stéphane leduc". URL: https://vimeo.com/179989629, 2016.
23. S. Leduc, The mechanism of life, 1911.
24. O. Catts and I. Zurr, The ethics of experiential engagement with the manipulation of life. *Tactical biopolitics: Art, activism, and technoscience.* 125–142, 2008.
25. B. Hayes, Computing comes to life. *American Scientist* 2001.

https://doi.org/10.1142/9789811257155_0024

Chapter 24

The Cosmic Vineyard

Alessandro Chiolerio

Istituto Italiano di Tecnologia, Bioinspired Soft Robotics, Center for Converging Technologies, Via Morego 30, 16165 Genova, Italy

University of West of England, Unconventional Computing Lab, Coldharbour Lane, BS16 1QY Bristol, UK

alessandro.chiolerio@iit.it

Everything is so open, even to the rain pushed by the wind and the clear rays of the sun, everything is as silent as the gaze of the spectator, only a patient spectator could see a dusty tool, a very noisy remnant of war, or a bright modern tractor, a technological creature, walking through the ranks. Yet this same technology could help us see a hill in a different way, the vineyard resting harmoniously on it, the ordered plantation of the vines, their stubborn sugar production, finally cleverly transformed into liquid red gold. The vineyard is constantly subject to natural electromagnetic fields, to the radio signals that the stars send us. This electric whisper polarizes our vines, disperses itself into a symphony of impulses, determines their metabolism, perhaps their mood. And what if their mood is refined by us, with the help of enzymes and bacteria, to become wine? Wine is the result of this calculation of cosmic and holistic proportions, which ends up invading us and intoxicating us with sparks of stars.

1. The Complexity of a Vineyard

The design of a vine takes place during the seasons, fatigue, the work of the peasant and it follows the curvature of hills and flatlands, as masterfully evidenced by the wood grains in the lithography shown in Figure 1. But it hides a lot of invisible secrets. The vine has a fascinating history, which accompanies the history of mankind

Fig. 1. "They slope down on us slopes/of low vineyards", lithography by Gianni Verna, mm 1360 × 480, 1991.

and his migrations for millennia. Recent discoveries show that wine production is documented from the 6th millennium BC in places far apart from each other, such as the Caucasus,[1] Sicily,[2] and Sardinia[3] (Figure 2). Wine production was so important that when the *Phylloxera* epidemic destroyed most European vineyards, they learned to create chimeras with the most resistant American vines. And part of that story remains to be written, as new disease-resistant hybrids are being developed to ensure a sustainable future for our winemakers.[4]

A vineyard is an extremely complex system, in which the main species can live for tens or even hundreds of years, extending their roots up to 5 meters in the horizontal and 1 meter in the vertical direction, creating a symbiosis over the years with many species of bacteria and fungi, collaborating with herbs in the development of soil chemistry (Figure 3). Countless animal species then coexist with the vines, most of them insects and arachnids, which intertwine their life and death with the tangle of silk fabrics, iron threads, and dried vine leaves.

2. Influx Coming from the Stars

A vineyard is naturally exposed to the external environment and as such it absorbs electromagnetic radiation, the flow of charged particles, neutrinos, cosmic rays, which interact weakly with ordinary matter. This interaction is as sporadic as it is crucial for the ecology of our planet. A striking example is represented by the processes

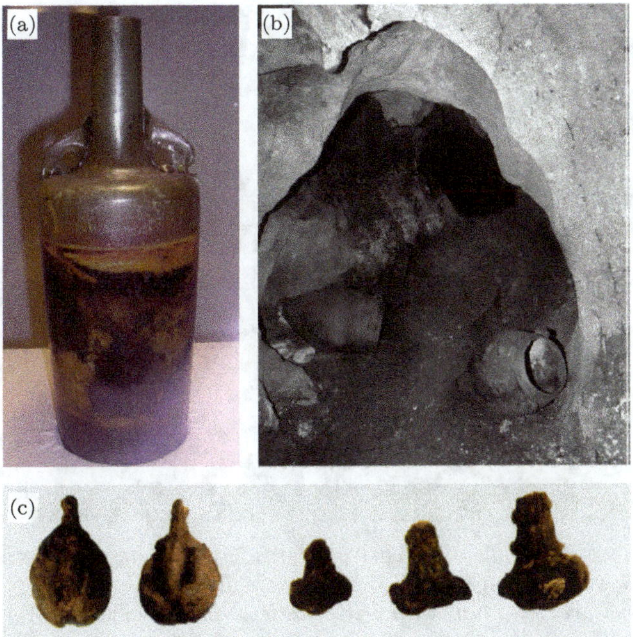

Fig. 2. (a) The speyer wine bottle, dated to about 325–359 AD, found during the excavations of a Roman patrician villa and preserved in the Historical Museum of the Palatinate, Germany (photo by Emmanuel Giel), is the oldest wine sample that has come down to us. (b) B1 and B2 jars found in the Bellitti tunnel at Monte Kronio, Sciacca (Sicily), containing traces of tartaric acid and sodium tartrate, date back to the early Copper Age, around year 4,000 BC. (c) Seeds of domestic grapes, *Vitis vinifera* subsp. *Vinifera* and pedicels, found in a Villanovan well near Bologna, circa 900 BC.

that govern the condensation of a cloud. The condensation nuclei carrying an electric charge undergo spontaneous growth, to form the water droplets that constitute a cloud. And there is no doubt that cosmic rays play a fundamental role in it.[5] A living organism, whether plant or animal, is sensitive to all external conditions to which it is subjected; it is possible to extrapolate information about it, for example, by monitoring the bioelectrical potential of this organism. The study and numerical analysis of action potentials could, among other things, make it possible to monitor the health of a crop, which is essential to prevent the appearance and spread of pathogens against which winegrowers are constantly fighting to safeguard agricultural production, such as late blight and powdery mildew. In addition,

Fig. 3. (a) Association of *Orchis purpurea* and *Ranunculus acris* with *Vitis vinifera*. (b) Association of *Taraxacum officinale* with *Vitis vinifera*. (c) Association of *Muscari neglectum* with *Vitis vinifera*. (d) Association of *Viola odorata* with *Vitis vinifera*. The vineyards in the photos are located on the territory of the municipality of Vigliano d'Asti (AT), Italy.

by "listening" to the thinking of the vines, we were able to collect enough data to determine their emotional well-being, and implement actions to make them feel better, so that the quality of the grapes, and therefore the processed wine, is positively affected. Examples of this are musical and sound experimentation.[6] And maybe we will discover that they behave like a collective organism![7]

3. Solar Wind put Inside a Bottle

Interplanetary scintillation is the variation of the intensity of a cosmic radio source, characterized by a very small emission diameter, induced by fluctuations in the refractive index of the turbulent interplanetary medium. The disturbances of the wavefront phase, which can be assumed to be parallel to the Earth's surface, come from diffraction that produces small inhomogeneities in the electron density of the medium, which are ultimately directly proportional to the density of the solar wind. This diffraction can in turn be correlated with sudden changes in solar wind speed, or particularly energetic events such as coronal mass ejection from the Sun. An instrument capable of measuring interplanetary scintillation along 24 hours is therefore an analyzer of the internal heliosphere that provides very useful information related to space weather,[8] useful to preserve our electronic and telecommunication systems that have now become fundamental, especially following the development of the pandemic. How is this instrument made? In all respects similar to a vineyard (Figure 4), consisting of long conductive metal wires,

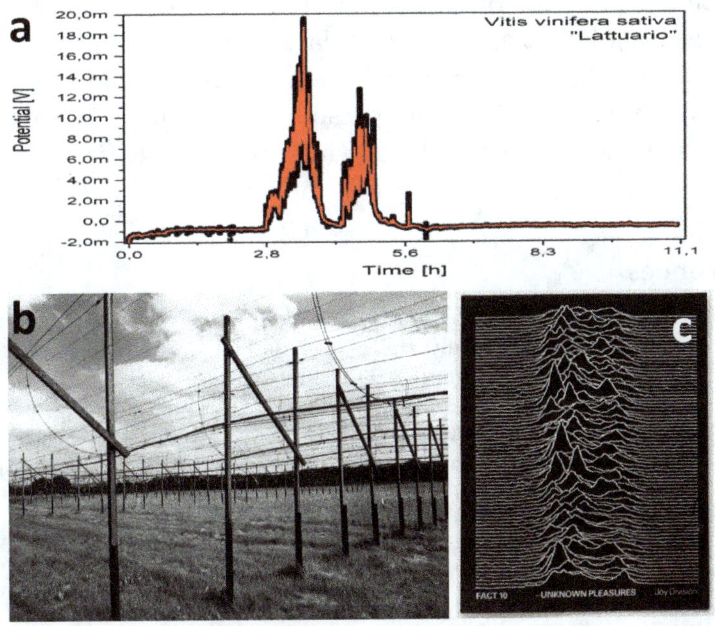

Fig. 4. (a) Example of a peak of bioelectric potential recorded during night hours from a Lattuario vine (Torino, A. Chiolerio, 2020). (b) Interplanetary Scintillation Array (ISA) at the Mullard Radio Astronomy Observatory, Department of Physics Photographic Archive, University of Cambridge (1967). (c) Cover of Joy Division's album *Unknown Pleasures* (1979). Although it is now part of the collective imagination, few know that it does not show sound traces, but the original radiogram recorded by Jocelyn Bell Burnell and Antony Hewish, two astronomers from Mullard, who discovered the first pulsar signal calling it LGM-1 (Little Green Man-1, today it is called CP1919) while looking for traces of quasars, represented by Harold Craft, a Cornell University student and pioneer in digital data visualization.

and possibly consisting of many rows placed in positions at different lengths,[9] which can map the sky as it flows before their eyes, to integrate local fluctuations in ionospheric density over time and cancel them out. Any better tool than a vineyard? The radio telescope would operate at frequencies between 50 and 500 MHz, depending on its construction characteristics: the length of the receiving wires determines only the tone on which you are tuned. And it would not be affected by bioelectrical signals from vines, which follow much slower physiological processes and closer to DC. In essence, the construction of an experimental plant would allow farmers to have a new tool for analyzing the health of their vineyards, scientists with a new tool for measuring interplanetary scintillation, and consumers to drink a traced wine, where modern blockchain and IoT techniques can converge to compile an electronic passport containing the notes of the cosmic symphony that shook the vineyard for an entire year.

References

1. D. Lordkipanidze, Le antichissime origini della viticoltura in Georgia in Vinum Nostrum, arte, scienza e miti del vino nelle civiltà del Mediterraneo antico Firenze 20 juillet 2010–30 avril 2011, Giunti editore, 32–33.
2. D. Tanasi, E. Greco, V. Di Tullio, D. Capitani, D. Gullì, and E. Ciliberto, 1H-1H NMR 2D-TOCSY, ATR FT-IR and SEM-EDX for the identification of organic residues on Sicilian prehistoric pottery. *Micromech. J.* 135:140–147, 2017.
3. M. Rubino, Sardo, rosso, di 3mila anni fa: Identikit del vino più antico del Mediterraneo. In *la Repubblica online*. https://www.repubblica.it/sapori/2016/12/19/news/sardegna_primo_vino_millenario-154444878/.
4. R. Testolin, E. Peterlunger, S. Collovini, S. Castellarin, G. Di Gaspero, F. Anaclerio, M. Colautti, M. De Candido, E. De Luca, A. Khafizova, et E. Sartori, Le varietà resistenti alle malattie. *Quaderni Tecnici VCR* 18 (3e édition), 2018.
5. J. Kirkby *et al.*, Role of sulphuric acid, ammonia and galactic cosmic rays in atmospheric aerosol nucleation. *Nature* 476:429, 2011.
6. Ş. Petrescu, R. Mustăţea, and I. Nicorici, The influence of music on seed germination of Beta vulgaris l. var. Cicla l. *J. Young Sci.* V:67–72, 2017.
7. V. Zappalà, Vini dell'altro mondo. Il mio libro, 2012.

8. J. A. González-Esparza, A. Carrillo, E. Andrade, R. Pérez Enríquez et S. Kurtz, The MEXART interplanetary scintillation array in Mexico. *Geofìsica Int.* 43(1):61–73, 2004.
9. M. Tokumaru, M. Kojima, K. Fujiki, K. Maruyama, Y. Maruyama, H. Ito, et T. Iju, A newly developed UHF radiotelescope for interplanetary scintillation observations: Solar Wind Imaging Facility. *Radio Sci.* 46:RS0F02, 2011.

https://doi.org/10.1142/9789811257155_0025

Chapter 25

Feral Automated System: Prototyping

Nikita M. Sazonov[*,‡], Ippolit V. Markelov[*,‡], Ekaterina B. Nikitina[*,‡], Dmitriy A. Lobanov[*], Natalia V. Alatortseva[*], Dmitriy E. Kadyrov[*], and Mikhail M. Petrov[†,§]

[*]*Posthuman Studies Lab, Moscow, Russia*
[†]*D. I. Mendeleev University of Chemical Technology of Russia Miusskaya sq. 9, 125047 Moscow, Russia*
[‡]*growinghumusitiesinrussia@gmail.com*
[§]*mikepetrovm@gmail.com*

The authors consider the technological and conceptual history behind the introduction of new plant species and the creation of the OGAS — their mutual dependence, intersections, and ecological significance for post-Soviet territories. Traveling from the past to the present, to the factories and fields of past experiments with industries and plants, they try to rework — both scientifically and theoretically — the unfortunate project of the "green" Soviet Internet.

The authors present an alternative scenario in which plants and machines actively collaborate to build bonds and connections between different creatures. The narrative elaborates on the idea of an information exchange system that uses phytomining and plant-based computing architectures along with microcircuits and modern digital protocols.

This text presents a conceptual and technological history of a kind of network prototype. Attempts to implement it were made in the Soviet Union back in the 1960s–1980s, when actively integrating cybernetics within the communist economy promised to radically modernize its systems of management. Back then computerization was paving the way for the boldest visions of the future of the key communist myths, such as the commonality of the communist body and the equality of its discreet agents, whether they be humans, machines, or other species. The prototype for this network was called the All-State Automated System of Management (OGAS), which represented "roiling streams of economic information, united by man with the help of computers and compressed into arrays of data".[1]

Here, the past and future are intertwined. That prototype was never actually implemented. We could say that, paradoxically, it was defeated by that which it was designed to combat — bureaucracy, since the project never went beyond ink and paper. Today we are trying to resume the prototyping of such a network, one whose purpose is to seek the globally common and to connect beings — but now within a posthuman condition and while facing environmental issues. We are working not with an all-state automated system, but a feral automated system because human federations actively coexist with non-human *ferations* — feral politics, the scale of which is incomparable to human scale, but whose interests overlap our own. Here, ferality harkens to the fundamental link between these overlapping and (to us) otherworldly worlds of ferations, and the human-sized world — as described in Ovid's *Fasti*,[2] with regards to the relationship between the living and the dead: the feral dead, the link with whom is broken, become a formidable threat.[a]

A multitude of people from different cities of the former Soviet territories are taking part in the creation of this prototype. By combining the stories of their past and that of their families with an archaeology of building networks and communicating with other species and eras, each participant creates their own point on the map of the Feral Automated System, simultaneously proposing their

[a]Today, the adjective feral is typically used to denote wildness — either biological or cultural. By contrast, we have introduced the concept of feration in our work, positing that ferality is a positive ecological status for all of Earth's inhabitants (those robbed of territories, heritage, touchstones) and thus aspires to serve as an alternative to human-sized politics and its "federations". Such an understanding of ferality is mostly rooted in its etymology, which refers to funeral feasts in Ovid — the feral here is situated at the boundary between the otherworldly (the dead) and the dangerous (a restless spirit) and here is already established as politics.[2,3]

own means of connection to others, no matter how unstable these connections may be.

1. Prototype Description

At its core, the Feral Automated System (FAS) is more than just the internet. It is an information-economic model inspired by the Soviet dream of automated management systems. Within the FAS, the main productive elements, or labor units, are called hubs (Fig. 1). Each hub is a Wi-Fi microchip that tracks the activity of various agents: people, plants, industrial objects (see paragraph 1, p. 408). The network is powered by a redox flow battery that uses vanadium which has been partially extracted from plant agents. Based on the process of degradation of the vanadium in the battery, we modeled a labor infrastructure where each hub employs its agents to mine the virtual equivalent of vanadium, thus making it a kind of cryptocurrency (see paragraph 2, p. 411). The main agent of automated task allocation that "replenishes" vanadium in the network is called the plant automation manager (see paragraph 3, p. 418).

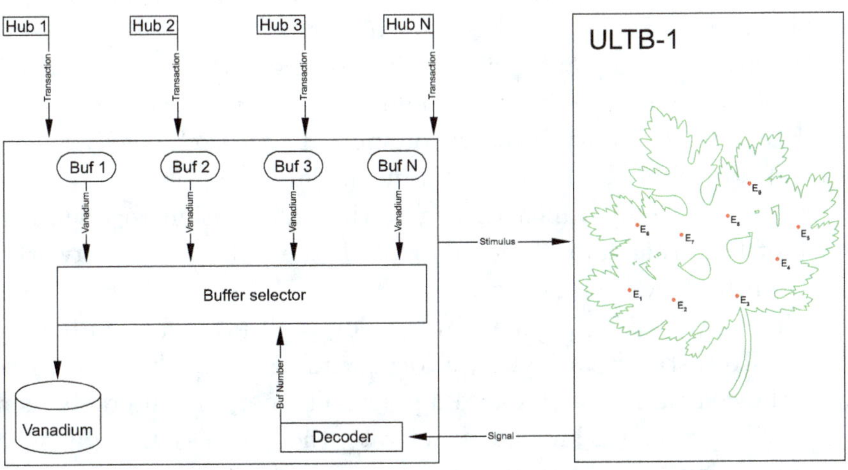

Fig. 1. Structure of the central hub.

The following paragraphs describe the individual constituent parts of the FAS. Each item is split into its scientific-technological and theoretical aspects.

2. Hubs

Hubs are objects equipped with Wi-Fi microchips (ESP8266) that are connected to form a special network architecture. All the hubs were built by artists from different parts of Russia as part of the Unstable Compounds summer school that was organized by the Posthuman Studies Lab and hosted by the Vyksa Air art residency in the Nizhny Novgorod region of Russia in 2021.

Each hub is tied to a certain area in the country, including the Moscow, Leningrad, and Yaroslavl oblasts, Stavropol Krai, Yekaterinburg, Omsk, Dzhankoi (Crimea), Vladimir, Petropavlovsk-Kamchatsky, Northern Kazakhstan, and Taganrog. These locations were chosen for a variety of reasons, such as their industrial histories, the biological activity of the endemic vascular plants capable of metal hyperaccumulation, and the participating artists' ties to these areas, industries, and local non-human agents.

As an economic unit, each hub determines certain activity, its so-called "labor", and records data about its work. This labor is a combination of the plants' vital activities, the participants' research, the labor costs of building hubs, and performative practices.

Each hub's activity is recorded, and data are exchanged with the central control hub. The data contain the labor that is then recorded by the central hub, which then uses this information to adjust the work of the other hubs. The labor of all the hubs in the network is thus synchronized.

The network is built in a star topology, with the main hub at the center. To install it in Vyksa, a local wireless network was created, with the central hub as its access point. The network interacts using the TCP protocol. The hubs activate the wireless module when necessary, join the network using a prewritten config, and exchange data with the central hub. Once the exchange is completed, the wireless module is again switched off to offload the central hub's bandwidth. Interaction with the central hub uses finite state-machine logic, where each message sent by a satellite hub contains its current state and parameters. It's worth noting that the states themselves are stored on the satellite hub, but switching between them is controlled by the main hub via reply messages.

A total of six types of messages and five possible machine states are declared. See Tables 1 and 2.

2.1. *Connection 1*

In the Soviet Union, labor united beings, both human and non-human. Various historical events — wars, crises, revolutions — made

Table 1. Types of messages during exchange.

TYPE_REGISTER	Message from a satellite hub, requesting registration on the central hub
TYPE_REGISTER_OK	Reply from the central hub that registration was successful
TYPE_SET_COEF	Request for configuration coefficients from the central hub
TYPE_SET_TASK	Request for a new vanadium mining task from the central hub
TYPE_GET_Q	Request from central hub to transfer mined vanadium
TYPE_PRODUCED_Q	Message with data on mined vanadium from satellite hub

Table 2. Possible machine states.

STATE_INITED	Hub initialized, registering
STATE_REGISTERED	Hub registered, getting configuration
STATE_CONFIGURED	Hub configured, awaiting vanadium mining task
STATE_TASK_SET	Task received, mining
STATE_NEED_GET_Q	Vanadium transfer request received, sending according to current task

a purely "human" development of communism impossible. Back when the bolshevik dictatorship was still being established, the fuel crisis formed a special attitude toward the non-human terrestrial past. This is when Lenin's slogan "for bread and oil" is born, where a union of fossilized and living energies of the Earth is seen as, "the correct economic basis for the relations between workers and peasantry", Ref. 4, p. 362. It can be interpreted as an act of communion, as a kind of revolutionary alliance of Earthly proletariat, no matter what time period they're from. Even the labor of the previous generations of Earth — now a dead, compressed labor — is exploited just like any "living" labor. The revolution takes on a planetary scale.

Alexander Bogdanov's science fiction works and sketches from his "Tektology"[5,6] are no less relevant here. Bogdanov outlines the organizational principles of uniting and building communities. Communism must go beyond the default organic individualism of human bodies by seeking points of connection at a more fundamental level. The most illustrative metaphor for "red" communism is the idea of a shared vascular system — a concept for a time made reality in the Soviet Blood Institute. A community built around blood transfusion is tied by leninist communion, where the private boundaries of biological bodies are, quite literally, reshaped into a shared communist body, a communal organism.

On the other hand, the Soviet Union during its post-war period is also worth considering. During the war and afterwards, wild plants were a key agent in designing and developing the economy of the USSR, turning into new kinds of crops in agriculture as they were introduced. This so-called "green proletariat" toiled on terms equal to those of the human workers: they served as a universal resource, as a living protection from harmful compounds, and as teachers promoting "love for communism".[7]

The idea of a posthuman labor system eventually culminated in the concept of an All-State Automated System (OGAS). Seeking to automate economic management, OGAS combined the labor power of earth (silicon and other chemical elements found in computer machines) and the living labor of collective farms and factories and their people, plants, and animals. The automating agency of the network was thus supposed to establish and maintain total connectedness between the disparate labor systems of the Soviet Union. The interspecies carbon-silicon assemblies were supposed to become interconnected. This was a large-scale renovation of the Soviet planned economy: the human-sized (more like superhuman-sized) 5-year plan (typically overachieved within three) was to be superseded by a dynamic one, which, according to OGAS mastermind Viktor Glushkov, ought to "live and develop", and serve as a "flexible and dynamic" planning system;[1] more than just a network or a data exchange system, it was supposed to become an organic way toward communion.

3. Vanadium

Redox flow batteries are a promising type of electrochemical energy storage, in which electricity is stored and stocked up on as chemical energy, that is, as bonds between certain electroactive chemical compounds that are capable of undergoing redox reactions. Electrolytes may be considered as fuel in redox flow batteries, and this fuel is not consumed in the process.[8] The best examples of flow batteries can maintain an acceptable energy capacity even after 100,000 charge–discharge cycles. For this and many other reasons, redox batteries are considered particularly promising storage devices for large amounts of energy — from tens of kWh to tens and hundreds of MWh.

The most common electrolytes used in flow batteries are sulfuric acid solutions of vanadium salts. Vanadium is a trace metal: its overall abundance in the Earth's crust is quite high, but it's concentration is very spread out. There are almost no rocks with a high vanadium content (10% and higher), but it is contained in minor impregnations in a variety of deposits. Today, significant amounts of vanadium are obtained not from mined ore, but from secondary raw materials — mainly from metallurgical solid waste. However, there is an additional theoretical way to extract vanadium. There exist hyperaccumulator plants which are capable of accumulating particular metals in their tissues in concentrations hundreds or even thousands of times higher than their naturally occurring content. Such plants are already being employed in the phytoremediation of contaminated areas,[9] and experimental plant farms are set up for the purpose of extracting certain metals such as nickel. According to the literature, there also exist plants capable of accumulating significant concentrations of vanadium[10] — especially from secondary raw materials, such as sludge from oil refineries and thermal power plants and spent catalyst waste used in sulfuric acid synthesis.

One of the more utilitarian goals behind designing the FAS was to test the hypothesis of whether plants can be used to extract vanadium, with the resulting compounds then used to synthesize the electrolytes in redox flow batteries. Today, the problem of vanadium phytomining from raw plant materials for subsequent use in a redox battery remains unsolved.

Using citizen science approaches and with the help of the 20 participants of our Unstable Compounds project, we selected samples of hyperaccumulator plants from different areas around Russia. The samples were then ashed in a laboratory at the D. Mendeleev University of Chemical Technology of Russia, Moscow, after which total chemical analysis was carried out with X-ray fluorescence spectroscopy.

When choosing plant species for further analysis, we considered two key parameters:

- the intensity of vanadium accumulation by the plant
- the proximity of the plant to the points of interest (i.e., metallurgical sludge waste containing vanadium in a form available for phytomining).

The highest amounts of vanadium can be contained in the roots of the plants of the Asparagus family (*Asparagaceae*), the Figwort family (*Scrophulariaceae*), the Rose family (*Rosaceae*), Cereals (*Poaceae*), Sedges (*Cyperaceae*), Legumes (*Fabaceae*), the aerial parts of the Rose (*Rosales*), Sedge (*Cyperaceae*), Buttercup (*Ranunculaceae*), Aster (*Asteraceae*), Madder (*Rubiaceae*), Mulberry (*Moraceae*), Horsetail (*Equisetidae*), Asparagus (*Asparagaceae*), and Buckwheat families (*Polygonaceae*).[11–13]

We carried out studies to determine the total of chemical elements content in the ashed plant samples with X-ray fluorescence spectroscopy (XRF) on an ORBIS PC X-ray fluorescence spectrometer by EDAX and comparing the data on the elemental composition of the area and the signal level on the X-ray detector. To verify the presence of vanadium in the samples, an Al filter (250 μm in thickness) was used. Accelerating voltage 35 kV; the emission current, 500 μA; magnification $\times$ 10; analysis types: analysis by area; the X-ray spot diameter on the sample, 2 mm; the working chamber operating mode was vacuum.

When observing an optically magnified area of the sample on the XRF spectra of secondary X-ray radiation, we identified peaks of intensity corresponding to the tabular positions of the characteristic X-ray emission lines of chemical elements. Then, using the Orbis PC software, we calculated the elemental composition of each sample.

The geography of sample distribution is shown in Figure 2. Tables 3 and 4 show general data on the composition of ash residues of collected plant samples.

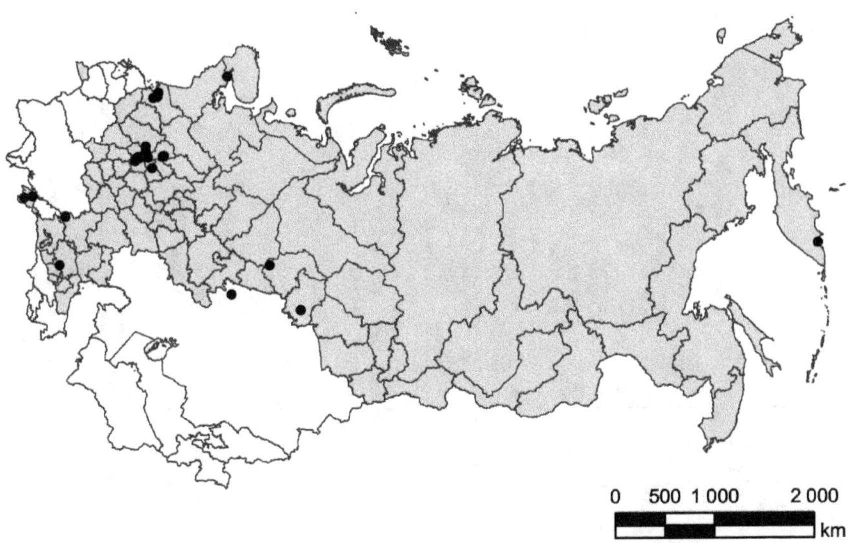

Fig. 2. The geography of sample distribution.

Table 3. Gross elemental composition of plant samples.

Plant sample description	Part of plant	Location	Ash content, %	Mg, wt %	Al, wt %	Si, wt %	P, wt %	S, wt %	Cl, wt %	K, wt %	Ca, wt %	Ti, wt %	Mn, wt %	Fe, wt %	Ni, wt %	Cu, wt %	Zn, wt %
Common yarrow, *Achillea millefolium L.*	Leaves	Russia, MoscowOblast, town of Roshal, Ash dump at the sulfuric acid production facility at the Roshal Chemical Plant	20	–	–	3.1	6.1	2.8	–	24.4	28	–	0.9	27.2	–	0.8	6.6
Wild saltbush, *Atriplexfera sp.*	Leaves and stems	Russia, Stavropol Krai, city of Georgievsk, area between rebar and nail factories	30.6	0.6	0.2	0.8	4	2.1	4.2	61.3	25.2	–	0.2	0.9	–	0.1	0.4
Wild chamomile, *Matricariadiscoidea sp.*	All plant		–	0.8	–	0.5	7.8	3.9	6.4	52.3	25.8	–	–	1.3	–	0.3	1
Horse sorrel, *Rumexconfertus*	Leaves	Russia, YaroslavlOblast, village Dubki	7.2	3.1	0.4	1.1	4.3	3.1	16.8	29.9	35.9	0.1	0.4	3.8	–	0.3	0.8
Knotwood, *Reynoutria sp.*	Stems	Russia, Yaroslavl region, city of Yaroslavl, the "Lakokrasa" ponds and sludge reservoir	3.8	3.1	–	0.1	6.1	2.3	10.1	48.8	27.9	–	0.3	0.4	–	–	0.9
	Roots		13.9	1.3	0.7	2.3	7.7	3.3	2.4	28.2	45.8	–	–	4.3	–	1.2	2.8
	Leaves		–	–	–	0.5	3.6	3.2	8.1	26	50.4	–	3.9	2.4	–	–	1.8
Parsnip, *Pastinaca sativa L.*	Leaves and stems	Russia, Omsk Oblast, city of Omsk, sludge and ash disposal site at CHP 5	7.8	3.9	–	0.4	8	3.4	6.7	44.6	31.8	–	0.3	0.5	–	0.1	0.3
Curlycup gumweed, *Grindeliasquarrosa sp.*	Stems	Crimea, city of Dzhankoy, the ruins of a foundry	5	0.4	–	0.2	4.6	2.2	7.8	54,9	28.5	–	0.1	0.6	–	0.2	0.6
	Leaves and flowers		10.5	2	0.1	1	4.3	3.6	10	40.8	35.5	–	0.2	1.3	–	0.3	0.9
Giant goldenrod, *Solidago gigantea*	Leaves	Russia, Moscow Oblast, town of Zaprudnya, Zaprudnensky Glass Plant	2.6	–	2.6	5.5	0.9	6.2	53.1	27.3	–	0.2	0.8	–	0.2	0.6	–

Table 4. Gross elemental composition of plant samples, continuation.

Plant sample description	Part of plant	Location	Ash content, %	Mg, wt %	Al, wt %	Si, wt %	P, wt %	S, wt %	Cl, wt %	K, wt %	Ca, wt %	Ti, wt %	Mn, wt %	Fe, wt %	Ni, wt %	Cu, wt %	Zn, wt %
Curlycup gumweed, *Grindeliasquarrosa sp.*	Roots	Crimea, city of Dzhankoy, the ruins of a foundry	12.1	1	0.5	1.9	3.8	2	7.3	44	33.1	0.1	0.2	4.6	–	0.4	1.1
	Stems		5	0.4	–	0.2	4.6	2.2	7.8	54,9	28.5	–	0.1	0.6	–	0.2	0.6
	Leaves and flowers		10.5	2	0.1	1	4.3	3.6	10	40.8	35.5	–	0.2	1.3	–	0.3	0.9
Giant goldenrod, *Solidago gigantea*	Leaves	Russia, Moscow Oblast, town of Zaprudnya, Zaprudnensky Glass Plant	2.6	–	2.6	5.5	0.9	6.2	53.1	27.3	–	0.2	0.8	–	0.2	0.6	–
Honey clover, *Melilotus albus*	Stems	Russia, Moscow Oblast, city of Dzerzhinsky, Zolotvaly Dzerzhinsky CHP 22	3.7	5.4	–	0.3	3.5	2.4	9	62.2	15.8	–	0.2	0.8	0.1	0.2	0.2
	Roots		7.4	4.4	0.7	1.1	5	3.1	13.8	53.5	15.6	0.1	0.2	2.1	–	0.2	0.3
	Leaves and seeds		5.9	2.4	0.9	1.7	2.9	4.4	5.2	45.9	31.7	0.1	0.1	3.1	0.1	0.6	1
White goosefoot, *Chenopodiumalbum sp.*	Leaves	Russia, Vladimir Oblast, city of Vladimir, sludge disposal site at the Vladimir Tractor Plant	61.1	2.6	–	0.7	3.4	2.7	3.2	56.3	28.5	–	0.1	1.6	–	0.2	0.6
Red clover, *Trifolium pratense*	All plant	Russia, Kamchatka Krai, city of Petropavlovsk-Kamchatsky, premises of the abandoned Concrete Plant 116	–	–	–	–	1.8	1	1.3	35.7	56.9	0.2	0.2	1	–	0.5	1.4
Wild perennial lupine, *Lupinus perennis*	All plant		–	–	–	0.4	3	1.5	1.3	25.2	61.8	0.2	0.5	2.5	–	0.2	3.5
Wild buckwheat, *Fallopia convolvulus*	Leaves and stems	Kazakhstan, Kostanay Oblast, town of Rudny, dumps in the vicinity of the Sokolov–Sarbay *mining and processing production association	–	4.3	1.1	2.9	4.4	3.4	1	53.1	22.6	0.3	0.3	5.2	0.3	0.4	0.9

As we can see, some plants have displayed an obvious capacity for phytoremediation, i.e., the capacity to accumulate heavy metals. For instance, the leaves of Knotwood (*Reynoutria sp.*) contained up to 3.9 wt.% manganese, the leaves of giant Goldenrod (*Solidago gigantea*) had up to 0.2 wt.% titanium and nickel, and the leaves and stems of wild Buckwheat (*Fallopia convolvulus*) up to 0.3 wt.% of the same titanium and nickel. Of special interest among these samples is the common Yarrow (*Achillea millefolium L*), gathered near the ash dump of a sulfuric acid production site at the Roshal chemical plant (Russia, Moscow Oblast, the town of Roshal) — its leaves contained 27.2 wt.% iron, 6.6 wt.% zinc, and 0.8 wt.% copper. Considering the sample's ash content (20%), its concentrations of iron, copper, and zinc in solid samples is at 50,000 mg/kg, 1,600 mg/kg, and 13,200 mg/kg, respectively. Achillea millefolium can thus be considered a hyperaccumulator of copper (its concentration must exceed 1,000 mg/kg) and zinc (concentration must be above 10,000 mg/kg).

Unfortunately, vanadium was not detected in any of the plant samples, at least within the limits of the sensitivity of the chosen method of composition analysis. This is probably due to low vanadium content in the soils where the plants had grown. Therefore, we are now conducting an additional experiment involving controlled plant growing in a soil artificially enriched with vanadium in the form of ammonium metavanadate. If this experiment is successful and the possibility of vanadium accumulation is demonstrated, the search for wild-growing plants can be reattempted by specifying the criteria for locations and collected plant species. In addition, to create the electrolytes in redox flow batteries, other elements can be used from those that have already been found in significant quantities in the composition of the studied plants — primarily zinc and iron, whose compounds also possess pronounced electroactive properties.[14,15]

3.1. *Connection 2*

"Technology" is lagging not too far behind "nature". They are intertwined, both metaphorically and biologically. Cryptocurrency farms resonate with phytomining processes. Within the modern hi-tech economy represented by, e.g., blockchain technology, labor is incorporated into currency. Cryptocurrency is mined the way gold used to be, as a resource. On the other hand, the ability of plant species for phytoremediation allows us to see them as independent economic agents, ones already equipped with all the necessary (bio)technologies for resource extraction. Phytoremediation reveals the technical nature of the plants themselves and their capacity for networking. As proto-economic agents, plants are tied to different ecological layers of the Earth and are vigorously active within these layers. By dynamically altering those they connect to with their vegetative organs, plants form systems of exchange and feedback.

On its own, housekeeping as a practice is closely linked to *oikos*, what Aristotle described as: "the art of household management", Ref. 16, p. 38. Aristotle contrasts the art of profit and the art of management in a fascinating way. The art of management is dictated by the logic of the *oikos* whose purpose does not lie in: "the unlimited acquisition of money", Ref. 16, p. 44. Instead, its purpose is in forming connections, inventing various new ways of using the same thing. That is, forming one's own household is fundamentally a deeply ecological activity. By the same token, phytoremediation is a way for plants to make use of that which enters their *oikos* — in this case, industrial waste and other byproducts. This is co-existing with industrial toxins, an attempt at ecological reconciliation — a practice where economy and ecology are most closely intertwined.

The attempt to include plant labor in the FAS is an attempt to include our own waste into our *oikos* instead of seeing it as something unecological. We thus acknowledge our ties to this waste, work though it by expanding our economic assemblages with the help of plants. We domesticate waste — not in terms of taming it, but by insisting that they are included in our landscapes as full fledged inhabitants. Such posthuman economic labor is tied together with the help of the cryptocurrency of vanadium, with "crypto-" taking on the additional meaning of "hidden". The purpose of the exchange occurring within the FAS prototype is not accumulating wealth, but optimizing the use of a domesticated toxin. The cryptocurrency-ness of vanadium serves as an oikonomical foundation of a shared ecology being built. Thus the feral, often perceived as dangerous and thus repressed in human economy, ecology, and consciousness, becomes an organizing basis for building a common network.

4. The Leaf Switchboard Operator

The FAS is controlled by a central hub, a plant automation manager. The central hub is an external device — a plant computing system that implements the operational logic of the FAS.

The structure of the central hub consists of:

(1) an input–output device (electronic oscillator, electrodes, analog-to-digital converter, amplifier, and a server communication system);
(2) a computational non-human biological agent (ULTB-1).

The operation principle of ULTB-1 is based on spike shifts. A stimulus is applied to the electrodes and the response is recorded. Depending on form, duration, and number of registered "response" spikes, a map of the distribution of logical operators in the leaf is formed.[17]

The central hub functions as a dispatch and resource management center for the FAS. Each satellite hub thus performs work mining virtual vanadium. The amount of vanadium mined is stored in a buffer installed on each hub in the network. The central hub receives data on the amount of vanadium produced by each satellite hub. This may be a single line with encrypted parameters of vanadium representing a single transaction from a satellite hub. This information undergoes certain modulation and is then transmitted as an electric impulse via electrodes to the leaf. After processing inside the leaf, the system registers a response. This response is a special data signal from a living plant leaf which is then translated into a buffer number from where the "upper" transaction is subsequently taken and transferred into the battery resource of the main hub.

4.1. *Connection 3*

The idea of an automaton — an autonomously existing artificial unit — corresponds to two dialectically related tendencies in culture: God and blasphemy (heresy). Hermann Diels, the author of *Antike technik*,[18] links the automata of late Greece to ritual purposes, and Walter Benjamin, when reflecting on a puppet and a hunchback sitting inside a chess machine in his *Theses on the Philosophy of History*,[19] points to the bond between theology and technology. Such ideas are indicative of a connection to the technological sublime inherent in automation. What is the technological sublime, if not a manifestation of life and divinity in artificially created objects? There is, however, a different trend which, using the words of Bruno Schulz, can be designated as "the great heresy".[20] This describes the human aspiration to create something in the image and likeness of man (and nature more widely) — that is, something artificial, but as alive and autonomous as organic life. It seems that this undisguised aspiration can be one of the defining features of what we call human activity, labor, art, creation, as well as what man is by nature — this is confirmed by the stories of alchemists, the legend of the golem, and Dr. Faust.

Umberto Maturana and Francesco Varela, when looking at systems, paid special attention to autopoiesis as the most important property of a system's self-reproduction, the construction of its own "vision" of the environment based on the analysis of said environment.[21] The automated technological assembly is thus co-constituted with its environment. A machine is never just a machine, the concept of autopoiesis vivifies it, turns it into a constant hybrid, whose coming into being involves everything the system comes into contact with.

Poiesis is bringing into existence. As Aristotle writes, "All Art has to do with Production, and contrivance, and seeing how any of those things may be produced which may either be or not be, and the origination of which rests with the maker and not with the thing made", Ref. 22, p. 167. When we talk about autopoiesis, through "auto" we connect the creator and the created. We shine a light on the creative self-sufficiency of that-which-is in being, that which the Greeks called *physis*. *Physis* is the capacity of something to be itself and not something else, to be oneself, to insist on its existence as such.[23] By and large, *physis* is autopoiesis. By saying this, we couple *techne* to *physis* — something brings itself into existence, creates itself.

The computational logic of the leaf switchboard operator harkens to the processes concentrated at the very core of its *physis*. By making these processes into an agent of automation, we turn the FAS into a living system. The natural processes inside the plant constitute that which brings existence to our network, extends and sustains it. We synchronize the life cycle of the leaf into an unconventional computer system — its growth and decay — with the economic cycle of our network. At different moments in the biological life of the leaf, its "fusiological" response will be different. Thus, the FAS lives and develops (and withers), but, unlike the OGAS, it now does so almost literally.

References

1. A. Adamatzky, Plant leaf computing. *BioSystems*. 182:59–64, 2019. ISSN 0303-2647.
2. A. V. Akhutin, *Ponyatie "priroda" v antichnosti i novoye vremya (The concept of "nature" in antiquity and modernity)*. Nauka, Moscow, 1988.
3. D. S. Antal, C. A. Dehelean, C. M. Canciu and M. Anke, Vanadium in medicinal plants: new data on the occurrence of an element both essential and toxic to plants and man, *Analele Universităţii din Oradea: Fascicula Biologie* XVI/2:5–10, 2009. ISSN 1224-5119.
4. Aristotle, *Politics*, trans. B. Jowett. Clarendon Press, Oxford, 1908.
5. W. Benjamin, Theses on the philosophy of history. In *Illuminations. Essays and Reflections*, H. Arendt, editor, translated by H. Zohn. Schocken Books, New York, 2007, 253–264.
6. A. Bogdanov, *Essays in Tektology: The General Science of Organization*, translated by G. Gorelik. Intersystems Publications, Seaside, 1984.
7. A. Bogdanov, *Red Star*, translated by C. Rougle. Indiana University Press, Bloomington, 1908.
8. H. Diels, *Antike technik (The Ancient Technics)*. Verlag B. G. Teubner, Leipzig and Berlin, 1914.
9. A. Dinesh, S. Olivera, K. Venkatesh, M. S. Santosh, M. G. Priya, Inamuddin, A. M. Asiri and H. B. Muralidhara, Iron-based flow batteries to store renewable energies, *Environmental Chemistry Letters* 16:683–694, 2018. ISSN 1610-3653.

10. V. Y. Glushkov, *Chto takoe OGAS? (What is OGAS?)*. Nauka, Moscow, 1981.
11. M. V. Katalymov, *Mikroelementy i mikroudobreniya (Microelements and Fertilizers)*. Khimia, Moscow, 1965.
12. C. T. Lewis and C. Short, *A Latin Dictionary*. Clarendon Press, Oxford, 1879.
13. H. R. Maturana and F. J. Varela, *Autopoiesis and Cognition: The Realization of the Living*. Springer, Boston and London, 1980.
14. V. L. Mindovskiy, *"Zelenoye Stroitelstvo" ("Green Bulding")*. Molotovgiz, Berezniki-Molotov, 1951.
15. E. Nikulina, A. Makarova, V. Meshalkin, V. Chelnokov, A. Matasov, and T. Avdeenkova, Integrated chemo-phyto-ecological process for the treatment of polymetal contamination in landfill sites and the consequent soil recovery, *Process Saf Environ*, 152:719–729, 2021. ISSN 1744-3598.
16. J. Noack, N. Roznyatovskaya, T. Herr, and P. Fisher, The Chemistry of Redox-Flow Batteries. *Angew. Chem. Int. Ed.* 54(34):9776–9809, 2015. ISSN 1433-7851.
17. Ovid, *Fasti*, translated by J. G. Frazer. W. Heinemann, London, 1931.
18. M. M. Petrov, A. D. Modestov, D. V. Konev, A. E. Antipov, P. A. Loktionov, R. D. Pichugov, N. V. Kartashova, A. T. Glazkov, L. Z. Abunaeva and V. N. Andreev, Redox flow batteries: Role in modern electric power industry and comparative characteristics of the main types. *Russ. Chem. Rev.* 90(6):677, 2021. ISSN 0036-021X.
19. B. Schulz, *Traktat o manekinach albo wtóra Księga Rodzaju. In idem Sklepy cynamonowe (The Cinnamon Shops)*. Wydawnictwo MG, Cracow, 2013, 83–95.
20. A. R. Sibirkina, S. F. Likhachev, and A. V. Kocherov, Biogeochemical Properties of Vanadium and Beryllium in the "Pinery Sands–Herbaceous Plants" Semipalatinsk Irtysh Forest of the Republic of Kazakhstan, Proceedings of Voronezh State University. Series: Chemistry. *Biology Pharmacy*, 1:141–151, 2018. ISSN 1609-0675.
21. L. Tian, J. Yang, C. Alewell, and J.-H. Huang, Speciation of vanadium in Chinese cabbage (*Brassica rapa L.*) and soils in response to different levels of vanadium in soils and cabbage growth. *Chemosphere* 111:89, 2014. ISSN 1879-1298.
22. V. I. Lenin. *Sochineniya. Tom 40. (Complete Works, Vol. 40)*. Izdatelstvo politicheskoy literatury, Moscow, 1974.
23. Aristotle. *Nicomachean Ethics*, translated by D. P. Chase. George Routledge, London, 1910.

https://doi.org/10.1142/9789811257155_0026

Chapter 26

Fungal Gray Matter

Andrew Adamatzky[*,‡], Irina Petrova[*], and Antoni Gandia[†]

[*] *Unconventional Computing Lab,*
University of the West of England, Bristol, UK
[†] *Institute for Plant Molecular and Cell Biology,*
CSIC-UPV, Valencia, Spain
[‡] *andrew.adamatzky@uwe.ac.uk*

Fungi are creatures with remarkably pronounced protocognition abilities. They reportedly control "thinking" of trees, and can produce mind-altering experiences in humans. They help us to live in the world and to see the invisible. Recently, we discovered that the electrical activity of fungi is similar to neurons. Fungi communicate with trains of spikes of electrical potential. Fungi respond to stimulation by changing the electrical properties and patterns of their electrical activity. We briefly overview our discoveries on sensing and computing with fungi.

1. Introduction

Fungi may have been the first creatures which arrived on our planet and constitute a pivotal hub between the food webs of the ecosphere. They are considered creatures of magic and wonder (Figure 1). They populate a thin layer of soil, just under the surface, and implement chemical, and possibly electrical communication between trees and plants. There is a chance that when sending a message from one tree to another, fungi do actually alter the meaning of the messages, thus controlling "thinking" of trees and ultimately governing the Mind of the Forest. To some degree fungi also shaped what is known as the *Mind of the Noosphere*. From the beginning of civilizations fungi have been an essential component of the spiritual ceremonies, rituals,

Fig. 1. *Deep Down the Rabbit Hole*. Installation with *Amanita pantherina* fungi. Direct Print On Aluminium Backing, 40 × 60 cm. © Irina Petrova. www.irina-petrova.com

community building events, healing, mind opening inspirational trips, and mental healing. No one will ever forget their first trip with entheogenic mushrooms and will always be grateful to the mushrooms for showing the unseeable realms of consciousness. Here we discuss how electro-physiological properties of fungi can be used in sensing and information, in unconventional computing.

A vibrant field of unconventional computing aims to employ space-time dynamics of physical, chemical, and biological media to design novel computational techniques, architectures, and working prototypes of embedded computing substrates and devices. Interaction-based computing devices are one of the most diverse and promising families of unconventional computing structures.[1] They are based on interactions of fluid streams, signals propagating along conductors, or excitation wave-fronts. Typically, logical gates and their cascade implemented in an excitable medium are "handcrafted" to address exact timing and type of interactions between colliding wave-fronts. The artificial design of logical circuits might be suitable when chemical media or functional materials are used. However, the approach might not be feasible when embedding computation in living systems, where the architecture of conductive pathways may be difficult to alter or control. During the last decade we produced nearly 40 prototypes of sensing and computing devices from the slime mould *Physarum polycephalum*, including shortest path finders, computational geometry processors, hybrid electronic devices.[2] We found that the slime mould is a convenient substrate for unconventional computing, however geometry of the slime mould's protoplasmic networks is continuously changing, thus preventing fabrication of long-living devices, and the slime mould computing devices are confined to experimental laboratory setups. Fungi Basidiomycetes are now taxonomically distinct from the slime mould, however their development and behavior are phenomenologically similar: mycelium networks are analogous to the slime mould's protoplasmic networks, and the fruit bodies are analogous to the slime mould's stalks of sporangia. Basidiomycetes are less susceptible to infections when cultured indoors, especially the commercially available species, they are larger in size and more convenient to manipulate than slime mould, and they could be easily found and experimented with outdoors. This makes the fungi an ideal object for developing future living computing devices (Figure 2). Availability and scalability of fungi is yet another advantage. The fungi is the largest, widely distributed and the oldest group of living organisms. Smallest fungi are microscopic single cells. The largest fungi, *Armillaria bulbosa*, occupies 15 hectares and weighs 10 tons[3] and the largest fruit body belongs to *Fomitiporia ellipsoidea* which at 20 years old is 11 m long, 80 cm wide, 5 cm thick and has an estimated weight of nearly half-a-ton.[4]

2. Spiking Fungi

Not only neurons spike. Action potential-like spikes of electrical potential have been discovered using intracellular recording of mycelium of *Neurospora crassa*[5] and further confirmed in intracellular

Fig. 2. Fungi *P. ostreatus* explores space while geometrically constrained.

recordings of action potential in hypha of *Pleurotus ostreatus* and *Armillaria bulbosa*[6] and in extracellular recordings of fruit bodies of and substrates colonized by mycelium of *Pleurotus ostreatus*.[7] While the exact nature of the traveling spikes remains uncertain we can speculate, by drawing analogies with oscillations of electrical potential of slime mould,[2] that the spikes in fungi are triggered by calcium waves, reversing of cytoplasmic flow, translocation of

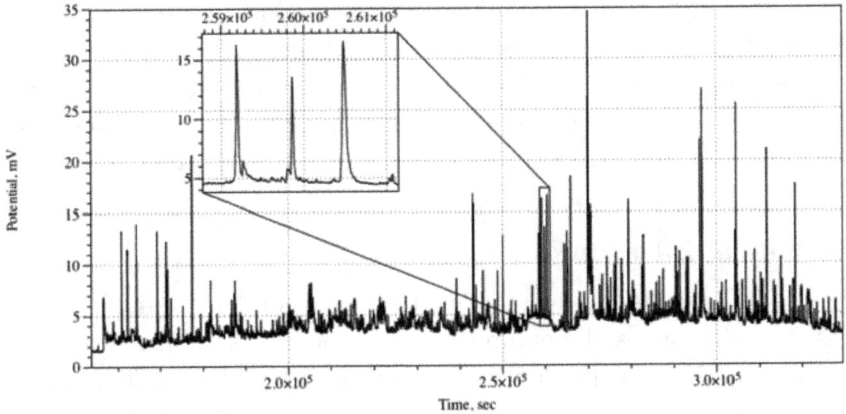

Fig. 3. Example of electrical spiking activity recorded from a hemp substrate colonized by mycelium of *P. ostreatus.*

nutrients and metabolites (Figure. 3). Studies of electrical activity of higher plants can bring us even more clues.[8] Thus, the plants use the electrical spikes for long-distance communication aimed to coordinate activities of their bodies. The spikes of electrical potential in plants relate to a motor activity, responses to changes in temperature, osmotic environment, and mechanical stimulation. In experiments with *Pleurotus ostreatus* 5, we demonstrated that fruit bodies of oyster fungi exhibit trains of action-like spikes of extracellularly recorded electrical potential. We observed two types of spikes: high-frequency spikes, duration nearly 3 min, and low-frequency spikes, duration nearly 14 min. The spikes are observed in trains of 10–30 spikes. The depolarization and repolarization rates of both types of spikes are the same. Refractory period of a high-frequency spike is one-sixth of the spike's period, and for a low-frequency spike, one-third of the spike's period. We showed that fruit bodies respond with spikes of electrical potential in response to physical, chemical, and thermal stimulation; not only a does simulated body respond with a spike but other fruit bodies of the cluster respond as well. We believe the spikes of electrical potential traveling in mycelium networks play the same roles of information carriers as action potential traveling along neural pathways in, e.g., human brains.

Thus it would be advantageous to discover what types of information processing devices we could make from fungi. To make information processing devices from fungi, we can either use fungi as electronic components of analog computers or employ internal dynamics of excitation in fungi directly. Both options are illustrated below.

3. Fungal Electronics

Fungi behave as memristors. The memristor is a device whose resistance changes depending on the polarity and magnitude of a voltage applied to the device's terminals. A memristor is a material logical implication. Therefore, any logical circuits can be made purely from the memristors. We have examined the conducted current for a given voltage applied as a function of the previous voltage to fungi fruiting bodies and substrates colonized by fungi and showed that the fungi exhibit remarkable memristive properties.[10] Indeed, demonstrating that a fruit body or a substrate colonized by fungi exhibits memristive properties is just a tiny step forward: cascading fungal memristors into function circuits will be the challenging task.

Despite being logically universal, memristors might not be enough to build fully functional computing circuits. We might need to store energy, implement digital memory, do signal coupling and decoupling, make high-pass and low-pass filters, suppress noise. All these can be done with capacitors. Fungi are capacitors, albeit rapidly discharging. In laboratory experiments we showed that the capacitance of mycelium is in the order of hundreds of pico-Farads and the charge density of the mycelium decays rapidly with increasing distance from the source probes.[10] Nevertheless, the mycelium can be used as part of a capacitors array. A fungal electronic circuit could have chemical, mechanical, or optical inputs.

A range of fungal responses to stimulation have been discovered in our experiments with fungal skin[9] — a thin flexible sheet of a living homogeneous mycelium made by a filamentous fungus. We demonstrated that a thin sheet of homogeneous living mycelium of *Ganoderma resinaceum* shows pronounced electrical responses to

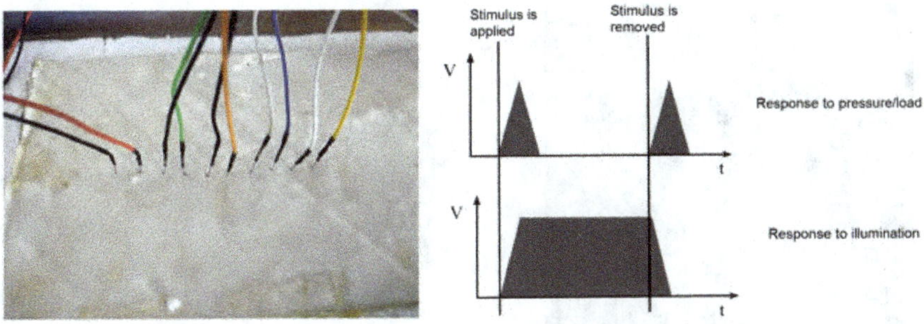

Fig. 4. Responsive fungal skin. Left, pairs of differential electrodes inserted in the fungal skin of *G. resinaceum*. Right, response of the fungal skin to pressure and illumination.[9]

mechanical and optical stimulation (Figure 4). It is possible to differentiate between the fungal skin's response to mechanical and optical stimulation. The fungal skin responds to mechanical stimulation with a 15 min spike of electrical potential, which diminishes even if the applied pressure on the skin remains. The skin responds to optical stimulation by raising its electrical potential and keeping it raised till the light is switched off. We can even differentiate the responses to loading and removal of the weight. While amplitudes of "loading" and "removal" of spikes are the same (0.4 mV in average), the fungal skin average reaction time to removal of the weight is 2.4 times shorter than the reaction to loading of the weight (385 s vs. 911 s). Also "loading" spikes are 1.6 times wider than "removal" spikes (1261 s vs. 774 s).

4. Fungal Computing

Mycelium networks are disorganized from the point of view of a computation architect and difficult to program at a fine-grained level. Thus, direct design of computing circuits might be impossible at this point of development. In such situations an opportunistic approach to outsourcing computation can be adopted (Figure 5). The system is perturbed via two or more input loci and its dynamics is recorded at one or more output loci. A spike appearing at one of

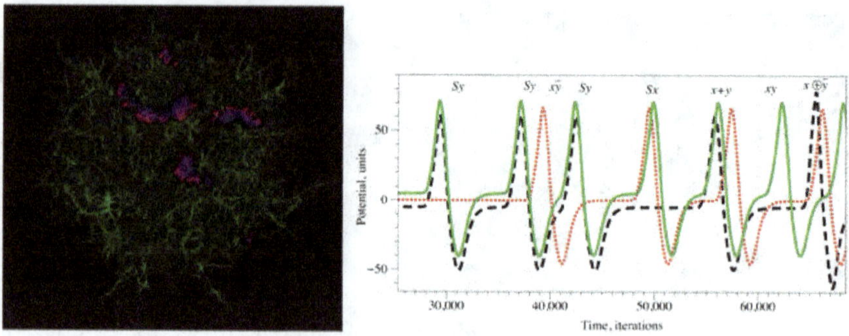

Fig. 5. Computing with traveling spikes in mycelium networks. Left, a snapshot of excitation waves propagating in the network. Right, encoding of the spiking response to Boolean gates.

the output loci is interpreted as logical Truth or "1" and absence of the spike as logical False or "0". Thus, a system with relatively unknown structure implements a mapping $\{0,1\}^n \rightarrow \{0,1\}^m$, where n is the number of input loci and m is the number of output loci, $n, m > 0$. Using numerical modeling of excitation wave-fronts propagating on images of real colony of *Aspergillus niger*, we have demonstrated how sets of logical gates can be implemented in single colony mycelium networks via initiation of electrical impulses.[9] The impulses travel in the network, interact with each other (annihilate, reflect, change their phase). Thus, for different combinations of input impulses and recording different combinations of output impulses, which in some cases can be interpreted as representing two-inputs-one-output functions, to estimate a speed of computation we refer to Olsson and Hansson's original study,[6] in which they proposed that electrical activity in fungi could be used for communication with message propagation speed 0.5 mm/s. Diameter of the colony, for which experimental laboratory images have been used to run the model, is c. 1.7 mm. Thus, it takes the excitation waves initiated at a boundary of the colony up to 3–4 s to span the whole mycelium network (this time is equivalent to c. 70 K iterations of the numerical integration model). In 3–4 s, the mycelium network can compute up to a hundred logical gates. This gives us the rate of a gate per 0.03 s, or, in terms of frequency this will be c. 30 Hz.

5. Programmability

To program fungal computers we must control the geometry of mycelium networks. The geometry of a mycelium network can be modified by varying nutritional and environmental conditions, such as temperature and light, especially the degree of branching is proportional to concentration of nutrients, and a wide range of chemical and physical stimuli. Also, we can geometrically constrain it. A feasibility of shaping similar networks has been demonstrated by us previously: high amplitude high frequency voltage applied between two electrodes in a network of protoplasmic tubes of slime mould *P. polycephalum* leads to abandonment of the stimulated protoplasmic tube without affecting the non-stimulated tubes, and low amplitude low frequency voltage applied between two electrodes in the network enhances the stimulated tube and encourages abandonment of other tubes.

6. Application Domain: Distributed Networks of Ecological Sensors

Likely application domains of the fungal devices could be large-scale networks of mycelium which collect and analyze information about the environment of soil, and possibly air, sending information to a collection network and executing some decision-making procedures. Fungi sense light, liquid chemicals and gases, gravity and electric fields. Fungi show a pronounced response to changes in a substrate pH, demonstrate mechanosensing; they sense toxic metals, CO_2, and direction of fluid flow. Fungi exhibit thigmotactic and thigmomorphogenetic responses, which might be reflected in dynamic patterns of their electrical activity. Fungi are also capable of sensing chemical cues, especially stress hormones, from other species, thus they might be used as reporters of health and well-being of other inhabitants of the forest. Thus, fungal computers can be made an essential part of distributed large-scale environmental sensor networks in ecological research to assess not just soil quality but an over health of the ecosystems. Interaction of voltage spikes, traveling along mycelium strands, at the junctions between strands is a key mechanism of

Fig. 6. *Post-Industrial Era.* Installation with *Amanita pantherina* fungi. Direct Print On Aluminium Backing, 60 × 40 cm. © Irina Petrova. www.irina-petrova.com.

the fungal computation. We can see each junction as an elementary processor of a distributed multiprocessor computing network. We assume the number of junctions are proportional to the number of hyphal tips. There are estimated 10–20 tips per 1.5–3 mm^3 of a substrate. Without knowing the depth of the mycelial network we go for a safest lower margin of 2D estimation: 50 tips/mm^2. Considering that the largest known fungus *Armillaria bulbosa* populates over 15 hectares, we could assume that there could be $75{\cdot}10^{11}$ branching points, that is nearly a trillion of elementary processing units. With regards to a speed of computation by fungal computers, electrical activity in fungi could be used for communication with message propagation speed 0.5 mm/s (this is several orders slower than speed of a typical action potential in plants: from 0.005 m/s to 0.2 m/s). Thus, it would take about half-an-hour for a signal in the fungal computer to propagate one meter. The low speed of signal propagation is not a critical disadvantage of potential fungal computers, because they never meant to compete with conventional silicon devices. The mycelium network computing cannot compete with existing silicon architecture, however its application domain can be a unique combination of living biosensors (a distribution of gates realized might be affected by environmental conditions) and computation embedded into structural elements where fungal materials are used[10] and fungal wearables.[11]

Acknowledgment

AA has received funding from the European Union's Horizon 2020 research and innovation programme FET OPEN "Challenging current thinking" under grant agreement No. 858132.

References

1. A. Adamatzky, *Advances in Unconventional Computing*. Springer, 2016.
2. A. Adamatzky, (editor), *Advances in Physarum Machines: Sensing and Computing with Slime Mould*. Springer, 2016.
3. M. L. Smith, J. N. Bruhn, and J. B. Anderson, The fungus Armillaria bulbosa is among the largest and oldest living organisms. *Nature* 356(6368):428, 1992.
4. Y.-C. Dai and B.-K. Cui, Fomitiporia ellipsoidea has the largest fruiting body among the fungi. *Fungal Biol.* 115(9):813–814, 2011.
5. C. L. Slayman, W. S. Long, and D. Gradmann, "Action potentials" in Neurospora crassa, a mycelial fungus. *Biochimica et Biophysica Acta (BBA)-Biomembranes* 426(4):732–744, 1976.
6. S. Olsson and B. Hansson, Action potential-like activity found in fungal mycelia is sensitive to stimulation. *Naturwissenschaften* 82(1):30–31, 1995.
7. A. Adamatzky, On spiking behaviour of oyster fungi Pleurotus djamor. *Sci. Rep.* 7873, 2018.
8. J. Fromm and S. Lautner, Electrical signals and their physiological significance in plants. *Plant, Cell Environ.* 30(3):249–257, 2007.
9. A. Adamatzky, E. Goles, G. J. Martinez, M.-A. Tsompanas, M. Tegelaar, and H. A. Wosten, Fungal automata, arXiv preprint arXiv:2003.08168, 2020.
10. A. Adamatzky, P. Ayres, G. Belotti, and H. Wösten, Fungal architecture position paper. *Int. J. Unconv. Comput.* 14, 2019.
11. A. Adamatzky, A. Nikolaidou, A. Gandia, A. Chiolerio, and M. M. Dehshib, Reactive fungal wearable, arXiv preprint arXiv:2009.05670, 2020.

https://doi.org/10.1142/9789811257155_0027

Chapter 27

The Chemical Machine

Preety Anand* and Grace Chung†
Architectural Association, School of Architecture
**preety_83@hotmail.com*
†graceland1125@gmail.com

1. Introduction

The Chemical Machine transgresses the notion of permanence of architecture and linear lifespan of its composing material as well as the centralized mode of control of building systems. The chemical machine, a cybernetic building system, is a space generator which transgresses the familiar models of space creation and autonomous regulation of micro and macro environments granted by chemically led information processing.

The Chemical Machine departs from an architectural critique of building services as accumulative systems without its spatial or architectural implications considered. Failing of a few subsystems has often resulted in the dysfunction of the whole. This is where the Chemical Machine differs as it incorporates what is categorized to be building services with space generation into a seamless whole. Rather than layered additions of separate systems over one another, the chemical Machine attempts to integrate technology, human needs, and environmental concerns in an architectural context.

This research Fig. 1 investigates an adaptive biological system based on environmental information processing for the purpose of self-regulation that prevents from systemic failure — the Physarum Polycephalum (Slime mold). These features form the base of adaptive

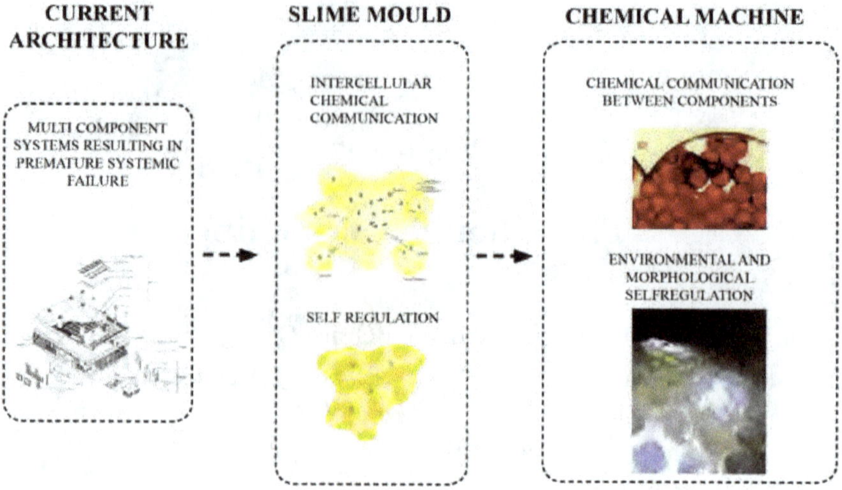

Fig. 1. Anand, P., Chung, G, *et al.* The Chemical Machine, 2013, illustration. Architectural Association, London.

building tectonic systems, which also entail critical re-evaluation of the role of the architect and their conventional attitude toward matter.

For demonstrating the mode of operation, the Chemical Machine is set in extreme, yet generative environments. In this context, the Chemical Machine fulfills a technological need for offering a non-human centric mode of inhabitation, which is given by its mode of operation.

The function of the chemical machine is not only that of generic architecture, but also a life support for both humans and the biology their harbor. The technicalities of this chapter have been developed by inter-scientific collaboration with computer scientists, material scientists, and chemical engineers. The project is situated within the emerging fields of Chemical Computation, Material Sciences, and cybernetics based on architecture-free computation.

> "Architects are suggesting how artificial machine-driven systems work around them, and are beginning to build spaces that can be constantly modified and adjusted."[1]

2. Architectural Motivation

The Chemical Machine is a critique of current multi-component building systems Fig. 2, which are based on separately fabricated parts with unsynchronized life spans. Their incapability to internally communicate and self-regulate prevents synchrony of individual life spans for the sake of overcoming premature systemic failures. The absence of communicative capabilities also implies a mode of centralized control and results in the absence of interaction with the user or the environment. Control in architecture as we know it is a result of a cause creating an effect, it is linear and almost never reversible. This linear system is translated into a static architectural outcome, which is given a certain time frame for its functional duration. In this model, all the irregularities that architecture faces, whether it

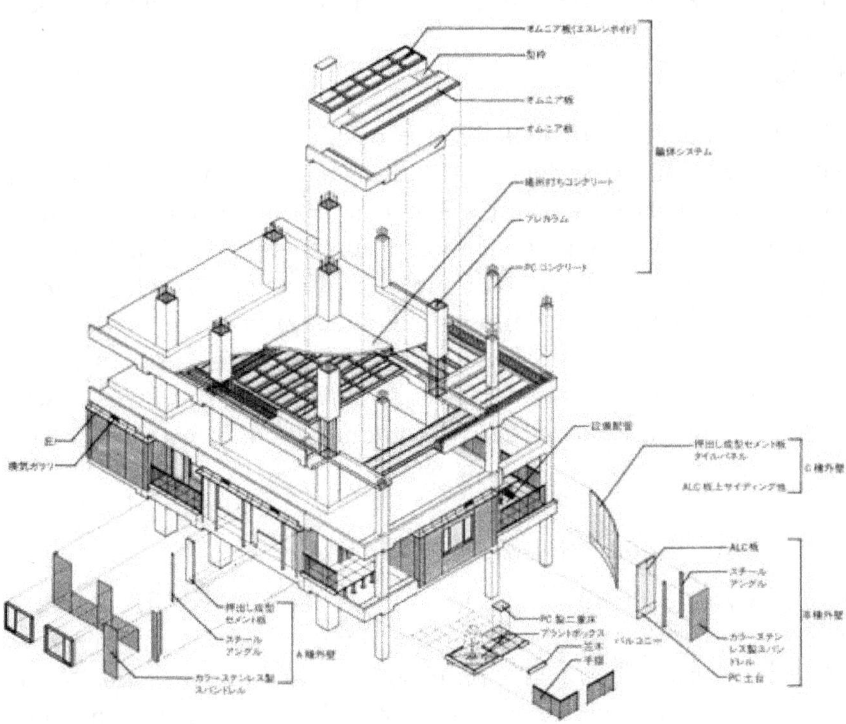

Fig. 2. Shu Koh Sha Architectural and Urban Design Studio. *Building System Diagram of NEXT21.1994*, illustration. Osaka, London.

be environmental or by inhabitants themselves, is computed by a central system and majority of the computational load is taken on by the center.

A precedent as well as an anti-model for our thesis has been Rayner Banhams's "Well-Tempered Environment". Banham discusses the issue of controlled interior environment of buildings with a vast array of machine-driven control systems which almost render the architecture as a carcass, an almost non-necessity in assistance of human inhabitation.[2]

Our project is also a response to the issue of buildings consisting of separate and unrelated mechanical systems, each with a singular or few functions to fulfill, all working without any orchestrated inter-systemic communication. Therefore, what needs to be translated into architecture are systemic models that are capable of functioning multiple tasks in autocatalytic perpetuation and parallel processing.

3. Research

The failing performance of highly specified and uncooperative disparate parts made us look at holistic and cooperative (biological) models, which are based on a small number of chemically reactive systemic members with an underspecified functionality. The investigation focused on adaptive biological system of self-regulation, decentralized control, and intercellular communication between parts enabling interaction with its environment. These systems excel in terms of a high fault tolerance and systemic robustness, which is granted by internal chemical communication and resulting self-regulation.

3.1. *Slime mold: Physarum Polycephalum*

Physarum polycephalum, more commonly known as slime mold, has been selected as a subject of study in its mode of operation and control. Physarum Polycephalum is an acellular slime mold which means despite appearances it is a single bag of fluid with numerous nuclei; hence the name "multi-headed" is also applied to this type of species. Within its life-cycle Figure 4 the plasmodium stage or

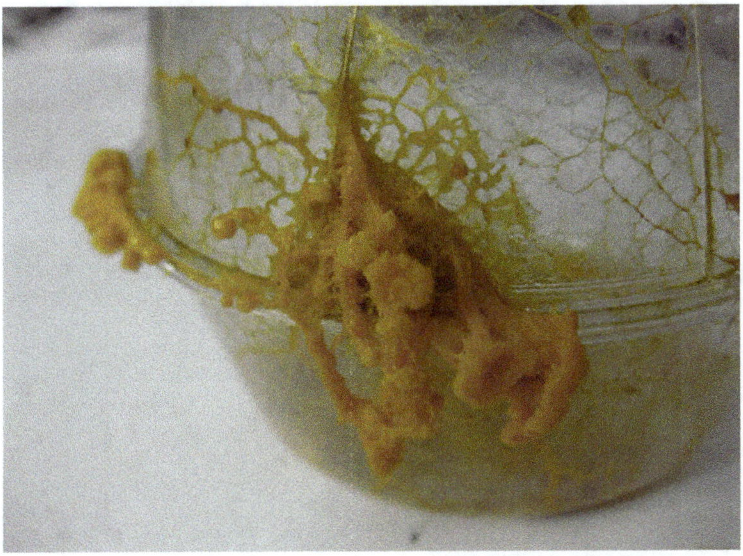

Fig. 3. Anand, P., Chung, G, *et al. Slime Mould lacking oxygen and nutrition*, 2013, photograph. Architectural Association, London.

the "vegetative" state of the organism exhibits behaviors that are described to be "intelligent" Figure 3 . During the plasmodium period it appears to be an amorphous yellow mass that develops a network of protoplasmic tubes to efficiently connect itself with several sources of nutrients and hydration. The morphology of the slime mold is highly environmentally driven and is computed in result of environmental interaction Figure 3.

The Chemical Machine learns from this adaptive biological system of self-regulation and intercellular communication between parts. This feature allows constant morphological computation for responding to ever changing environmental conditions and thus ensures the survival of the system. Chemical signaling allowing morphological computation as a form of self-regulation enables a high fault tolerance and systemic robustness.

Our research aims to situate the adaptive and information processing capabilities of the slime mold within an architectural context. The behavioral intelligence of the slime mold, including the communicative and regulative mechanisms, are mathematically

Fig. 4. Anand, P., Chung, G., *et al. End of life-cycle: Fruiting body of a slime mold*, 2013, photograph. Architectural Association, London.

expressed with the Two Variable Oregonator model. The biological modes of information processing can be retraced chemically, as the BZ Reaction and the slime mold share the same systemic logic. Therefore, we are implementing chemistry as a synthetic means within matter to embed slime mold intelligence into architecture.

The Chemical Machine learns from this adaptive biological system of self-regulation and intercellular communication between parts. This feature allows constant morphological computation for responding to ever changing environmental conditions and thus ensures the survival of the system. Chemical signaling allowing morphological computation as a form of self-regulation enables high fault tolerance and systemic robustness.

Our research aims to situate the adaptive and information processing capabilities of the slime mold within an architectural context. The behavioral intelligence of the slime mold, including the communicative and regulative mechanisms, are mathematically expressed with the Two Variable Oregonator model. The biological modes of information processing can be retraced chemically, as the BZ Reaction and the slime mold share the same systemic logic. Therefore, we are implementing chemistry as a synthetic means within matter to embed slime mold intelligence into architecture.

3.2. *Belousov-Zhabotinsky Reaction*

The BZ reaction is a chemical reaction which is a causal model for complex biological systems with coupled feedback loops. Despite the difference in constitution and composition comparative to its biological archetype, the reaction successfully produces phenomenological similarities Figure 5.

In biology feedback loops are often observed in communication and internal self-regulatory processes (of cells), and occur at indiscriminate levels of scale. The limited capabilities of the conventional digital mode of simulation are now beginning to be recognized, and this opens the possibilities of a direct jump from chemical computation to material actuation.

On the surface of the solution, patterns observable in nature and variations of the Turing pattern emerge. These patterns are manifested as a result of several operations running simultaneously — reaction and diffusion. This results in a system in which the concentrations of reactants and products oscillate temporally and spatially and in which this oscillation can result in ordered patterns.[4]

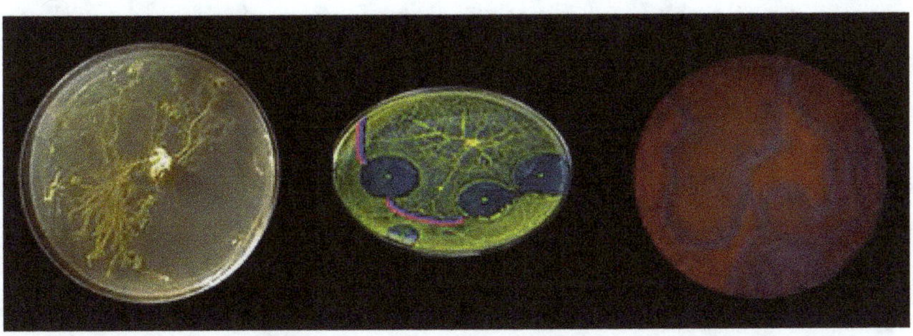

Fig. 5. Anand, P., Chung, G., *et al. Slime Mold — Belousov Zhabotinsky reaction,* 2013, photograph Architectural Association, London.
Source: Adamatzky (2010).[3]

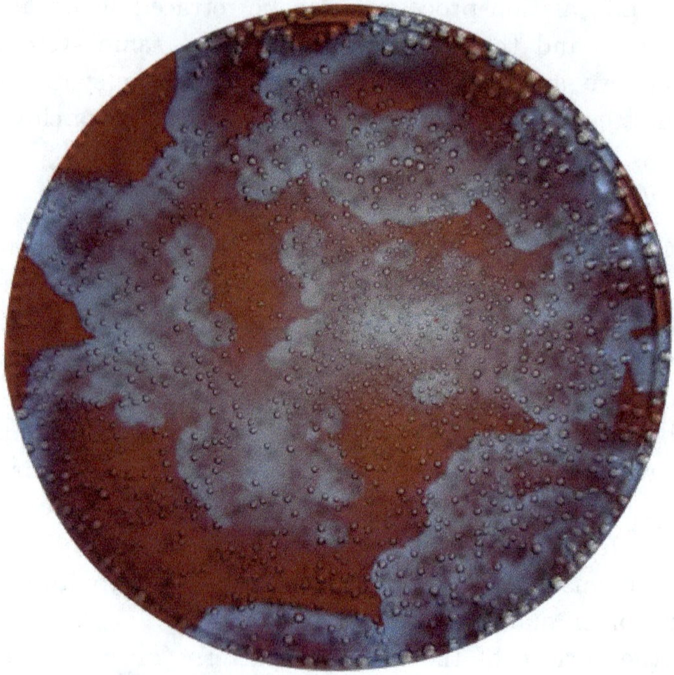

Fig. 6. Anand, P., Chung, G., *et al. Belousov Zhabotinsky Petri dish test*, 2013, photograph. Architectural Association, London.

These phenomena are based on perturbation of the homogeneous chemical system. The BZ Fig. 6 is a highly sensitive system which reacts to a wide range of perturbations from an intended mechanical perturbation to a dust particle in air settling on the reagent's surface.

The reaction is also responsive to a range magnitude of perturbation — and this is also a parameter for design implementation, as environmental, climatic, and inhabitant-based perturbations will approach our project in many different forms and magnitudes. The environmental implication on the BZ reaction has been simulated physically through the prototype, in order to emulate the possible material implication of the BZ encapsulated within hydrogel. There have already been attempts mainly in Japan and the US to directly materialize BZ reaction behaviors into "self-oscillating polymer gels", which are amorphous gels that engage in some characteristic behaviors of living organisms.

> "Polymeric hydro gels that exhibit autonomous, coupled chemical and mechanical oscillations are a unique example of synthetic, active soft matter."[5]

The precedents of such applications are currently limited to a small size, however with the prospects of possibly larger sized implications.

4. The Chemical Machine

We contextualize the mode of environmentally driven chemical control, the communicative and self-regulative capacities of the slime mold into the architectural context. This holistic approach to building tectonic systems enables a high fault tolerance and systemic robustness concluding in the proposal of The Chemical Machine. It is conceived as a cybernetic ecology capable of purposeful chemical and morphological regulation of the micro and macro environment by chemically led communication between parts and nonlinear feedback loops. The proposed system is sensitive to environmental conditions of light and carbon dioxide.

The chemically self-contained machine Fig. 7 comprises the environment, the inhabitant, microorganisms, environmental chemicals, and the BZ-components, whereas all are bound together by chemical inter-dependencies. It is a cybernetic ecology based on intercellular communication, signaling, and self-regulation. While the BZ-component-based habitat requires perturbations given by the chemical agencies and the environment in order to avoid the state of equilibrium, the chemical agencies need to be provided with oxygen and light in order to survive. In contrast to mechanically driven automation systems and sensors that clutter building services, the chemical machine proposes subtraction of all mechanical or deliberate control. In terms of architectural merit, architecture capable of self-regulation is capable of adapting to changing environmental conditions and occupancy rates.

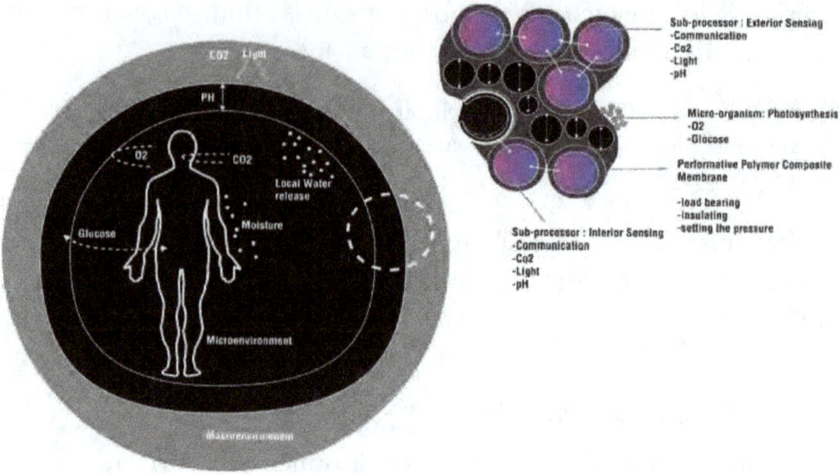

Fig. 7. Anand, P., Chung, G., *et al. The Chemical Machine,* 2013, illustration. Architectural Association, London.

4.1. *Design as a form of control*

The key design investigation aims to discover the morphology of control enabling the necessary environmental as well as morphological performances.

The approach to designing the physicality of the Chemical Machine is partly derived from the current research in the emerging fields of soft-robotics, which argues for compliance in design. This design intention requires a scientific instead of an intuitive design approach. In the context of the Chemical Machine, the conventional definition of architectural design in regards to a purely form-driven approach will no longer apply. For the Chemical Machine, design would be defined as material selection of individual components according to their intended performance and global arrangement. The Chemical Machine is not a mono-material system, as differentiated tissue can even be observed in simple biological systems as a necessity for performing diverse environmental and morphological tasks. Through time and continued repetition of environmental and habitual circumstances the Chemical Machine would also organically morph its design to better accommodate the inhabitants and adapt

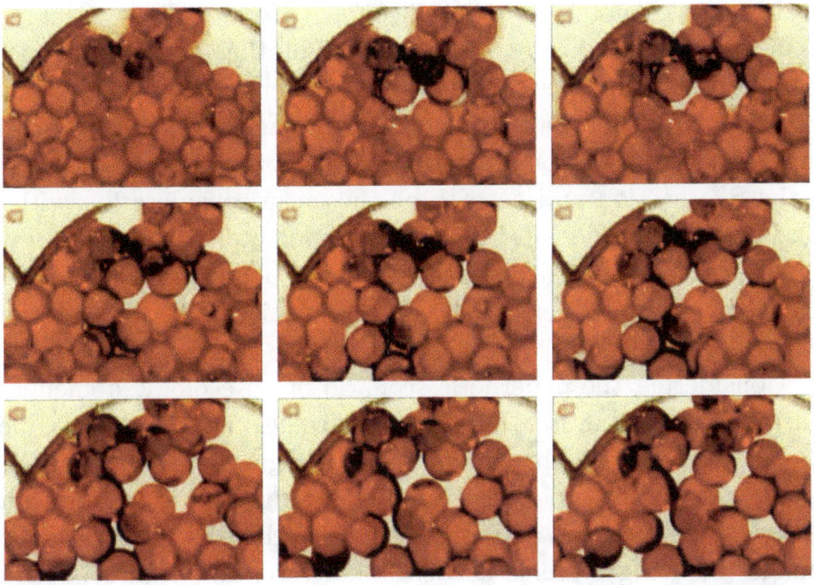

Fig. 8. Anand, P., Chung, G., *et al. BZ Components with ongoing chemical reaction*, 2013, photograph. Architectural Association, London.

to the ever changing environment throughout its own performative duration.

4.2. *Functionality of components*

In order to enable communication and inherent self-regulation within architecture, we fabricated components, which follow the functionality of the slime mold nuclei Figure 8 . Although there is no direct architectural translation, these components come closest to a building component. Each component is chemically independent and can be seen as the computing engine of The Chemical Machine.

4.3. *Materiality of components*

In order to fabricate synthetic nuclei, the self-perpetuating BZ chemical reaction has to be materialized to a more tangible solid matter. The synthetic nuclei are made of spherical silicone — hydrogel components, which absorb the BZ-chemical solution Figure 9 .

In terms of materiality, the silicone–hydrogel enables reversibility of morphological change. As the BZ Solution can be chemically preprogrammed, a specific environmental sensitivity such as to light and carbon-dioxide can be achieved at the same time. Chemical preprogramming can be achieved by the choice of reagents engaging in the BZ reaction and need to be chosen in consideration with the specific scenario and interrelated environmental necessities. Besides enabling chemical sensitivity, a even higher behavioral control can be achieved by embedding thresholds, e.g., excitability of the chemical reagents.

Considering the cybernetic scenario of an ecology, a chemically sensitive material system based on hydro-gels is appropriate due to their capability of being coupled to biological and chemical, e.g., human metabolic processes as well as to the BZ reaction. These materials can undergo reversible dynamic changes in accordance to changes in living systems as well. Therefore, materiality is capable of successfully executing self-regulative aspects of the Chemical Machine and does not have a secondary role.

The components of the Chemical Machine are able to receive, transmit, and process chemical signals in order to be engaged in self-regulation, which turns these into chemical transducers.

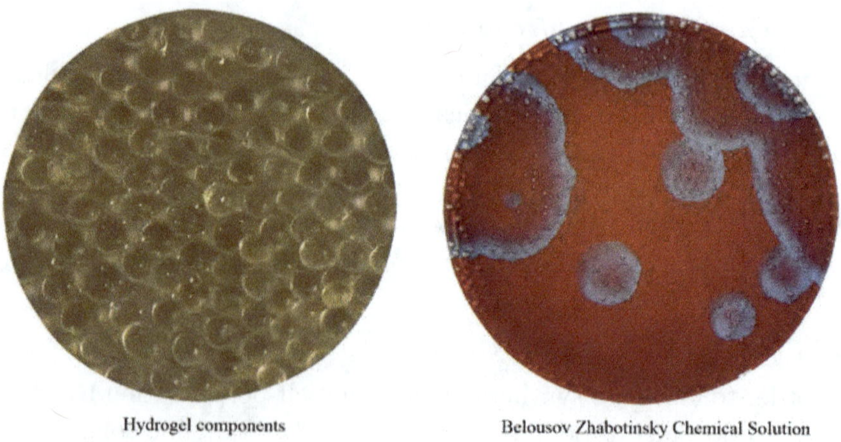

Fig. 9. Anand, P., Chung, G., *et al.* *Materiality of components*, 2013, photograph. Architectural Association, London.

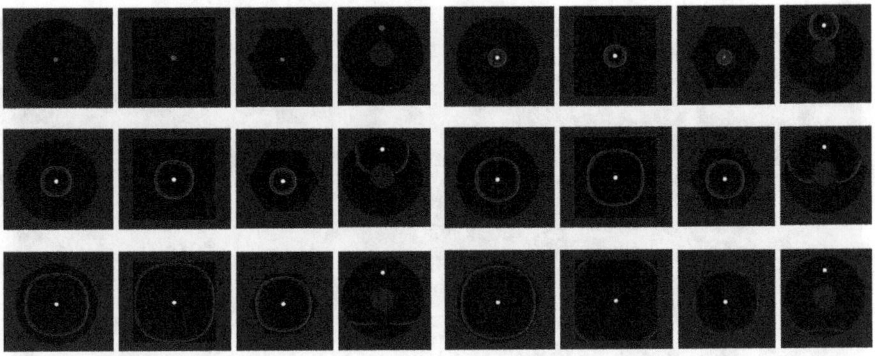

Fig. 10. Anand, P., Chung, G., *et al.* *Geometry of components*, 2013, photograph. Architectural Association, London.

Furthermore, silicone hydrogel is capable of absorbing the chemical reagents fully.

4.4. *Geometry of components*

As a result of digital simulations Figure 10 , the most suitable component geometry is spherical, as a wave hitting a sharp edge dissipates. Having spherical components also responds to the systemic need of compartmentalization. In this way, the spherical compartments serve the need of sustaining the propagating wave.

4.5. *Scale and productivity*

Digital simulations Figure 11 have proven that the most productive component diameter is 20 mm. As per the simulations, a decreasing size of the components results in a decreasing capacity to communicate among the components.

4.6. *Packing logic of components*

Digital simulations Figure 12 investigating on the packing logic of the spherical components have shown that a hexagonal packing is the most suitable form of arrangement. This allows the maximum number of neighbors Figure 13 meeting the inter-systemic need of communication while enabling the successful and nonlinear wave transmission.

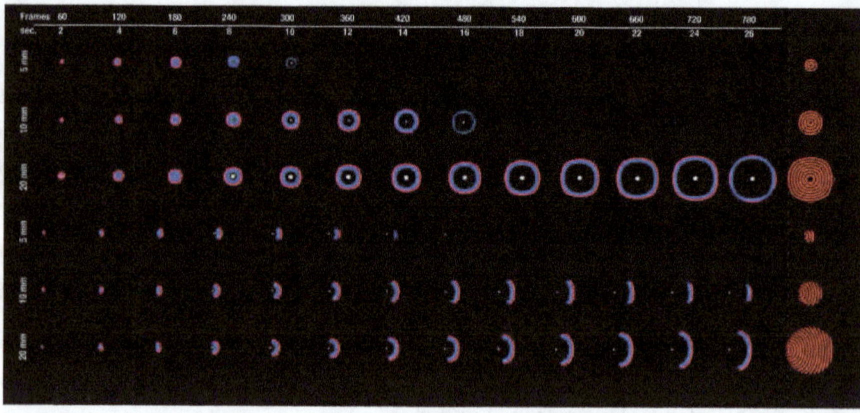

Fig. 11. Anand, P., Chung, G., *et al. Scale and productivity*, 2013, illustration. Architectural Association, London.

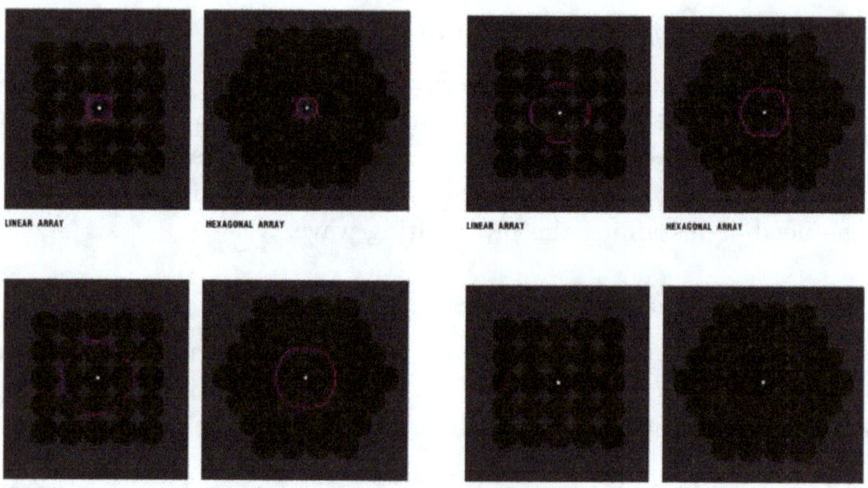

Fig. 12. Anand, P., Chung, G., *et al. Organizational logic of components*, 2013, illustration. Architectural Association, London.

4.7. *Encompassing membrane*

An encapsulating tensile membrane Fig. 14 prevents the premature dehydration of the components and the depletion of the chemical

Fig. 13. Anand, P., Chung, G., *et al. Maximum neighbors increase probability of wave transmission*, 2013, illustration. Architectural Association, London.

Fig. 14. Anand, P., Chung, G., *et al. Membrane with pores*, 2013, photograph. Architectural Association, London.

reagents. The membrane with its carefully distributed pores enables perturbation by environmental information and decentralized control.

The overall organization of the BZ components, the exterior skin, and the carefully positioned perturbation points allowing data exchange with the environment result in a prototypical material system Fig. 15 where organizational structure and mode of operation are interlinked.

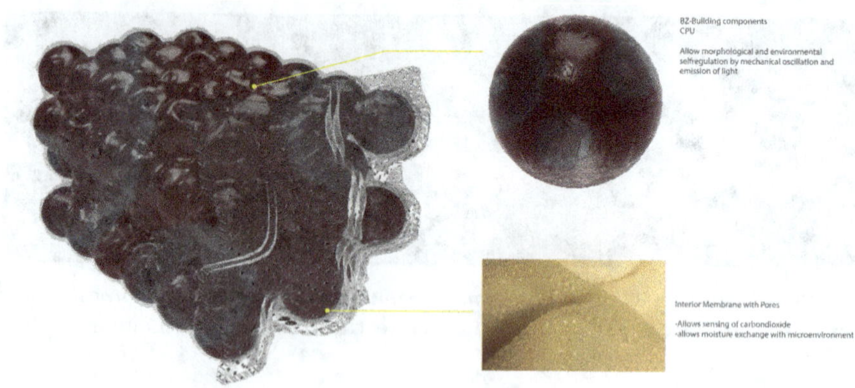

Fig. 15. Anand, P., Chung, G., *et al. Prototypical material organization,* 2013, illustration. Architectural Association, London.

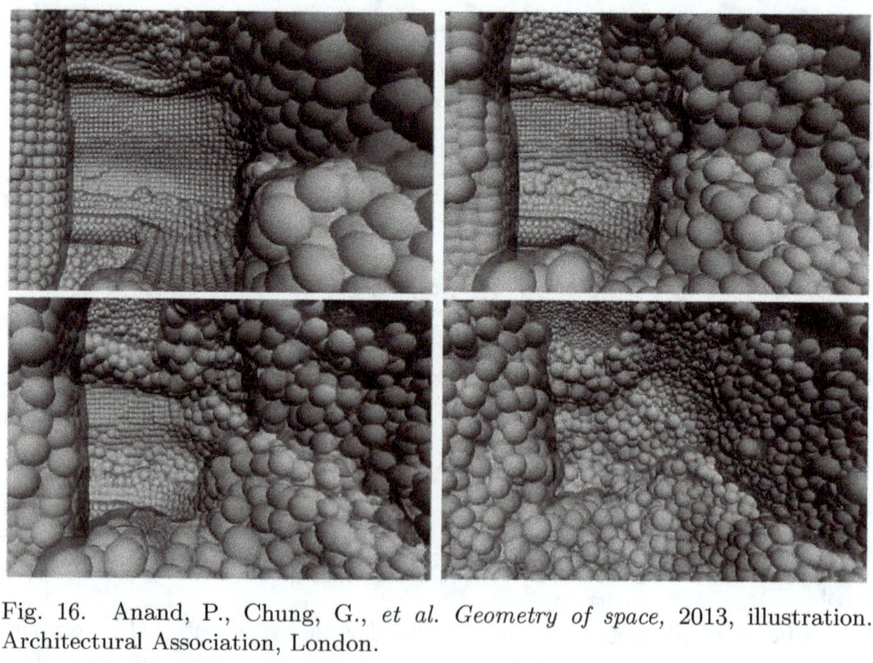

Fig. 16. Anand, P., Chung, G., *et al. Geometry of space,* 2013, illustration. Architectural Association, London.

4.8. *Geometry of space*

Besides matter distribution, spatial geometry Fig. 16 has a heavy impact on the capability of the system to pass information. Digital

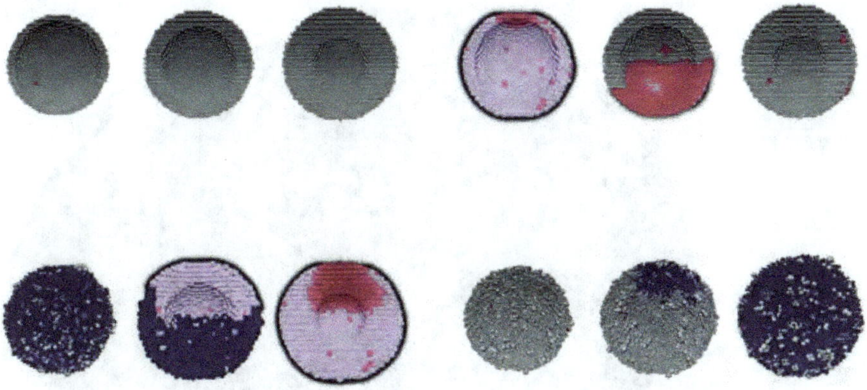

Fig. 17. Anand, P., Chung, G., *et al. Matter distribution and interlinked productivity,* 2013, illustration. Architectural Association, London.

simulations show that the most successful geometry of space is spherical. Chemical computation and communication inform form.

4.9. *Matter distribution*

Digital studies Fig. 17 have shown that the systemic capability of environmental and morphological self-regulation is closely related to the strategic quantitative distribution of the components. In this way, a thicker initial setup results in the wave propagation to be slow, but more stable, whereas a thinner section can be perturbed more easily and the wave propagates faster. Summing up, a larger component aggregation is more productive in terms of morphological and environmental self-regulation.

4.10. *Prototypical space*

As the chemical machine might have a higher performative need for the macro-environment, we need to strategically accumulate a larger share of BZ-components in the exterior facing part of the spherical spatial geometry.

The internal space Figs. 18–20 is generated by pressurization, whereas the innermost layer is built by a porous tensile membrane.

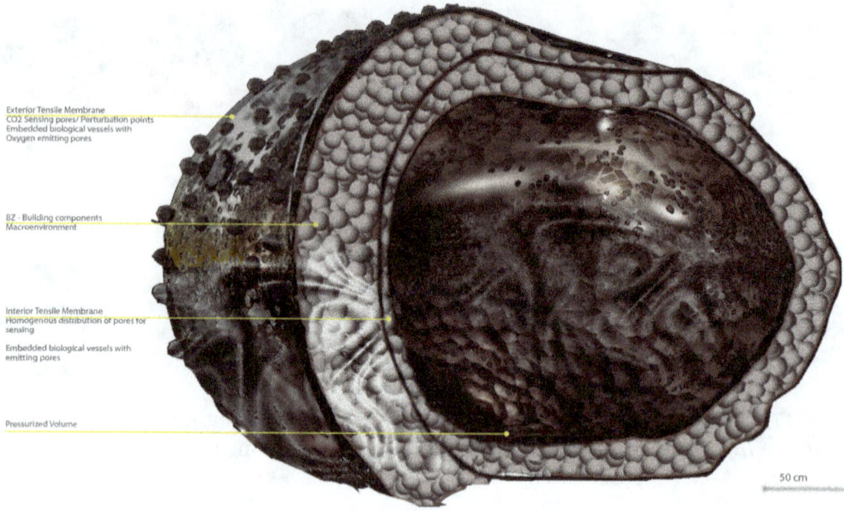

Fig. 18. Anand, P., Chung, G., *et al. Prototypical space*, 2013, illustration. Architectural Association, London.

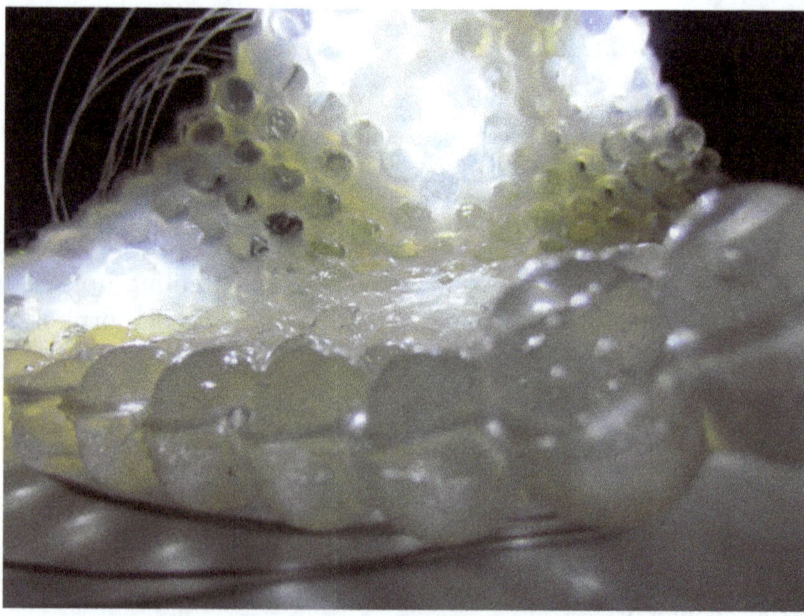

Fig. 19. Anand, P., Chung, G., *et al. Spatial speculation*, 2013, illustration. Architectural Association, London.

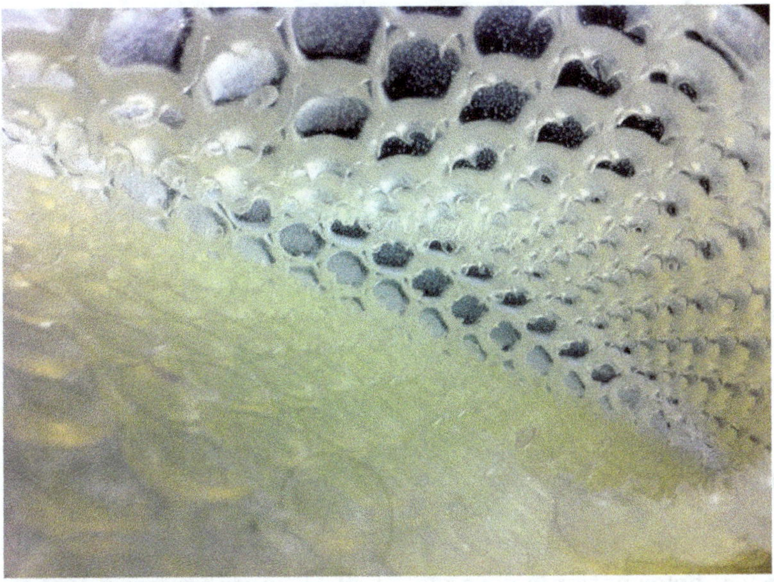

Fig. 20. Anand, P., Chung, G., *et al. Spatial speculation*, 2013, illustration. Architectural Association, London.

4.11. *Organizational logic of space*

Similarly to the arrangement logic of components, which responds to the design needs of chemical computation, the arrangement logic of individual spaces also needs to follow the logic of maximum neighbors Fig. 21. This packing logic globally results in cluster formation of spaces allowing an overall compact volume. This is how chemical computation informs form and design. We have attempted to link the observations made during our petri-dish scale experiments as well as digital simulations to design rules applicable to the larger scale.

4.12. *Architectural speculation*

There are three major determining factors that affect the physicality of the Chemical Machine beyond the stages of initial setup: Chemical perturbation, environmental perturbation, and neighborhood relationships. The local physical change will gradually lead to a much larger global change. Similarities can be drawn between the Chemical

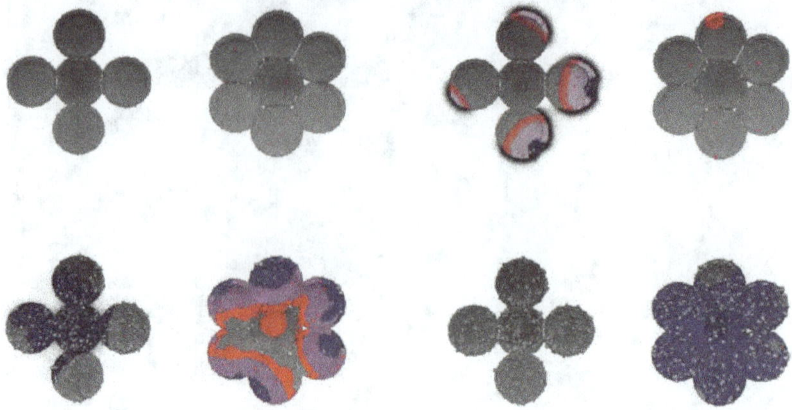

Fig. 21. Anand, P., Chung, G., *et al. Organizational logic of space*, 2013, illustration. Architectural Association, London.

Machine and the Physarum where local sensing and its sequential physical deformation have a global effect on its organism. In order to enable these transformations, numerous non-fixed routes of parallel sensing and computation are required.

The digital realm offers to examine architectural speculations by the use of simulations. On a more local level, the cyclic material behavior of swelling and de-swelling, which follows the chemical oscillation of the BZ solution, can be emulated. Considering the larger scale, scenario-based behavioral trajectories of the material system by varying chemical concentration ratios as well as environmental conditions can be simulated. These simulations can consider chemical agencies such as the inhabitant and varying locations for the perturbation points. The digital realm is the only way to examine spatial implications of this highly speculative proposal.

5. Architectural Merit

5.1. *Morphological self-regulation: Non-deterministic morphology*

Morphological self-regulation occurs in context with changing inhabitation and environmental conditions. The initial setup is populated by oxygen producing microorganisms. The absence of carbon dioxide is a perturbation to the system and results in the BZ components expanding, which in turn leads to the time delayed shrinkage of the respective spatial units. Once carbon dioxide producing inhabitants populate the Chemical Machine, the components shrink. Globally seen, this leads to the expansion of the inhabited units. As inhabitants start moving within the Chemical Machine, we can observe a time delayed and reversible morphological adaption to local chemical conditions.

Morphological self-regulation is expressed by the time delayed, localized, and reversible oscillatory swelling and de-swelling of the hydro gel components depending on pH-levels. This may lead to uninhabitable spaces emphasizing the overall non-human-centric approach. Morphological aspects of self-regulation can also influence spatial subdivisions as well as the closure or opening of passages and channels depending on the frequency and intensity of utilization by the chemical agencies.

Morphological self-regulation, depending on material behavior, cannot be correlated to existing building tectonic functions. Self-regulation within the Chemical Machine improves the surface-to-volume ratio globally and locally, as space and volume adapt in accordance to occupancy rate.

This is a purposeful semi-permanent form of adaptation to a specific environmental chemical condition or agency. Due to the reversibility of self-regulation, the past of a system, such as the interaction with a previous user, is inaccessible to current inhabitants.

5.2. *Regulation of light*

As the BZ solution can be sensitive to light, chemical preprogramming could allow the linkage to the chemical Luminol. In this way light could be generated chemically Figures 19, 20 in relation to environmental light conditions. This feature allows setting up a building system without the requirement of electricity.

5.3. *Deployment*

The Chemical Machine Figure 22 could be deployed in contexts with a real necessity for this technology such as harsh, remote and uninhabitable environments as well as for extraterrestrial exploration.

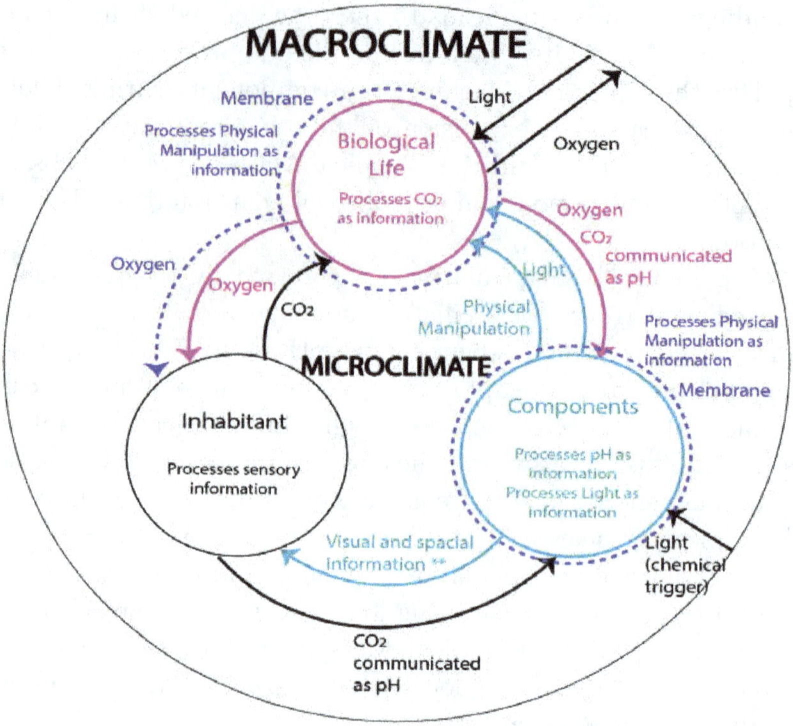

Fig. 22. Anand, P., Chung, G., *et al. Feedback loop in the Chemical Machine*, 2013, illustration. Architectural Association, London.

The initial small sized and light weight package is also beneficial for transportation to difficult deployment conditions. As the BZ solution can be chemically preprogrammed, components could be designed to perform certain environmental functions as per the requirements of the potential deployment scenario.

6. Conclusion

The Chemical Machine questions the role of materiality in general and the role of it within architecture. Instead of looking at inert multiple components with individual life cycles, materiality is critical for performing self-regulative tasks.

The non-human-centric proposal of the Chemical Machine is in general not a project of certainty and thus introduces aspects of speculation and provocation to the realm of architecture. Furthermore, the idea of animate architecture is introduced, which does not require a human agency in order to be operated as it is environmentally driven.

This project also stresses the need for inter-scientific research and scientific collaboration within the field of architecture, as this could potentially lead to new ways of solving architectural problems and yet unknown ways of working. Architectural projects might not be conducted as intuitive form-making exercises anymore, but as scientific research projects, whereas the traditional role of the architect will cease to exist.

Application of Slime mold functionality into building services is not only a critique of the contemporary, but an exploration in to the complete upheaval of how a building should be assembled and controlled through its lifespan. The aim of the project was to reflect the systematic functionality of the Slime Mold. Despite the intriguing patterns which the Slime Molds are capable of creating, the main focus and interest was not in replicating their physical patterns, but rather their intelligence and behaviors that would subsequently reach their functional aesthetics. Therefore, since the inception of the project it never had a preset physical form or preconceived aesthetics. Such an approach is contradictory to the generic architectural design process; however, it was a necessary approach to fully exploit the natural models and their subsequent productivity.

References

1. G. Pask, The architectural relevance of cybernetics. *Archit. Des.* 7(6):494–496, 1969.
2. R. Banham, *Architecture of the Well-tempered Environment.* (Architectural Press, London), 1969, 21.
3. A. Adamatzky, Routing Physarum with repellents. *Eur. Phys. J. E* 31(4):403–410, 2010.
4. G. M. Whitesides and R. F. Ismagilov, Complexity in chemistry. *Science* 284(5411):91, 1999.
5. I. C. Chen, O. Kuksenok, V. V. Yashin, R. M. Moslin, A. C. Balazs and K. J. Van Vliet, Shape-and size-dependent patterns in self-oscillating polymer gels. *Soft Matter*, 7(7):3141, 2011.

https://doi.org/10.1142/9789811257155_0028

Chapter 28

Collapsing the Wave Function on Postquantum Unconventional Computing

Richard Mayne
Unconventional Computing Lab, University of the West of England, Bristol BS16 1QY, UK
richard.mayne@uwe.ac.uk

Unconventional computing is the field that drives innovation and progress in computer science. One of its sub-fields, quantum computing, is a potentially breakthrough, disruptive technology in which there has been gradual but encouraging progress in recent years. If developed sufficiently, it is predicted that quantum computing will revolutionize a great many fields of human enquiry through laying bare cryptosystems, accelerating research in the natural sciences, and providing enormous speed and efficiency increases in machine learning, to name but a few applications. This chapter examines what quantum computing is, why we need to be aware of it, and whether there is a role for unconventional computing in a postquantum world.

1. Is Quantum Computing "Unconventional"?

Unconventional computing is the search for new materials, methods, and applications for computing technologies. This doesn't necessarily imply making smaller, faster, or more efficient general purpose computers, as is commonly considered: many research foci concern making new types of computers that do things which classical, silicon-based architectures can't do, or otherwise use inspiration from nature to program classical systems in novel ways.

This is an extremely wide remit for a field of enquiry and accordingly, advances therein are typically highly multidisciplinary, melding the expertise of the natural sciences, applied mathematics, and philosophy into this branch of computer science. This past decade has seen functional unconventional computing devices arising such as slime molds tackling problems of graph optimization (with better success rates than undergraduate mathematics students),[1] soldier crab logic gates,[2] neuromorphic "liquid marble" ballistic-reaction-diffusion circuits[3] and progress toward using intracellular protein networks used as nano-scale data buses[4] (Figure 1). The applications of these rich and varied prototypes clearly do not support general purpose computation, but rather suggest new routes to understanding the sciences, such as "reprogramming" of live cells for biomedical benefits or realizing true massively parallel processing to the scale of Avogadro's number. The creativity evident in the design of these devices speaks of the close interrelations between the arts and sciences, with unconventional computing at their nexus; manifold studies in modern art,[5] architecture, and wearable fashion,[6] etc. both inspire and emerge from the field.

Under this definition of unconventional computing, the emerging field of quantum computing would doubtlessly find a home. In spite of this, quantum computing is generally considered to be its own field, as it happens to be the more successful offspring of the original parent field — artificial intelligence which is arguably the most significant other example. Quantum computing's proponents argue that it offers routes toward enormous speed increases in computation of certain tasks, with database searches and factorization of prime products being among the most intensively researched upon applications in the field to date. Some go further to suggest that quantum computers will also reach the stage of general-purpose computation, although all purported future applications are the topics of much speculation and debate.

This raises an important question: if quantum computers are developed sufficiently that they will revolutionize computing and, by extension, every field of human endeavor, what will be the purpose of unconventional computing?

The purpose of this chapter is to briefly delineate what quantum computers are, evaluate whether they are likely to achieve such lofty

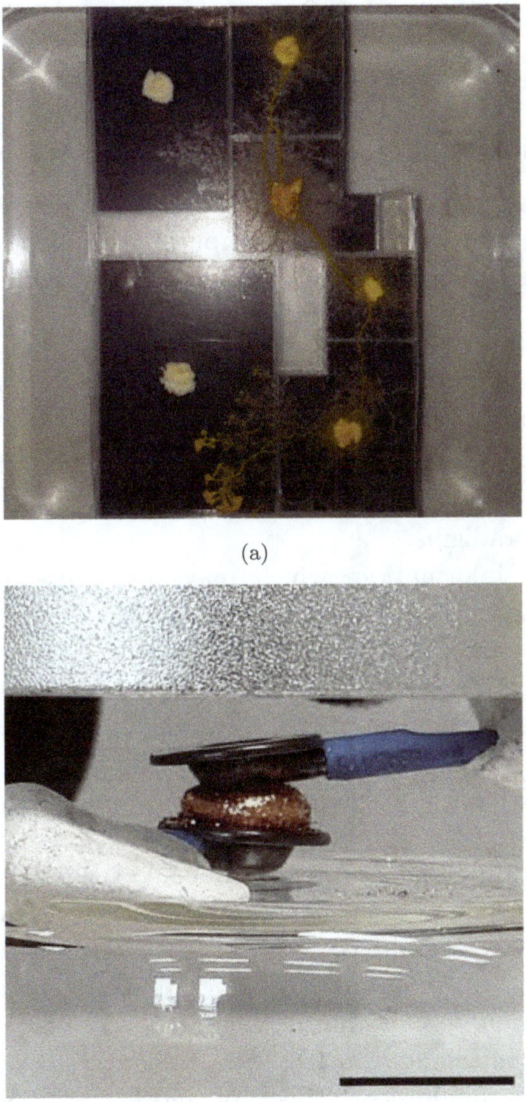

(a)

(b)

Fig. 1. Examples of unconventional computing devices. (a) A slime mold (Physarum polycephalum) navigates its way between distributed nutrient sources, optimizing its morphology in a manner that may be exploited to plan transport networks. (b) A "neuromorphic" liquid marble containing carbon nanotubes whose electrical resistance "remembers" past electrical activity in a manner similar to human neurons. Scale bar = 10 mm.

goals, and discuss what role unconventional computing will play in a postquantum world.

2. How Does/Will Quantum Computing Work?

Let us consider some fundamental information and theoretical concepts relevant to classical computer systems. Data exist as electrical charge distributed across silicon circuit components, where the presence of charge is a binaric bit representing "1" and its absence is a "0". We may observe these bits and store them: doing something to a bit won't alter anything else in the system and the process is generally lossy, i.e., one cannot deduce the input pattern from observing its output with no prior knowledge. None of these facts are true for quantum computing, or indeed many other unconventional computing paradigms.

Quantum information, which may be represented by any of a variety of properties of quantum matter, is stored as "qubits", which may assume a "1", a "0", or a linear combination of both of these (superposition). This chapter's title alludes to a concept in quantum science which states that when a qubit is observed, its manifold states will appear to collapse, at which point the qubit will resolve probabilistically as a binary bit. While it would be technically inaccurate to say that superposition enables a greater informational density in a qubit than a conventional bit, there are tools to allow for multiple calculations to be resolved on a single qubit. It should be noted that quantum computers are deterministic under most interpretations of quantum mechanics, but our inference of their output may not be. Another strange property of quantum matter is called entanglement, which is where changes in the state of one qubit will have simultaneous effects on another qubit, even if they are separated by great distances. Coupled with superposition, these are the primary two physical principles that are exploited in the design of quantum computers and underly the claims that these architectures will achieve enormous increases in efficiency over classical machines, at least at some tasks.

True to the ethos of general unconventional computing, quantum computers seek to exploit the natural transfers of state and energy by interpreting them as information processing. For example, the aforementioned probabilistic interpretation of a qubit's state arises from a process of overlying the wave functions of quantum information (as quantum matter assumes the properties of both particles and waves), observing and thresholding patterns of constructive and destructive interference therein. One would struggle to describe the appearance of contemporary quantum computer prototypes as anything other than the domain of the unconventional, as may be observed in the distinctly alien device shown in Figure 2.

Fig. 2. Photograph of the IBM Q 53 qubit quantum computer. Image courtesy of IBM.[10]

Quantum computing concepts were first devised in the 1970s, but there was arguably little commercial justification for their development until the mid-1990s, when two algorithms were described. First, Shor's algorithm described a method by which a quantum computer may factorize prime numbers efficiently,[7] something which is not possible on conventional computers: as several cryptosystems are based on the intractability of prime factorization, this algorithm understandably generated significant interest in academia, industry, and government cybersecurity divisions. Second was Grover's algorithm, which described methods to achieve quadratic increases in search for efficiency over classical systems.[8]

There have been a significant number of interesting developments in the field since these algorithms emerged (e.g., quantum teleportation as a basis for a new wireless internet[9]), but experimental progress in the physical engineering of quantum computers has not yet caught up. While both Shor's and Grover's algorithms have been experimentally realized, both have been done so on systems with extremely limited capabilities: only the numbers 15 and 21 have been successfully factorized by true quantum systems and Grover's searches haven't been implemented on systems utilizing more than a handful of qubits. In 2019, researchers at Google claimed to have experimentally demonstrated what they called "quantum supremacy": using a 54 qubit system called "sycamore", they implemented a non-useful function in 200s which would have apparently taken about 10,000 years on a classical supercomputer, although both these results and the estimations of their significance are contested.

The reader should now appreciate that quantum computers utilize unconventional media and algorithms to do certain tasks very well, but that we are yet to witness the advent of quantum computing at which it becomes a useful and widespread technology. Before we evaluate whether quantum computers will ever become useful and widespread, en route toward assessing the postquantum future of unconventional computing, let us briefly examine why we should be excited by the prospect of quantum computing, as it is perhaps difficult to be enthralled by discussing tedious and abstruse mathematical algorithms.

3. Why Should We Care about Quantum Computers?

An obvious application of quick searches is the ability to use brute force to crack encryption schemes; indeed, the emergence of Grover's algorithm contributed to the industry-wide adoption of 256-bit encryption in 2001. Similarly, Shor's algorithm is purported to be a route toward breaking RSA encryption schemes, thereby laying bare all internet-based communications and transactions. From the paucity of published progress, we must concede that these applications are a long way away; furthermore, in the author's opinion, these tasks are also an inelegant and restrictive application of such technologies, despite their obvious utility.

To reiterate a theme expounded upon in the previous section, unconventional computing concerns doing things that classical computers cannot, as opposed to competing with them. One intuitive application for quantum computers is, therefore, simulation of the quantum world: something which is extremely difficult to do on classical systems as the interaction environment needed to represent a quantity of quantum particles must necessarily be exponentially larger than the number of particles contained. This presents a problem as quantum physical experiments are notoriously complex, time-consuming, and expensive. Simulation of these events therefore allows us access to previously unfound tools for unlocking the secrets of the natural world. Recent results from IBM's quantum group, based on their previous numerical work, have already found uses in the design of next-generation rechargeable batteries.[11]

There have also been encouraging results in the field of quantum machine learning. Conventional machine learning refers to techniques which use computers to do classification or prediction, typically using very large datasets, where the experiment is repeated many times with minor variations in run parameters. The computer keeps a score of what works well and "learns" how to adjust itself to achieve optimal results, which can then be leveraged on new data. Incidentally, machine learning and its parent field, artificial intelligence, were once considered unconventional and derive from the biological sciences. Quantum machine learning in its purest sense, which is using quantum computers and their inherent mechanisms for doing efficient work on very large quantities of data to do machine learning, may potentially greatly accelerate the course of inquiry in all fields where data science is applied, which is to say, practically ever field of contemporary research. Most popular machine learning methods now have their own quantum implementations ready-and-waiting for the architectures that can run them, including fully scalable neural networks. Quantum machine learning may also refer to the use of conventional machine learning to interpret the output of a quantum computer which is, unfortunately, currently very noisy. There is an elegant self-perpetuation here in the use of unconventional computing to accelerate the progress of unconventional computing research.

The past decade has seen an enormous increase in the use of machine learning concomitant with the emergence of affordable computers capable of having arguments with large quantities of data. The field of data science has arisen from this and industry is only starting to realize the great many incentives that have emerged in the field including accelerated research, optimization, anomaly detection, and intuitive interpretation of natural languages. It is therefore no exaggeration to say that quantum computing will, through enhancing simulation of the natural world and enabling further advances in artificial intelligence, lead to momentous progress in every field of human enquiry.

The only caveat to these optimistic appraisals is that quantum computers may never reach these dizzy heights: therein will lie the answer to our main issue as to what role there is for the entirety of unconventional computing in this process, one way or the other.

4. The Postquantum Role of Unconventional Computing

It is impossible to predict the scale of investment in quantum computing to date, due to the nature of hidden business interests and state involvement in potentially disruptive technologies, but the size of government grants and magnitude of bespoke laboratories of tech giants (notably, IBM, Google, and Microsoft) indicate that tens of billions have been spent. Yet, the pace of experimental development after 20 years of work and 50 of academic interest is unequivocally glacial. Quantum matter is extremely difficult to manipulate due to its sensitivity to even minor environmental changes: a phenomenon called "decoherence" stalks every advance in the field, which introduces error into calculations once qubits experience the changes in entropy that are unavoidable during computation, for example resulting from self-examination, moving data, and performing calculations. Quantum computers must be isolated from all forms of radiation, vibration, temperature, etc. For these reasons, it's difficult to imagine miniaturized, affordable quantum chips being available during this century, although they may reach academic institutions and large businesses much sooner. Furthermore, several cloud computing services are planning to or have already begun to offer access to quantum computing clusters.

A significant number of dissenters argue that, due to the requirement for quantum computers to have vastly larger qubit capacities than their algorithms require to compensate for error correction mechanisms, we will never achieve the aforementioned quantum computing goals. These arguments, which are based on convincing quantum theory, are difficult to counter, although developments such as quantum machine learning seek to bridge the gap by resolving engineering issues through intelligent use of software.

We are currently at a historic fulcrum between which we must consider the putative future of unconventional computing as both existing and not existing (no pun intended): will quantum computation falls by the wayside and leaves room for other unconventional substrates to fill the gap, or will the future lie with quantum and leave us wondering how we may advance the field yet another "quantum" leap?

The most salient fact here is that we are already in a quantum era: the sparse experimental advances in the past 2 or 3 years have shown that the theoretical benefits of these architectures may be achieved on a small scale; it's unreasonable to assume that we won't realize significant improvements on this design. This cannot be overstated: we are already using quantum computers and they are already being applied to real problems, albeit on an extremely limited scale.

A key role for unconventional computing researchers is to recognize and understand the failings of quantum computers, rather than just their advantages. This chapter offers no specific mathematical or experimental method by which we can propose to exert control over a number of error-corrected qubits greater than the number of subatomic particles in the observable universe and hence enable the creation of "useful" quantum computers, but we have seen how a bioinspired unconventional computing tool (machine learning) has already been applied to clean the output of true quantum computers, which goes at least some way to bridging this gap.

Quantum computers are noisy, don't achieve true parallelism, and will demand comparatively strenuous engineering, but this doesn't detract from the magnificient vision with which they are being developed. Regardless of whether the future is quantum or unconventional computing, research feels both seek to improve their sister field and look for alternatives. We will still find enormous use in parameterizing biological mechanisms that afford us control over biomedical systems, designing chemical systems which realize true massive parallelism, and developing naturally-inspired algorithms for use on conventional systems that solve open problems on computability, to name but a few, limited examples. In an ideal world, we would have a different specialized computing substrate for each task: a biological CPU for neural networks, a quantum computer for database searches, and a conventional computer to compose text documents with, and so on. Until then, unconventional computing research will continue to find new methods, materials, and applications for novel computing materials.

References

1. A. Tero, S. Takagi, T. Saigusa, K. Ito, D. P. Bebber, M. D. Fricker, K. Yumiki, R. Kobayashi, and T. Nakagaki, Rules for biologically inspired adaptive network design. *Science* 327(5964):439–442, 2010. ISSN: 1095-9203. Doi: 10.1126/science.1177894.
2. Y. P. Gunji, Y. Nishiyama, and A. Adamatzky, Robust soldier crab ball gate. *AIP Conf. Proc.* 1389:995–998, 2011. ISSN: 0094243X. Doi: 10.1063/1.3637777.

3. R. Mayne, T. C. Draper, N. Phillips, J. G. H. Whiting, R. Weerasekera, C. Fullarton, B. P. J. de Lacy Costello, and A. Adamatzky, Neuromorphic liquid marbles with aqueous carbon nanotube cores. *Langmuir* 35(40):13182–13188, 2019. ISSN: 0743-7463. Doi: 10.1021/acs.langmuir.9b02552.
4. A. Chiolerio, T. C. Draper, R. Mayne, and A. Adamatzky, On resistance switching and oscillations in tubulin microtubule droplets. *J. Colloid Interface Sci.* 560:589–595, 2020. ISSN: 0021-9797. Doi: 10.1016/j.jcis.2019.10.065.
5. A. Adamatzky, *The Silence of Slime Mould.* Luniver Press, Bristol, UK, 2019.
6. A. Adamatzky, (editor), *Slime Mould in Arts and Architecture.* River Publishers, Gistrup, Denmark, 2020.
7. P. W. Shor, Polynomial-time algorithms for prime factorization and discrete logarithms on a quantum computer. *SIAM J. Comput.* 26(5):1484–1509, 1997. ISSN: 0097-5397. Doi: 10.1137/S0097539795293172.
8. L. K. Grover, A fast quantum mechanical algorithm for database search. In *Proceedings of the Twenty-Eighth Annual ACM Symposium on Theory of Computing*, STOC '96. Association for Computing Machinery, New York, NY, USA, 1996, 212–219.ISBN: 0897917855. Doi: 10.1145/237814.237866.
9. W. Pfaff, B. J. Hensen, H. Bernien, S. B. van Dam, M. S. Blok, T. H. Taminiau, M. J. Tiggelman, R. N. Schouten, M. Markham, D. J. Twitchen, and R. Hanson, Unconditional quantum teleportation between distant solid-state quantum bits. *Science* 345(6196):532–535, 2014. ISSN: 0036-8075. Doi: 10.1126/science.1253512.
10. IBM. Ibm newsroom gallery. Available at https://newsroom.ibm.com/image-gallery-research, 2020. Accessed: October 22, 2020.
11. J. E. Rice, T. P. Gujarati, T. Y. Takeshita, J. Latone, M. Motta, A. Hintennach, and J. M. Garcia, Quantum chemistry simulations of dominant products in lithium-sulfur batteries, *J. Chem. Phys.* 154:134115, 2021.

https://doi.org/10.1142/9789811257155_0029

Chapter 29

Tactile Computing Interaction and its Application for Arts

Yasuhiro Suzuki
Department of Informatics, Nagoya University, Furocho Chikusa Nagoya, Japan
ysuzuki@i.nagoya-u.ac.jp

Nature has been calculating with algorithms since before the beginning of humankind. Tactile interaction is the basis of nature's computation. We have proposed the Tactile Score to describe the algorithm of tactile calculation. Tactile interaction can be divided into direct touch and indirect interaction, which is conveyed by sound as vibrotactile. In this chapter, after an overview of the algorithm, we present a work of art that uses direct interaction with Tactile Score and indirect interaction with vibrotactile technology.

1. Algorithm

A principle of computing is an algorithm. A father of algorithm in the numerical calculation was al-Khwarizmi, and in the logical calculation, it was Ramon Llull. G. W. Leibniz integrated numerical and logical calculations of algorithms. He found that logical calculation is possible by numerical calculation and proposed a universal mathematical framework based on an algorithm (Dissertatio de arte combinatoria).[1,a]

[a]Leibniz produced a computer by using gears by improving Pascal's computer and also realized a digital computer. Leibniz developed a decimal to the binary converter and a binary to decimal converter. Hence, he had produced an "unplugged" digital computer, whose architecture was the same as electronic digital computers nowadays.[2]

In a dictionary, the definition of an algorithm is *a set of mathematical instructions or rules that, especially if given to a computer, will help to calculate an answer to a problem* (Cambridge Advanced Learner's Dictionary & Thesaurus); however, the algorithm has existed before computers emerged. Hence, *a list of instructions for solving a problem* (Cambridge Business dictionary) would be suitable for the definition. On this definition of algorithm, we may say that Nature computes with algorithms. Nature uses algorithms. How can humans use algorithms to communicate with Nature? When we interact with a person, we usually use "natural language". Language is composed of "a list of instructions" to deliver information, where "the order of word" is an algorithm. "You are a student" and "Are you a student" have the exact words. Since "the order of words" are different, their information is also different. So, we can communicate with Nature by an algorithm to choose a common language in Nature.

1.1. *What is "a common language" in Nature?*

A common language is a tactile sense. For example, a DNA molecule finds a pairing partner (base-pair) with tactile sense; the DNA molecule "binds" to the other molecule if it "fits". Whether or not it will fit the other molecule depends on the shape of the molecule. It is similar to doing a jigsaw puzzle with our eyes closed. If we do a jigsaw puzzle blindfold, we will touch the puzzle pieces, recognize the shape, and try to fit them together. Since the DNA molecule has no sense of sight, recognizing the molecule's shape is based on the "sense of touch." Tactile interactions are not only molecular but also multi-scale, including cells, organisms, and societies.

Animals, including humans, strengthen their emotional bonds through tactile interactions such as grooming. In other words, the sense of touch is an information medium of sensitivity. The sense of touch as a sensory medium has long been considered.[3] The art historian Alois Riegl used the metaphor "haptic" to define the fact that in late Roman art and crafts, the surface of the work was perceived as if it were traced by sight.[4] We proposed Tactile Score as an algorithm to describe the sense of touch as a sensory medium.[5]

2. Tactile Score

We live near a kindergarten, and in the morning, we hear the cheerful voice of the children: "Good morning!"; Then I turn on the radio and hear a calm announcer's voice saying, "Good morning". The children's "Good morning" and the announcer's both have the same "meaning". However, I felt "energy" from the children and "calmness" from the announcer. In other words, the meaning is the same, but the sensation is different.

Why do we feel "energy" from the children's "Good morning"? What makes it different from the "Good morning" of a skilled announcer? The difference is in the way we say it. We say it is characterized by the height and volume of our voice, but for the sense of touch, it is the volume of our voice that matters.

It has been investigated whether frequency or amplitude, in this case, the loudness of the voice, is more relevant for tactile perception. The results confirmed that tactile sensitivity is correlated with the product of the amplitude power and the frequency.[6] The volume of the voice more influences tactile perception than the height of the voice.

Thinking back to the preschool children's "Good morning", their voices gradually became louder. Therefore, we focus on the change of "loudness of voice". Then, let us describe the change in loudness of the voice by appropriating the music notation. In staff notation (musical notation), the higher the line, the higher the note, and the lower the line, the lower the note. So the middle line (the third line) is the "normal" volume, and the lower the line, the higher the volume, and the higher the volume, the lower the line. The "rhythm of speech" is based on the notes themselves (Figure 2).

We can now describe the "vigorous" tone by notating the change in loudness of the voice. This Score is called the Tactile Score.[1] Using the "Genki notation" shown in Figure 1, we can make other ways of saying things more cheerful. For example, try saying "Welcome" and "Hello" in a "gradually increasing volume" manner. On the other hand, how about "gradually lowering the volume of your voice"? A muffled "Good morning" or "Welcome" gives the impression of being

Fig. 1. Tactile score of cheerful Good morning.

"listless". The tactile representation of "lack of energy" is shown in Figure 1, reversed.

2.1. *Tactile qualities*

The sense of touch has two essential properties, called tactile qualities. The Tactile Score describes tactile quality. Imagine a thick and stiff spring. The hardness of the spring is described by a constant k, called the spring constant. The spring's restoring force is expressed by the product of the spring constant and the amount of deformation, δx, $f = k\delta x$.

Under the same force of pushing, if the spring is stiff, δx is smaller, while if it is soft, δx is large. When δx is small, the changing of f is also tiny, while δx is large, its change is significant. Hence, we can denote hardness as the abrupt change in a Tactile Score. The tactile quality can be judged from the Tactile Score alone: if the vertical variation of the Tactile Score is slight, the spring is "hard", and if the vertical variation is significant, the spring is "soft".

Next, let us look at roughness as expressed by the Tactile Score. A well-maintained lawn has very little change in hardness when we walk on it. Whether the grass is hard or soft, we perceive a smooth lawn if the change in firmness is constant. How would we compare the case where the lawn is cut in a particular pattern, and the firmness varies regularly and significantly with the case where the lawn is cut randomly, and the firmness varies randomly and significantly? To illustrate, a constant change in hardness is when we are riding on a railway, while an unexpected change in hardness is when we are driving a jeep on a gravel road. In this case, the more random the change in hardness, the rougher we feel it to be. Therefore, the simpler the combination of tactile patterns, the smoother it is, and the more random the combination, the rougher it is. When driving a luxury car on a smoothly paved road, the change in hardness is slight. In this case, the Tactile Score is combined with a slight change in hardness. On the other hand, several Tactile Scores are combined in the same order if we are on a train or a boat. If we are driving a jeep on a forest road, the different Tactile Scores are randomly combined. The Tactile Scores now represent the hardness and roughness of the important tactile qualities.

2.2. *Tactile Score as a partial differentiation of sensitivity/tactile sensation*

Tactile Score is proposed for the description of spatiotemporally varying tactile sensation. Since the tactile sensation is a medium of sensitivity, it is possible to describe spatiotemporally varying sensations.

Our senses perceive differences. It is usually unrecognizable if we write something on a white wall with paint of the same whiteness. However, if we use optical instruments that can detect minute changes, we can recognize the pattern of white paint on a white wall. The size of the recognizable difference is called the "sensitivity". Some professionals can recognize minimal differences; for example, a skilled worker in a car company can recognize microscopic irregularities just by touching them.

The Tactile Score describes the spatiotemporal change of the recognized difference. The difference between things depends on the senses. For example, the visual difference of a white wall is the visual pattern, the tactile difference of a white wall is the feeling of touch. We denote the difference of sensing by browning the concept of partial differentiation. Partial differentiation is a mathematical description to denote the change of a particular variable in a multi-valued function. By appropriating the concept of partial differentiation, we call the extraction of differences according to specific senses "partial differentiation of sensitivity", PDS.[7]

We use this notation to clarify "which senses and which differences" of things we are focused on, which sensitivity. In the following, a "thing" is denoted by Ψ_i. The subscript labels the sort of the thing; for example, "white wall" and "sound in the train" are labeled as Ψ_i, Ψ_j, respectively. Furthermore, the senses that perceive a thing, such as visual, hearing, touching, etc. are denoted as $s_j, s_k, s_l, \ldots$. Differences of Ψ_i in the sense of s_j are denoted as $\frac{\partial \Psi_i}{\partial s_j} = \dot{\Psi}_i^{s_j}$. For example, let Ψ_w be a white wall and s_v be a visual sense, $\frac{\partial \Psi_w}{\partial s_v} = \dot{\Psi}_i^{s_v}$ denotes the visual difference of the wall.

In order to illustrate the visual difference of the white wall obtained, it is necessary to extract the difference of the spatial variation of this visual difference. There are various definitions of space, but for the sake of simplicity, we will denote the partial derivative in space as ∂sp. Hence, the pattern of wall is denoted as $\frac{\partial \dot{\Psi}_w^{s_v}}{\partial Sp} = (\dot{\Psi}_w^{s_v})^{s_{sp}}$. We denote the partial derivative of time as ∂t; If the room of wall keeps constant brightness $\frac{\partial \dot{\Psi}_w^{s_v}}{\partial t} = (\dot{\Psi}_w^{s_v})^{s_t} = 0$, while if it changes, $(\dot{\Psi}_w^{s_v})^{s_t} \neq 0$.

For simplicity, we denote $\frac{\partial \Psi_i}{\partial s_j}$ as $PDS\Psi_i^{s_j}$, thus, the visual difference of the wall is denoted by $PDS\Psi_w^{s_v}$, the spatial variation of the visual difference, by $PDS(\Psi_w^{s_v})^{s_{sp}}$, and the temporal variation, by $PDS(\Psi_w^{s_v})^{s_t}$.

Tactile Score is a discretized notation of PDS. Tactile Score of "Good morning" is obtained from discretizing $(PDS_{\Psi\text{voice}}^{s_{\text{volume}}})^{s_t}$ into n equal parts; in case $n = 12$, it is equal to one octave. The larger the number of divisions (n), the more detailed the description of the volume change; the smaller ∂t, the shorter the description of the time change.

3. Tactile Sense of Tale of the Heike

The Tale of the Heike (Heike Monogatari) is an epic account compiled prior to 1330 of the struggle between the Taira clan and Minamoto

Table 1. The Tales of Heike; (left column) Original text in Romanized Japanese, (right column), English translation.

Gion Syozya no Kaneno-koe,	*(the sound of Gion Syozya Temple,)*
Syogyo Muzyo no Hibbiki ari,	*(it is saying, everything is unstable,)*
Sarasohjyu no hana no iro,	*(the color of Sarasohjyu flower,)*
Jyousya Hissui no kotowari wo arawasu,	*(it expresses, flourish means collapse,)*
Ogoreru mono mo hisasi karazu,	*(pride goes before a fall,)*
Tada haruno yono yumeno gotoshi,	*(just like a dream on a spring night,)*
Takeki monomo tsuiniha horobinu,	*(even the fiercest of men will perish at last,)*
Tada kaze no mae no chiri ni onazi,	*(it is the same as a dust in the wind.)*

clan for control of Japan at the end of the 12th century in the Genpei War (1180–1185).[8] The beginning of the story (Table 1).

Tale of the Heike is a long story; Minstrels have handed it down as a chant. We thought that the *Tale of the Heike* should have its characteristics as a chant. Therefore, we converted the *Tale of the Heike* into a Tactile Score.

We have converted the *Tale of the Heike* into a Tactile Score using the PDS of vowel counts (Figure 2). As the beginning of the *Tale of the Heike* is eight sentences long, we have taken each sentence as one measure. We carried out a morphological analysis of the sentences. Moreover, we divided them into semantic units. Then we examined the number of vowels in each semantic unit. We divided the maximum and the minimum number of vowels into 8. The average number of vowels was considered as "normal intensity". If the number of vowels was more significant than 8, it was considered "strong"; if it was less, it was considered "weak".

The *Tale of the Heike* has been translated into English. We have used the same method of PDS to convert the English translation of the *Tale of the Heike* into a Tactile Score. We made a video work of chanting and massage using the Tactile Score of the *Tale of the Heike* (Figure 3[9]). The Tactile Score of the English translated version shows a disruption of the tactile rhythm compared with the original one. It would be due to the faithfulness of the translation to the original.

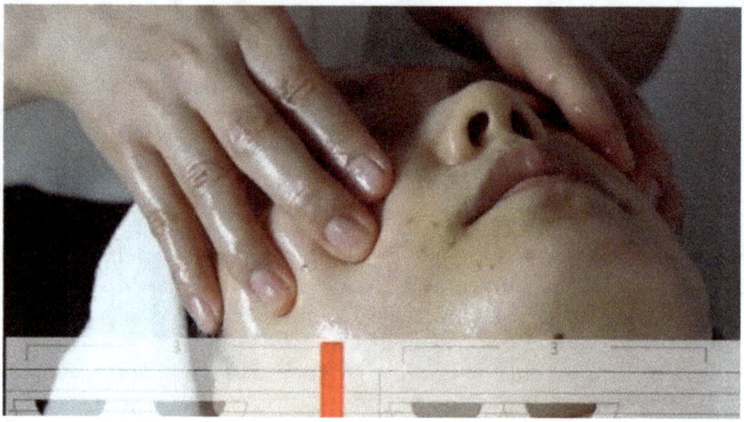

Fig. 2. Massage extracted from the *Tale of the Heike*; the pattern at the bottom of the figure is a simplified Tactile Score, the red bar is the position.

Fig. 3. (Above) Tactile Score of the original Japanese version of the *Tale of the Heike*, (Below) Tactile Score of the English translation (the first 6 bars).

3.1. *Mass media of the sense of touch*

In the work of *The Tale of the Heike's* Tactile Score, we noticed the particularity of the sense of touch as a medium. Unlike music or graphics, the sense of touch cannot be shared by many people. The tactile sensation of our message cannot be shared with others. We have made it possible to share the sense of touch with others following three different ways:

(1) Converting it into a medium that can be shared with others (such as music, graphics, ...),
(2) Converting it into vibratory tactile sensations,

(3) Converting it into non-contact vibratory tactile sensations, low frequency sounds: Deep Micro Vibrotactile, DMV.[10]

This chapter mainly addresses works that transform Tactile Score into sounds and transform Tactile Score into a non-contact tactile sensation using low-frequency acoustics.

4. Sounds, Translated from Tactile Score

We converted the various things into Tactile Scores using the PDS. The converted Tactile Score was further converted to Music Instruments Digital Interface, MIDI code. We can create a sound from a Tactile Score by using Digital Audio Workstation, DAW; Although MIDI codes transformed from Tactile Scores were unsuitable for music, we did not change the MIDI code. Because to change the MIDI code is to change the underlying tactile sensation. Hence, to compose a sound, we used several MIDI tracks and shifting times or changing pitches. Steve Reich [1936-] is one of the leading composers of Minimal music, compositions that combine sound patterns. For example, his "Clapping music" was composed of eight notes shifting the same rhythmic pattern. We referred to his style of composing.

We have created 81 sounds from various things; for example, in our work "Keiba" (*horse race* in Japanese), we converted the sound of a live broadcast of a horse race into a Tactile Score. In this work, the sound of the live broadcast is included, but it is played backwards to eliminate the meaning of the language; in "early spring", we converted the sound of Stravinsky's *The rite of spring* into a Tactile Score. We converted a part of the sound of *Augurs of spring* into Tactile Score. After converting it to MIDI, we composed it as a multi-piano serial piece. All created sounds are available on Soundcloud (https://soundcloud.com/hapticahaptico). Most of them were composed via Tactile Score.

We named this project as *Lost in Translation.* Music, for example, is the art of the pitch, and the intensity of sound is secondary. The pitch of the sound is perceived as auditory, the intensity of the sound is tactile. The composer expresses something by changing the sound pitch; when something is translated into pitch, the intensity of the sound, or the sense of touch, is lost. The *Lost in Translation* project complements the sense of touch that is lost in the translation of something into sound.

5. Deep Micro Vibrotactile, DMV

We cannot hear low-frequency sound waves below 20 Hz and higher than 20,000 Hz. Even if we cannot "hear" the waves, they exist. We know little about low-frequency sound; it has focused on noise pollution, and low-frequency sound with large amplitude harms health. In Nature, low-frequency sounds are produced by a variety of natural phenomena. For example, low-frequency sounds are included in tectonic movements, such as volcanoes, celestial motions, such as asteroids, meteorites, meteors, and fireballs entering the atmosphere, meteorological phenomena, such as typhoons and lightning, and water movement such as waterfalls and rivers. Since low-frequency components distort the sound in the music industry, such sound elements are removed in the production process. However, low-frequency sounds cannot be removed entirely from music. Instead, it is an integral part of the music. In the Middle Ages, low-frequency sounds were used in music. Acoustic studies have shown that pipe organs can be generated from 10 Hz to 10,000 Hz. Pipe organs generate a very low-frequency sound with a 32 feet long pipe.

Inaudible low-frequency waves can stimulate the sense of touching; it has proved that we perceive amplitude of waves rather than frequency.[6] Low-frequency sound creates a sound field. We have confirmed that low-frequency sound changes the sound field.[11] What happens when low-frequency sounds surround a place? In a living room or a concert hall, there are many different frequencies of sound. When low-frequencies sounds are played strongly, these sounds mix with every different frequency in the room. We have confirmed that the mixing of low-frequency sound changes the frequency spectrum. For example, if a music recording is of poor quality and has a weak high-frequency component, the high-frequency component can be enhanced by mixing it with low-frequency sounds.

We have used this discovery to create a sound art piece that uses inaudible sounds to create a sound field. This work is called "inaudible sound art", in which the structure of audible sound is transformed by inaudible sound. Although we are not aware of it, low-frequency sound is abundant in Nature, as mentioned above. We have discovered that there are places where low-frequency sounds are extreme. We have analyzed tactile qualities of such low-frequency sounds and confirmed that the tactile quality of the low-frequency sound in each spot is different (Fig. 4).

We participated in the "Performative concert" of the filmmaker Yoshimasa Ishibashi with the City of Kyoto Symphony orchestra (January 17, 2021, at Rohm Theatre, Kyoto, Figure 5). Ishibashi has interested in "invisible expression" through the sense of touch and DMV. The tactile artist Rieko Suzuki designed a vibrating device and DMV for Ishibashi's "Celebration — Japanese Polish Contemporary Art Exhibition, 2019".

Generally, in an orchestral concert, there is an echo board at the back of the stage to allow the sound to reach the audience, but Ishibashi removed the echo board and moved the orchestra to the back of the stage to make room for the dance performance. The problem arose when the orchestra played without the reverberation board, as the orchestra's sound was muffled.

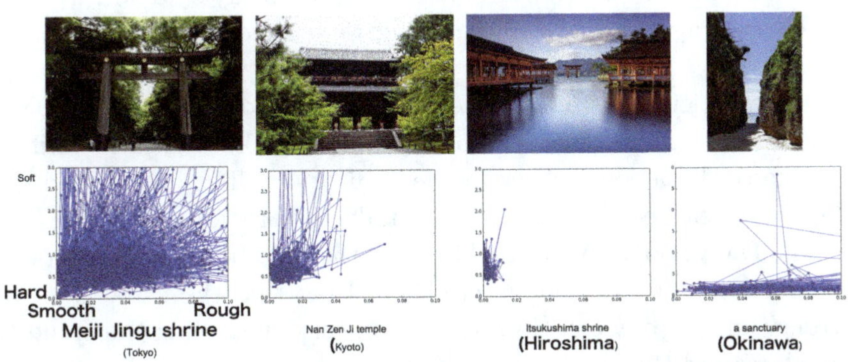

Fig. 4. (Top row From left to right; Tactile quality of Meiji jingu shrine (Tokyo), Nanzenji temple (Kyoto), Itsukushima shrine (Hiroshima), a sanctuary in Okinawa island (southern part of Japan), Bottom row) Tactile quality of low-frequency sounds of each space, the vertical axis illustrates hardness and horizontal axis illustrates roughness.

Fig. 5. From *Preformative Concert* at Rohm Theatre Kyoto, Yoshimasa Ishibashi, City of Kyoto Symphony Orchestra (January 17, 2021), Photo: Kenta Yamaji.

For this piece, we created a new DMV based on natural sound; we used the large woofers in the upper and lower sections of the main hall of the Rhom theatre Kyoto and produced the DMV. Since the final piece of the performance was Stravinsky's "The Firebird", we created a vital DMV piece using the sound based on Hawaiian volcanoes.

We participated in the general probe and coordinated the DMV. Before the general probe, some members of the orchestra and the conductor did a soundcheck in the auditorium. They confirmed that the sound reached the auditorium, unlike the previous day when there was no DMV. So, they decided to leave the reverberation board off for the performance. The sound crew and I felt that the DMV was "too strong" in the preview. However, no one else, including the orchestra members and the conductor, felt that way.

The concert began with Stravinsky's Symphonic Fantasy "Fireworks", followed by quieter pieces and songs, such as Debussy's "Prelude to the Afternoon of a Pastor" and Ravel's song cycle

"Scheherazade". Therefore, we prepared several DMVs with a gentle touch and several DMVs with an intense touch for "Fireworks" and "Firebird".

On the day of the performance, the DMV was played before the doors opened, we investigated the effect of the DMV's absorption by the audience. The absorption by the audience was not strong. In the finale, we used various DMV simultaneously at maximum volume.

Since we were in the sound booth, we could not know the "atmosphere" of the auditorium. Later, Ishibashi told us that he could not stop crying at the finale of the Firebird. The backstage staff and audience feedback confirmed the same. We did not have any complaints from the audience about the DMV.

6. Sound Creature

If tactile language is the common language of Nature, then tactile language should allow us to integrate the PC with Nature. We are trying to integrate the PC with Nature through haptic interaction.

If we want to integrate the PC with Nature, what do we need to "transmit"? Unlike the Internet, the natural system is not the PC, and the Internet cannot directly connect the PC and the natural system. PCs, in general, do not have a sense of touch, so tactile interaction requires tactile sensors and output devices.

One way to connect the PC to Nature is through sound: the PC has a microphone and a speaker, and through the microphone, the PC can convert sound into electrical signals, process the information, and output the sound through the speaker.

What is sound? A sound is a form of energy. Energy is converted into waves, which travel through a medium. The propagation of the waves depends on the Nature of the catalyst. If we can sense the waves as they travel through the medium, we can convert them back into energy. In other words, by connecting a natural system to a PC via waves, energy can be exchanged between the natural system and the PC.

In the natural system, the energy is converted into waves, which travel through space and reach the PC's microphone. The energy is converted into an electrical signal in the microphone, processed in the PC, and converted back into an audio signal. The audio signal is then sent out through the speakers, allowing the PC to transmit energy in waves to the natural world.

6.1. *Concept of the Sound Creature, SC*

To connect Nature and PC via waves, we created the Sound Creature, SC Fig. 6. This creature obtains energy from outside via waves and metabolizes them, and generate another wave. In other words, this artificial creature is an energy transformer. SC consists of a transformation from sound to two-dimensional patterns and vice versa and the game of life.

John Conway, who proposed the Game of life, strongly rejected the deterministic position in an invited talk at ALife, the international conference on artificial life in 2014. He emphasized contingency

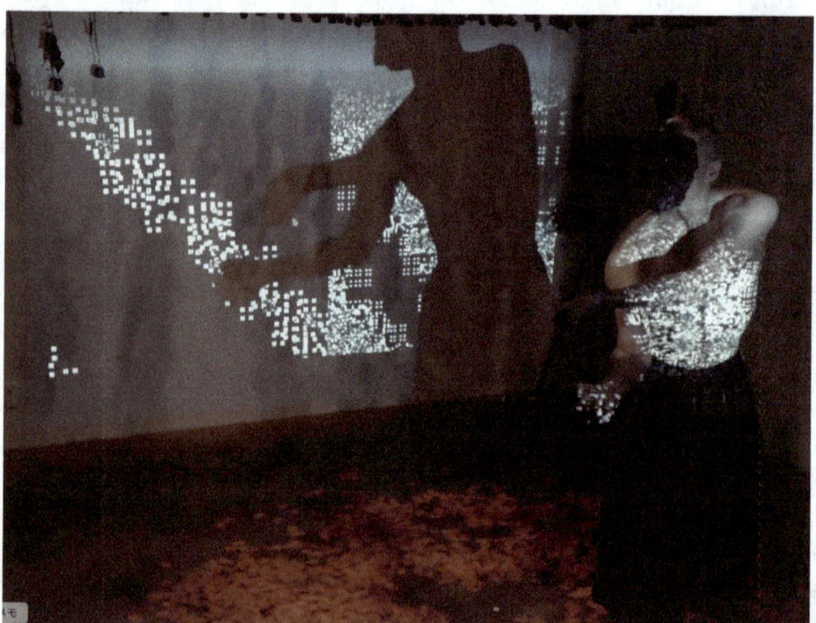

Fig. 6. From *Garandoh no Niwa* (empty garden) with Norihito Ishii (Sankai Jyuku company) (July 30, 2021 at gallery shell102, Tokyo).

is much more critical than determinism; he said that, "*there is full of contingencies in the world*".

6.2. *Game of Life*

The universe of the Game of Life is an infinite, two-dimensional orthogonal grid of square cells, each of which is in one of two possible states, alive or dead (or populated and unpolluted, respectively). Every cell interacts with its eight neighbors, which are the cells that are horizontally, vertically, or diagonally adjacent. At each step in time, the following transitions occur:

(1) Any live cell with two or three live neighbors survives,
(2) Any dead cell with three live neighbors becomes a live cell,
(3) All other live cells die in the next generation. Similarly, all other dead cells stay dead.

The initial pattern constitutes the seed of the system. The first generation is created by applying the above rules simultaneously to every cell in the seed, alive or dead; births and deaths coincide, and the discrete moment at which this happens is sometimes called a tick. Each generation is a pure function of the preceding one. The rules continue to be applied repeatedly to create further generations.[12] The game of life is entirely deterministic; there is no room for "contingency", So we integrate the Game of life and the sound surrounding a PC, where the Game of Life is working. The PDS for volume (amplitude) converts the ambient sound of the PC into a two-dimensional pattern. Since the amplitude of the input sound is between -1 and 1, we divide this into N equal parts, where N is the sensitivity of the SC. The average amplitude of the input sound during δt (about 0.01 seconds) is measured and converted into a Tactile Score. Then, the input speech is converted into a two-dimensional pattern based on the Tactile Score at time t and $t + \delta t$.

This two-dimensional pattern is the volume change pattern of the input speech, and the Game of Life converts the two-dimensional pattern into other patterns. Recall the episode of the introduction of the Tactile Score. What was the difference between a cheerful good morning and a cheerless good morning? The difference was the way they said it, that is, the pattern of the volume change.

To change the pattern of the volume change is to change the way we say things. In other words, when the Game of Life converts a volume change into a two-dimensional pattern into a different volume change, it is equivalent to changing the wording. The SC outputs the input environmental sound differently. In order to output sound, we need vocal cords. So the Game of Life converted the 2D pattern into a MIDI code. This MIDI code was then used as input to a Digital Audio Workstation, DAW. We used Ableton Live as the DAW. We then generated "musical noise" from the continuous MIDI input.

7. Garandoh no Niwa (empty garden)

We worked with Butoh dancer Norihito Ishii (Sankai jyuku company) to create a dance piece using SC. *Butoh (Butō) is a form of Japanese dance theatre that encompasses a diverse range of activities, techniques, and motivations for dance, performance, or movement.*[13] *A founder of Butoh, Hijikata explored the transmutation of the human body into other forms, such as those of animals. He also developed a poetic and surreal choreographic language, butoh-fu (fu means "score" in Japanese), to help the dancer transform into other states of being.*

Butoh Score is a choreographic pair of counts and poetic statements. For example, it can be described as follows:

```
A body lies in a heap of flesh, 40 counts.
The corpse comes to life. 20 counts.
```

7.1. *A fusion of Butoh Score and Tactile Score*

We read and recorded the Butoh Score according to the count. The recorded sound was then subjected to a low-pass filter to extract very low-frequency sounds below 20 Hz. We created a sound art by layering natural sounds recorded in the forest, based on the very low-frequency sound.

The length of this work follows the count of Butoh Score, so it is almost the same as the dance length. This sound art piece was played at the beginning of the dance and was used as an input to Sound Creature during the performance. The Butoh Score was transformed into the Tactile Score via voice and fed into the Sound Creature, thus creating a fusion of Butoh Score and Tactile Score.

References

1. G. W. Leibniz, (Berlin: Akademie Verlag, 1923), Dissertatio de arte combinatoria, 1666, Sämtliche Schriften und Briefe A VI 1, p. 163; Philosophische Schriften (edited by Carl Immanuel Gerhardt), Bd. IV, 30.
2. Y. Sasaki, *Leibniz jyutsu* (in Japanese), Kohsaku sha, 2002.
3. A. Riegl, (Vienna, 1901 und 1923), Die spätrömische Kunstindustrie nach den Funden in Österreich-Ungarn.
4. Ryo Sawayama, srtscape, (vol. October 01, 2001), DNP Museum Information Japan.
5. R. Suzuki and Y. Suzuki, *Tactile Score — A Knowledge Media for Tactile Sense.* Springer Verlag, 2014.
6. M. Prsa, D. Kilicel, A. Nourizonoz, *et al.* A common computational principle for vibrotactile pitch perception in mouse and human. *Nat. Commun.* 12:5336, 2021. https://doi.org/10.1038/s41467-021-25476-9.
7. Y. Suzuki, Partial differentiation of sensibilities and its applications. *J. Rob., Net. Art. Life* 7(4):258–261, 2020.
8. *The Tale of the Heike.* Penguin, 2012.
9. Y. Suzuki, R. Suzuki and J. Watanabe, Translation from text to touch: touching a "Japanese old tale", In CHI '14 *Extended Abstracts on Human Factors in Computing Systems* (CHI EA '14). Association for Computing Machinery, New York, NY, USA, 2014, 685–694, 2014. Doi: https://doi.org/10.1145/2559206.2578877.
10. Y. Suzuki, Method of extracting sensibility from time series data and converting it to vibrotactile. *J. Rob. Net. Art. Life*, 7(10), 2008.
11. Y. Suzuki, Effects of Mixing low-frequency waves — Deep micro vibrotactile with sounds, *J. Rob. Netwo. Art. Life*, 8(4), 2022.
12. https://en.wikipedia.org/wiki/Conway%27s_Game_of_Life.
13. https://en.wikipedia.org/wiki/Butoh.

https://doi.org/10.1142/9789811257155_0030

Chapter 30

Complex Excitable Media: Activators Design, While Inhibitors Embellish

Sergio Alonso Muñoz
Department of Physics, Universitat Politecnica de Catalunya, Barcelona, Spain
s.alonso@upc.edu

A system composed by coupled reaction–diffusion equations is one of the most classical scenarios of pattern formation. There are also other mechanisms of pattern formation,[1] however, all have in common the interplay between a mechanism of transport and the nonlinearities of the system.[2] Probably, the simplest example of pattern formation involving reaction–diffusion equations is a bistable front moving with constant velocity. A certain concentration U experiences an autocatalytic nonlinear growth and the corresponding increase of U travels along the media by diffusion.[3] Such a process is simply modeled by a single reaction–diffusion equation:

$$\frac{\partial U}{\partial t} = D\nabla^2 U + F(U), \tag{1}$$

where $F(U)$ is a nonlinear function, typically a cubic function on U, giving rise to two stable solutions, i.e., bistable system. Under an adequate initial condition a front changes the stability from the metastable to the more stable solution. In such a case, the concentration U activates his own production and can be noted as *activator*. Such activation propagates by diffusion through the medium forming a traveling front with a fixed velocity which depends on the parameters on Eq. (1). The resulting fronts are robust and very stable waves which propagate through the system.

We can change the stability of the stable solutions by the addition of an extra variable V which inhibits the production of U. The new variable V is usually noted as *inhibitor*. Such an extra variable is produced due to the presence of U, and it gives rise to a typical activator–inhibitor system[4] of two coupled equations:

$$\begin{aligned} \frac{\partial U}{\partial t} &= D\nabla^2 U + F(U) - V, \\ \frac{\partial V}{\partial t} &= D_v \nabla^2 V + \epsilon(U - V). \end{aligned} \tag{2}$$

The effect of the inhibitor is to return to the initial stable state after the increase due to the autocatalytic part of the activator, resulting in an excitation, a large excursion in the variable space, instead of a definitive change on the variable value. This type of dynamics corresponds to excitable media.[5] This effect produces the generation of a traveling wave[6,7] instead of the traveling front typical in bistable media. The perturbation of the neighboring region corresponds to an excitation which propagates forming the wave. The temporal scale of the return to the rest state is given by the parameter ϵ in Eq. (2). While the velocity and curvature are designed by the dynamics of the activator U, the inhibitor V outlines the thickness and shape of the wave.

In two dimensions, the breakup of the traveling wave generates a free end. The resulting wave rotates around this free end, giving rise to a rotating spiral wave, see an example in Figure 1. The beauty of spirals is out of discussion, it has been a typical motive for human ornaments forever and ever. Spiral waves in excitable media have been observed in chemical systems like the Belousov-Zhabotinsky reaction,[8] CO Oxidation in catalytic surfaces;[9] in biological systems like calcium waves inside Xenopus laevis oocytes,[10] and action potential propagation in cardiac tissue.[11]

Fig. 1. Rotating spiral wave in a generic model of excitable media. A rotating spiral wave is the most typical spatio-temporal structure appearing in excitable media. It rotates with constant velocity around the free end of the spiral giving rise to a continuous emission of traveling waves. The spiral wave is obtained using a FitzHugh–Nagumo[6,7] model consisting in an activator and an inhibitor, see Eqs. (2). We plot the value of the variable U, but a similar pattern is obtained if we plot the variable V.

Under random initial condition excitable media can produce multiple spiral waves, see Figure 2, a set of interacting rotating spiral waves which keep the relative distances among the centers of rotation.

The complexity of the spatio-temporal patterns produced in excitable media is increased with additional inhibitors. Following the

Fig. 2. Multiple spiral waves formation under random initial conditions in an excitable model. Several spiral waves rotate simultaneously with the same frequency in a homogeneous extended system generated by a set of random initial conditions. Waves are obtained in a model of chemical waves in microemulsions, however, similar behavior can be obtained in simpler models with adequate initial conditions.

structure of Eqs. (2) we add an additional inhibitor:

$$\begin{aligned} \frac{\partial U}{\partial t} &= D\nabla^2 U + F(U, V, W) - V - W, \\ \frac{\partial V}{\partial t} &= D_v \nabla^2 V + \epsilon_1 G_1(U, V), \\ \frac{\partial W}{\partial t} &= D_w \nabla^2 W + \epsilon_2 G_2(U, W), \end{aligned} \tag{3}$$

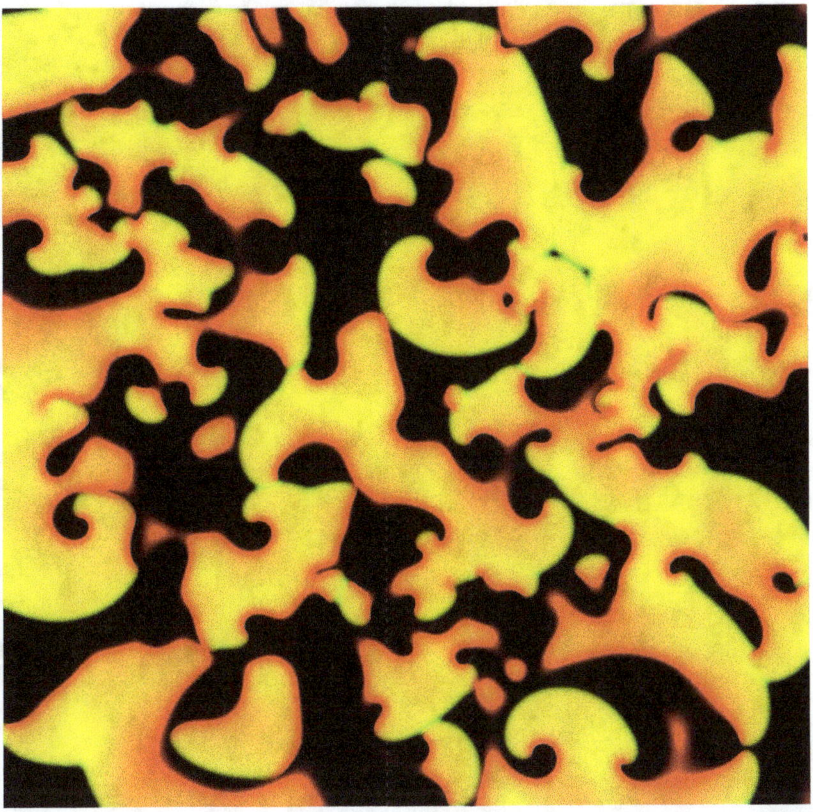

Fig. 3. Spatio-temporal chaos from an alternan instability in a model of action potential propagation in cardiac tissue. Single spiral waves may be unstable and may breakup because of the interaction between consecutive waves. The rotation of the spiral wave naturally produces a periodic pacing of the tissue which can produce a bifurcation to long-short waves. If the interaction is strong enough, it produces the local decay of the wave, conduction block, and the production of two new spiral waves. Repeating the same mechanism and due to interaction among spirals, complex wave patterns are found.

where ϵ_1 and ϵ_2 are two temporal scales which may be very different.[12] Therefore, each inhibitor controls the dynamics at different temporal scales. The two temporal scales of the inhibitors in Eqs. (3) can be tuned to produce very different patterns. The inclusion of such a second inhibitor produces complex excitable media, which have been employed in the context of computational modeling

Fig. 4. Spatio-temporal chaos from an interaction of Turing and wave instabilities, a model of chemical waves in microemulsions. The use of two inhibitors permits the simultaneous induction of two different instabilities which under certain windows of parameters permit the coexistence of spiral waves with Turing patterns. The resulting pattern is a complex mix of spatio-temporal structures.

of cardiac tissue[13] and on the modeling of chemical reactions in microemulsions.[12,14]

The interaction between consecutive waves of a rotating spiral wave can induce an alternation among the thickness of the waves. Finally, it gives rise to unstable spiral waves and produces a turbulence of waves which reminds of the cardiac fibrillation,[15] see Figure 3. On the other hand, spiral waves can be unstable and produce multiple spiral waves due to interaction with other kind of instabilities typical of reaction–diffusion systems, see in Figure 4 the

Fig. 5. Turing patterns in reaction–diffusion systems. The increase of the diffusion of the inhibitor in two coupled reaction-diffusion equations like Eqs. (2) can produce the formation of stable and static patterns. It is actually not a typical pattern for excitable media, however, its formation together with the presence of spiral waves in more complex systems, for example Eqs. (3), can give rise to the patterns obtained in Figure 4.

interplay between the rotating spiral waves, for example see Figure 2 and a Turing instability, see Figure 5. Turing patterns result from large values of the diffusion coefficient of one of the inhibitors in comparison with the diffusion coefficient of the activator.[4]

The dynamics is more complex in three-dimensional systems. Scroll waves are the three-dimensional equivalents to spiral waves.[16] In its cross section, a scroll wave has the shape of a spiral.[17] Such spirals are stacked one upon another and rotate around a filament

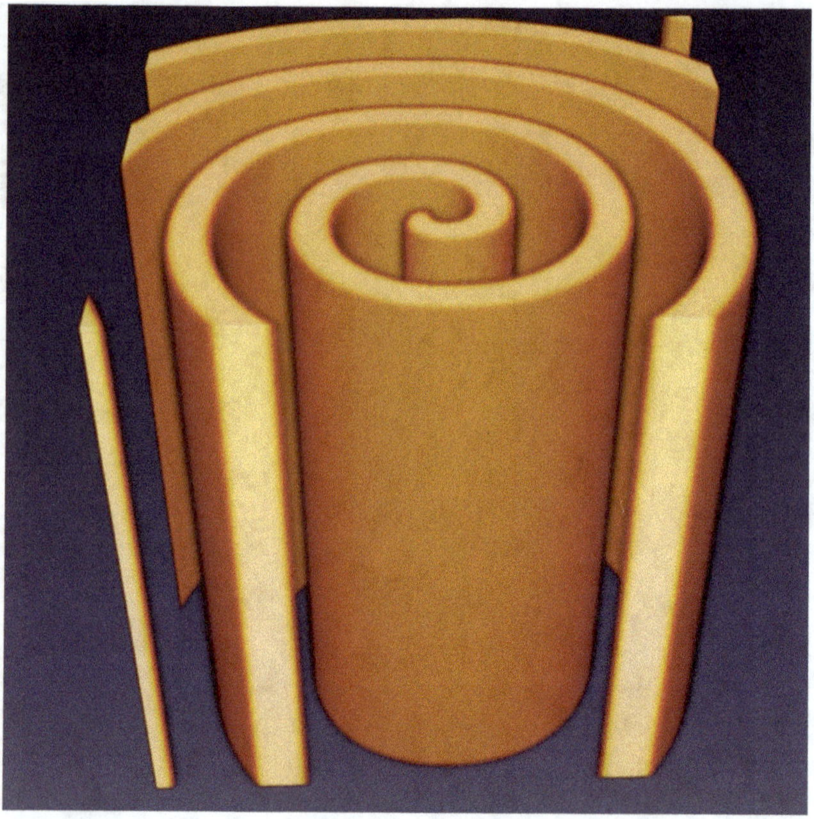

Fig. 6. Rotating scroll wave in a generic model of excitable media. Scroll waves are the three-dimensional analogs to the two-dimensional spiral waves. They rotate around a filament analogously as the spiral waves typically rotate around a circular core. Somehow, a scroll wave can be considered as a stack of rotating spiral waves that are organized by the filament, a line connecting the cores. A straight scroll wave is stable if the filament has positive tension and there are no strong interactions among consecutive waves.

that occupies the centre of the scroll.[18] The scroll may be straight, see Figure 6, or it may also be closed into a ring, form knots, or even generate a complex turbulent state,[19,20] see an example in Figure 7.

Finally, there are several computational tools to solve the reaction–diffusion equations in a restricted region of the space. The use of a phase field facilitates the reconstruction of the borders of the irregular domains in the heart[18] or the cylindrical shape of a

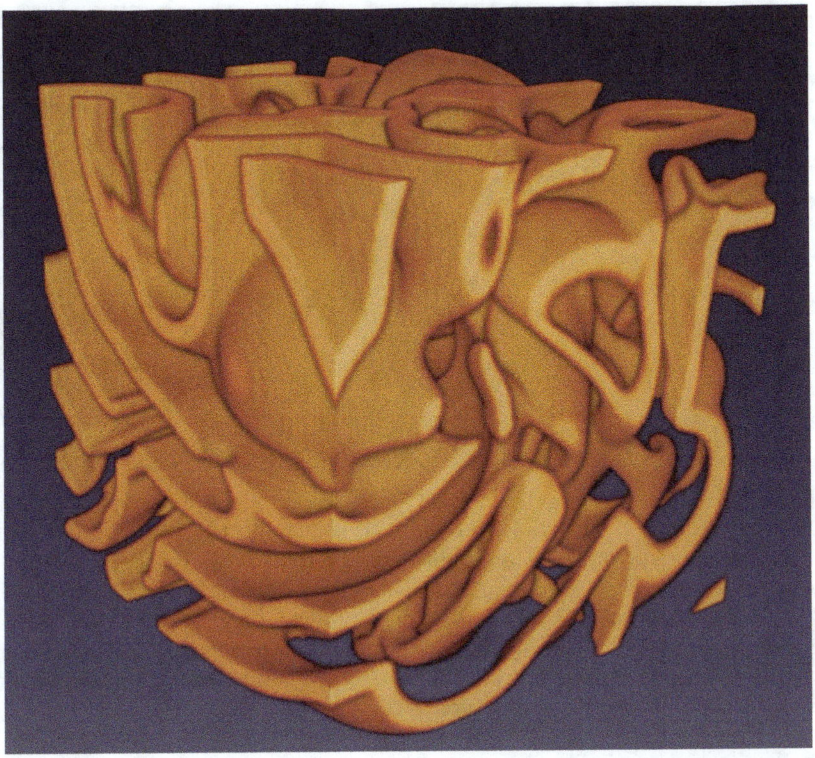

Fig. 7. Negative filament tension of the filament of scroll waves in a generic model of excitable media. If the filament tension of the scroll wave is negative, then the filament will tend to increase its length and the straight scroll wave shown in Figure 6 becomes unstable. The filament tends to stretch, bend, loop, and produce an expanding tangle that fills up the volume. Through the process of local stretching and bending of a filament, very complex and potentially chaotic wave patterns were expected to develop.

three-dimensional chemical reactor.[21] For example, we can fit the complex dynamics depicted by the excitable waves inside regular or irregular domains, see, for example, the turbulent waves inside two neighboring spheres in Figure 8.

In summary, several pretty patterns obtained in complex excitable media have been shown in this chapter. While the activator marks the direction of the propagation and even the velocity, the inhibitor can produce important interactions responsible for the formation of complex patterns. The increase of the number of

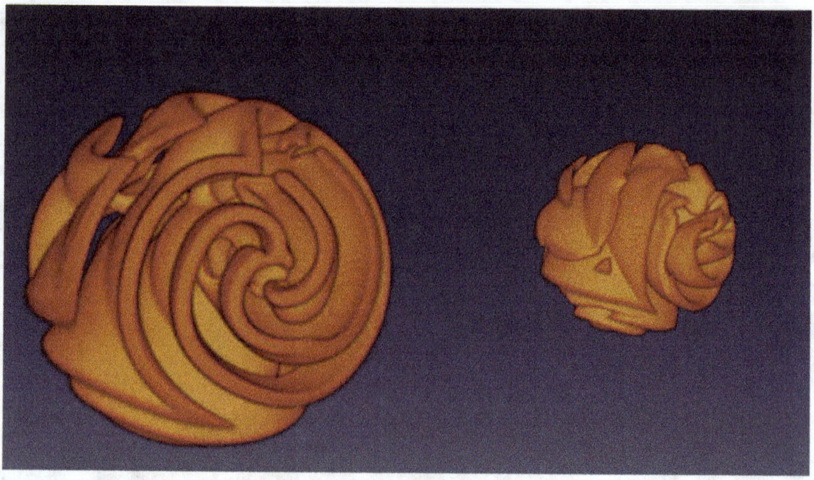

Fig. 8. Negative filament tension of the filament of scroll waves in a generic model of excitable media under spatial confinement in two spheres. We employ a phase field model which is an additional variable whose value goes from 1 in the interior of a certain domain to 0 outside. The reaction–diffusion variables U and V are multiplied by the additional phase field. The transition between the two values is smooth and keeps the non-flux boundary conditions in the irregular border. Two disconnected spheres are generated by the phase field and the dynamics of the turbulent behavior computed in their interior.

inhibitors can directly increase the complexity of the patterns. We have also shown that the pass to three dimensions unveils hidden instabilities.

I hope the reader will enjoy the beauty of the patterns obtained in complex excitable media as much as I have enjoyed with their production.

References

1. M. C. Cross and P. C. Hohenberg, Pattern formation outside of equilibrium. *Rev. Mod. Phys.* 65(3):851, 1993.
2. R. C. Desai and R. Kapral, *Dynamics of Self-organized and Self-assembled Structures.* Cambridge University Press, 2009.
3. J. P. Keener and J. Sneyd, *Mathematical Physiology.* Springer, vol. 1, 1998.
4. J. D. Murray, *Mathematical biology. II Spatial models and biomedical applications {Interdisciplinary Applied Mathematics V. 18}*. Springer-Verlag, New York, Incorporated New York, 2001.

5. E. Meron, Pattern formation in excitable media. *Phys. Repo.* 218(1):1–66, 1992.
6. R. FitzHugh, Impulses and physiological states in theoretical models of nerve membrane. *Biophys. J.* 1(6):445–466, 1961.
7. J. Nagumo, S. Arimoto and S. Yoshizawa, An active pulse transmission line simulating nerve axon. *Proc. IREE.* 50(10):2061–2070, 1962.
8. A. T. Winfree, Spiral waves of chemical activity. *Science* 175(4022):634–636, 1972.
9. S. Jakubith, H. Rotermund, W. Engel, A. Von Oertzen, and G. Ertl, Spatiotemporal concentration patterns in a surface reaction: Propagating and standing waves, rotating spirals, and turbulence. *Phys. Rev. Lett.* 65(24):3013, 1990.
10. J. Lechleiter, S. Girard, E. Peralta, and D. Clapham, Spiral calcium wave propagation and annihilation in xenopus laevis oocytes. *Science* 252(5002):123–126, 1991.
11. J. M. Davidenko, A. V. Pertsov, R. Salomonsz, W. Baxter, and J. Jalife, Stationary and drifting spiral waves of excitation in isolated cardiac muscle. *Nature* 355(6358):349, 1992.
12. S. Alonso, K. John, and M. Bär, Complex wave patterns in an effective reaction–diffusion model for chemical reactions in microemulsions. *J. Chem. Phys.* 134(9):094117, 2011.
13. F. Fenton and A. Karma, Vortex dynamics in three-dimensional continuous myocardium with fiber rotation: Filament instability and fibrillation. *Chaos* 8(1):20–47, 1998.
14. I. Berenstein and J. Carballido-Landeira, Spatiotemporal chaos involving wave instability. *Chaos* 27(1):013116, 2017.
15. S. Alonso, M. Bär, and B. Echebarria, Nonlinear physics of electrical wave propagation in the heart: A review. *Rep. Prog. Phys.* 79(9):096601, 2016.
16. A. T. Winfree, Scroll-shaped waves of chemical activity in three dimensions. *Science* 181(4103):937–939, 1973.
17. S. Alonso, R. Kähler, A. S. Mikhailov, and F. Sagués, Expanding scroll rings and negative tension turbulence in a model of excitable media. *Phys. Rev. E.* 70(5):056201, 2004.
18. E. M. Cherry and F. H. Fenton, Visualization of spiral and scroll waves in simulated and experimental cardiac tissue. *New J. Phys.* 10(12):125016, 2008.
19. S. Alonso and A. V. Panfilov, Negative filament tension in the luo-rudy model of cardiac tissue. *Chaos* 17(1):015102, 2007.
20. S. Alonso, F. Sagués, and A. S. Mikhailov, Taming winfree turbulence of scroll waves in excitable media. *Science* 299(5613):1722–1725, 2003.
21. P. Dähmlow, S. Alonso, M. Bär, and M. J. Hauser, Twists of opposite handedness on a scroll wave. *Phys. Rev. Lett.* 110(23):234102, 2013.

https://doi.org/10.1142/9789811257155_0031

Chapter 31

Designing Microchemical Systems Like Living Matters Induced by Interfacial Phenomena

Daigo Yamamoto[*,‡], Erika Okita-Nawa[†], and Akihisa Shioi[*]

[*] *Doshisha University, Kyoto, Japan*

[†] *Osaka Prefecture University, Osaka, Japan*

[‡] *dyamamot@mail.doshisha.ac.jp*

The details of the mechanism of the phenomenon of spatio-temporal self-organization and autonomous motion, which is a mundane occurrence in the living matter systems, are yet to be elucidated due to its complexity, but it functions follow the laws of physical chemistry. Our goal is to express the phenomena, which occur, in nature, in a non-living system and to investigate their mechanisms. In the chapter, we show the five interesting phenomena, induced by a variety of interfacial phenomena, with micro objects (<1 mm).

Our research at the Molecular Chemical Engineering Laboratory in Doshisha University, Japan, is aimed at designing chemical systems similar to living matter. This means that the systems change autonomously in response to variations in their surrounding environment and alter their temporal or spatial patterns to maintain their essential functions. Living matter on Earth exhibits non-equilibrium phenomena, such as self-motility and self-organization, with the expense of chemical energy under isothermal conditions (i.e., body temperature). Such systems are only feasible as non-equilibrium open systems.

We mainly targeted micro objects (<1 mm) as the chemical systems. The chemical systems introduced in this chapter were accomplished by simple macro-operations, and therefore the same phenomena can be easily reproduced in any laboratory possessing an optical microscope. One of the most remarkable points in the difference between micro- and macrosystems is that in the former, the viscosity dominates the inertia, i.e., an ultralow Reynolds number ($Re \ll 1$).[1] In this situation, the strategy used for macroscopic motions (e.g., direct current macromotors) can never be applicable. To understand the harsh environment surrounding a microsystem, let us consider the scenario in which a football player kicks a ball at a speed of 100 km/h. The high-speed ball is hardly decelerated until it reaches a goal net. Let us also imagine that the player kicks a micron-sized ball at the same speed. Its motion is immediately damped by the environmental viscosity, dissipated to the thermal fluctuation, and converted to Brownian motion, within a time range of microseconds to milliseconds (if human beings were micro-sized organisms, the strategy of football could not be pass-and-shoot but only dribbling!!). Therefore, a microsystem requires a continuous driving force to maintain non-equilibrium phenomena, and it is important to find a different principle to successfully fabricate such a microsystem. The key is that a small object has a large specific surface area, indicating that the sustainable driving force arising from "interfacial phenomena" is useful with a decrease in the scale of a system. The moving object can be an open system in the presence of an interface: The interface itself is a real open system that can maintain the system non-equilibrium. In this chapter, we introduce the following five interesting phenomena induced by a variety of interfacial phenomena.

- Ex. 1: Catalytic reaction (Figure 1)
  Autonomous motion of catalytic particle[2,3]
- Ex. 2: Diffusion and reaction (Figure 2)
  Deformation of a vesicle[4–6]
- Ex. 3: Dewetting transition and evaporation (Figure 3)
  Dynamic self-organization of oil droplets[7]
- Ex. 4: Drying (Figure 4)
  Spiral patterning of carbon nanotubes[8]
- Ex. 5: Diffusion and crystal growth (Figure 5)
  Micro-patterning of Barite and Celestite particles[9]

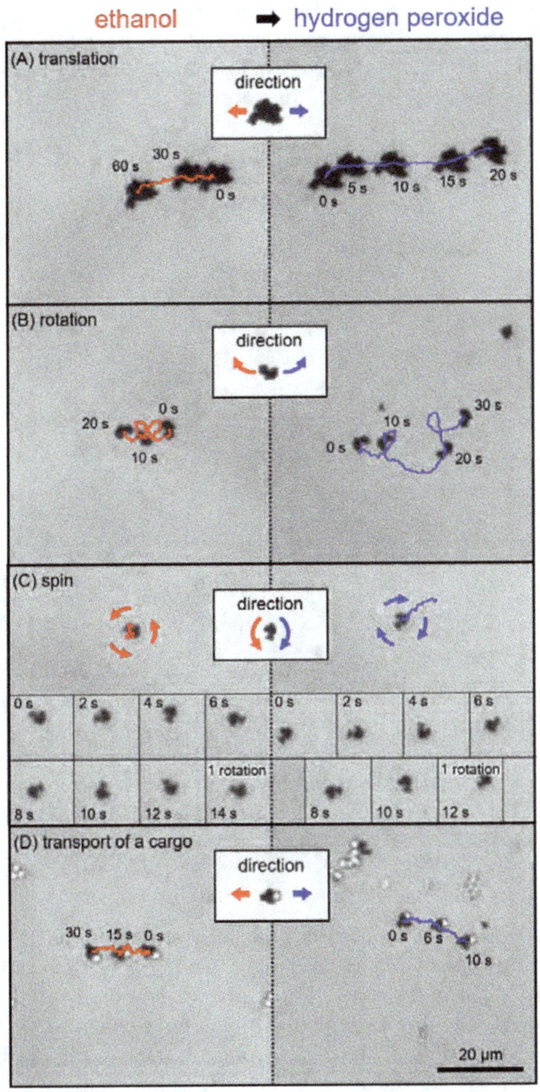

Fig. 1. Autonomous motions of simple catalytic particles without compositional asymmetry propelled by catalytic reactions. How do catalytic particles behave in solutions containing reactants? We have observed the behavior of commercial Pt powders in aqueous solutions containing hydrogen peroxide[2] and ethanol.[3] Spherical Pt particles show more vigorous, random motion than Brownian motion. However, their aggregates show various regulated motions, such as translation, rotation, and spin, depending on the particle shape. Interestingly, motile direction is reversed by exchanging ethanol with hydrogen peroxide. The trajectory and snapshots of the (a) translational, (b) rotational, and (c) spin-regulated motions

Fig. 1. (*Continued*) for Pt aggregates, and (d) accompanying silica cargo (diameter: 1.5 μm) are shown above before and after switching from 1 vol% ethanol to 1 wt% hydrogen peroxide in water.

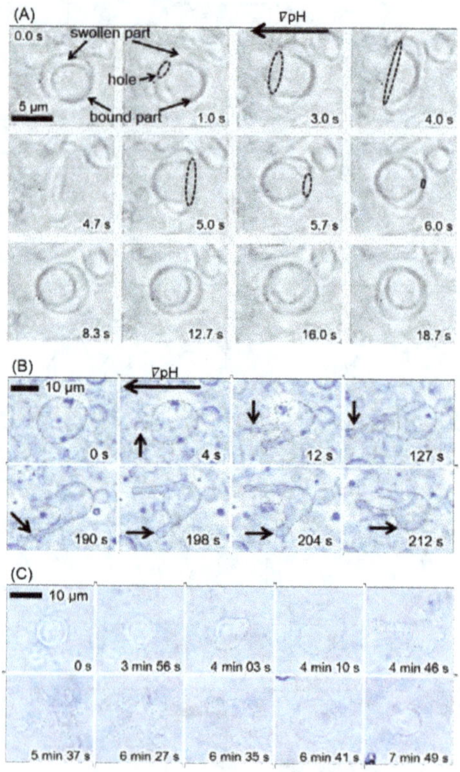

Fig. 2. Various transformations of a vesicle by surrounding pH variation. These vesicles were composed of oleate and oleic acid. (a) The vesicle structure was similar to that of a stomatocyte.[10] The inner water was located on one side (the swollen part), in the center of which a small hole exists. The opposite side of the vesicle was nearly dehydrated (the bound part). A pH gradient around the vesicle was formed by the diffusion of NaOH. When there was no pH gradient, the hole faced various directions randomly. However, after a pH gradient had formed, the swollen part was located at the higher pH side. Then, the hole opened and the vesicle reversed it inside out until the hole was located at the lower pH side. After this motion, the vesicle rotated to turn the hole toward the higher pH side again. This rhythmic shape change was repeated several tens of times.[4] (b) The vesicle had a disk-like structure. A pH gradient around the vesicle was formed by the diffusion of a diluted NaOH solution. The inner water leaked out by osmotic pressure formed by the NaOH; hence, a pseudopod was extruded from the surplus membrane. Then, the osmotic pressure was eliminated

Fig. 2. (*Continued*) by a transport of H^+ across the membrane (flip-flop[11]). The pseudopod was retracted by membrane elasticity, and the vesicle shape returned to the initial state. The osmotic pressure continued to increase by the diffusion of NaOH. All the pseudopods extruded toward the higher pH side. This amoeboid-like shape change was repeated.[6] (c) Vesicle transformation in water containing glucose oxidase and glucose. A pH change around the vesicle was induced by the oxidation of glucose. The stomatocyte vesicle opened the hole and changed to a disk-like structure. Then, several pseudopods were formed and retracted. The vesicle returned to a stomatocyte structure again. The reversible transformation was repeated two times.[5]

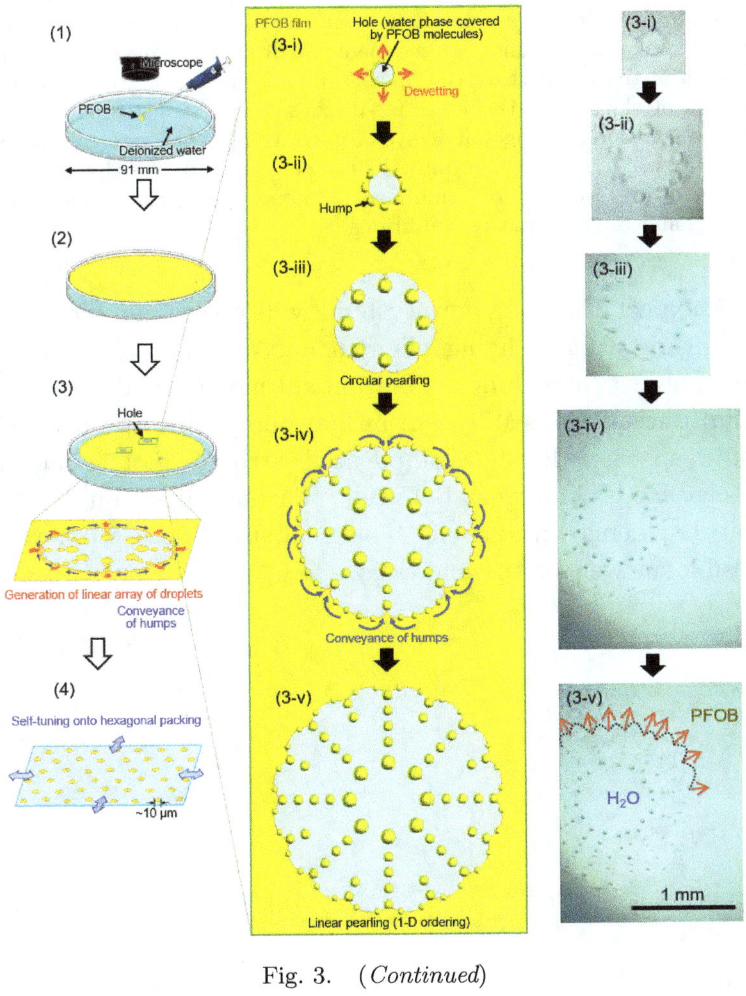

Fig. 3. (*Continued*)

Fig. 3. Dynamic self-organization of oil droplets spreading on water. We show a unique cascade in the self-ordering of droplets accompanied by a dewetting transition: from circular pearling, through the appearance of linearly arranged pearling (one-dimensional ordering), to transformation into a hexagonal pattern on a two-dimensional surface.[7] Such dynamic behavior of self-emergent droplets was observed when a thin liquid layer (film) of an immiscible fluorocarbon oil (perfluoro octylbromide (PFOB)) was placed on a water surface. The sequence of events was as follows. (1) PFOB droplet (8 μL) was pipetted onto a water surface. (2) Immediately after contact with the water, it spread to form a thin film. (3-i and ii) As the PFOB gradually evaporated, a hole appeared spontaneously and gradually grew as a result of a local dewetting transition. Interestingly, a circular pearling structure at the rim generated spontaneously with the growth of the dewetting hole. (3-iii, iv, and v) At the next stage, linear arrays, or a one-dimensional arrangement, of droplets formed accompanied by the growth of the hole, where the linear arrays formed in a radial manner from the center of the hole. (4) These one-dimensional arrangements then evolved into two-dimensional hexagonal arrays of microdroplets through collective rhythmical shrinking/expanding motions. The emergence of such dynamic patterns is discussed in terms of the nonlinear kinetics of wetting transition under thermodynamically dissipative conditions.

We expect that our new systems will give a significant insight regarding mysterious living phenomena. Why can a motor protein carry an object toward its destination autonomously despite vigorous thermal fluctuations? Why can an ion pump pick up only a certain cationic species? Is the complicated structure of living matter necessary for its emergent behavior? Through designing "artificial" chemical systems without any biological substances, we may obtain a possible answer for these questions.

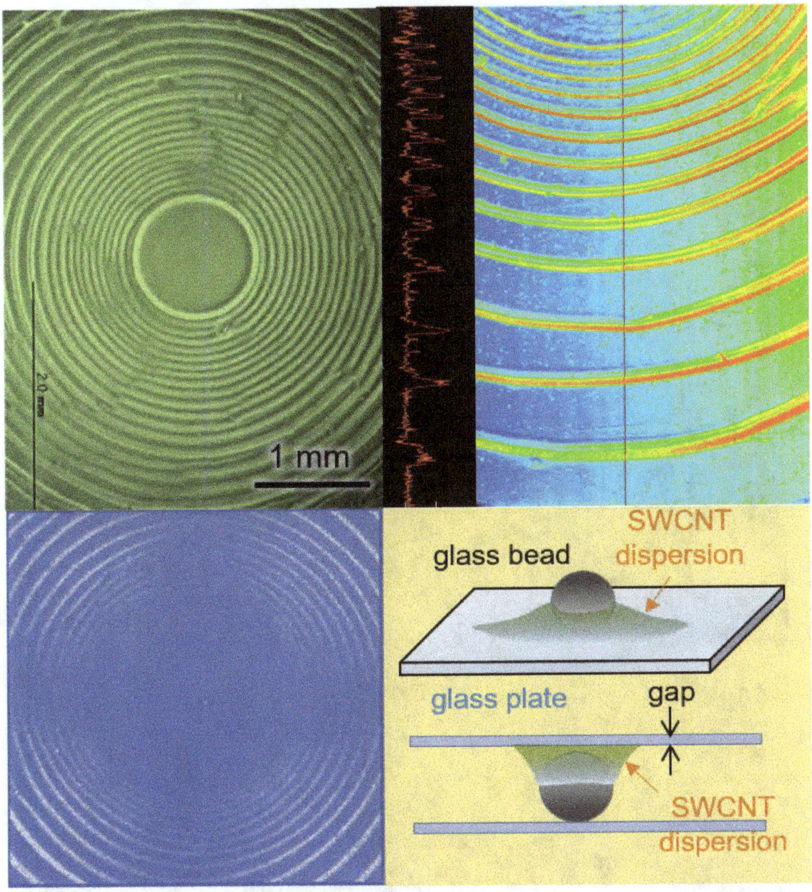

Fig. 4. Microspiral patterning of single walled carbon nanotube (SWCNT) via Drying. What will be expected when a coffee-ring pattern formation is applied to SWCNT dispersion? In a designed setup, SWCNTs were spontaneously aligned in a stripe that formed a single spiral.[8] A glass bead played the role of tuning the drying pattern (bottom-right).[12] During the drying, the contact line of the dispersion retreated, depositing SWCNTs and the dispersant (polyfluorene). The deposited line drew a microspiral. (Central hole of top-left photo is 1.1 mm.) The height of the line was almost constant, around 0.5 mm by laser microscopy (top-right). As the deposited SWCNTs aligned along the stripe, the microspiral was optically anisotropic under cross-Nicol polarized microscopy (bottom-left). This technique may be applied to various types of nanotubes and nanorods for tuning the optical and electromagnetic properties of microscale objects.

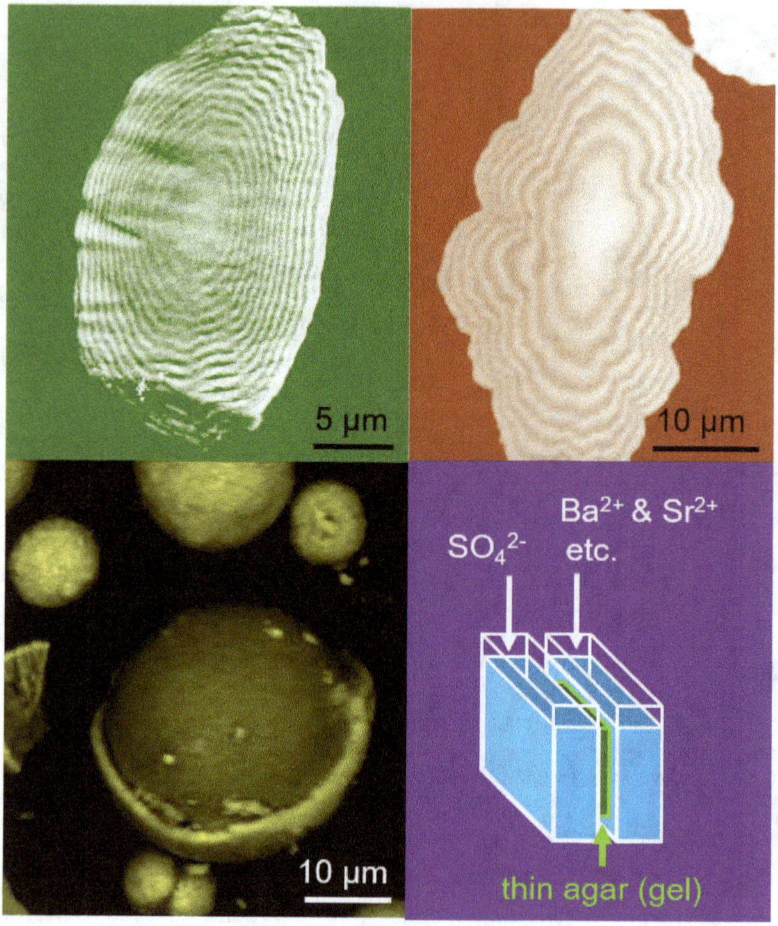

Fig. 5. Micro-patterning of Barite and Celestite within laboratory time scale. A patterned mineral such as Agate is prized as a treasure as it takes an extremely long time to form and its reserve in Earth's crust is limited.[13] Materials with chemical and physical patterns in them are useful for various fields in modern technology. If the acceleration of natural mineral patterning is realized, the technique may become a part of modern technology. As can be seen in the above photograph. The ions that form barite ($BaSO_4$) and celestite ($SrSO_4$) diffused into a thin gel (agar) from both the side reservoirs (bottom-right).[9] The cations and anion form small particles in the agar within a short time. One cannot obtain any characteristic particles via the continuous diffusion. However, micropatterned particles are obtained by intermittent diffusion, meaning that the reservoirs are periodically replaced by empty boxes (top). This technique and its analogue can be utilized to design other types of particles, such as a core-shell type particle (bottom-left).

References

1. D. Yamamoto and A. Shioi, Self-propelled nano/micromotors with a chemical reaction: Underlying physics and strategies of motion control. *KONA Powder and Particle Journal.* 32:2, 2015.
2. D. Yamamoto, A. Mukai, N. Okita, K. Yoshikawa, and A. Shioi, Catalytic micromotor generating self-propelled regular motion through random fluctuation. *J. Chem. Phys.* 139(3):034705, 2013.
3. D. Yamamoto, T. Takada, M. Tachibana, Y. Iijima, A. Shioi, and K. Yoshikawa, Micromotors working in water through artificial aerobic metabolism. *Nanoscale.* 7(31):13186, 2015.
4. E. Nawa, Y. Nishigaki, D. Yamamoto, and A. Shioi, Rhythmic shape change of a vesicle under a pH gradient. *Soft Matter* 9(32):7832, 2013.
5. E. Nawa, D. Sakashita, K. Owaki, D. Yamamoto, and A. Shioi, Reversible transformation of an acid/soap vesicle by glucose oxidation. *Colloids Interface Sci. Commun.* 8:10, 2015.
6. E. Nawa, D. Yamamoto, and A. Shioi, Chemotactic amoeboid-like shape change of a vesicle under a pH gradient. *Bull. Chem. Soc. Japan* 88(11):1536, 2015.
7. D. Yamamoto, C. Nakajima, A. Shioi, M. P. Krafft, and K. Yoshikawa, The evolution of spatial ordering of oil drops fast spreading on a water surface. *Nat. Commun.* 6:7189, 2015.
8. K. Mae, H. Toyama, E. Nawa-Okita, D. Yamamoto, Y.-J. Chen, K. Yoshikawa, F. Toshimitsu, N. Nakashima, K. Matsuda, and A. Shioi, Self-organized micro-spiral of single-walled carbon nanotubes. *Sci. Rep.* 7(1):5267, 2017.
9. K. Suzuki, A. Okazaki, and A. Shioi, Control of periodicity and growth rate of oscillatory zoning in particle by chemical process. *J. Chem. Eng. Japan* 45(3):174, 2012.
10. U. Seifert, Configurations of fluid membranes and vesicles. *Adv. Phys.* 46(1):13, 1997.
11. F. Kamp and J. A. Hamilton, pH gradients across phospholipid membranes caused by fast flip-flop of un-ionized fatty acids. *Proc. Nation. Acad. Sci. the United States of America* 89(23):11367, 1992.
12. Y.-J. Chen, K. Suzuki, H. Mahara, K. Yoshikawa, and T. Yamaguchi, Self-organized Archimedean spiral pattern: Regular bundling of fullerene through solvent evaporation. *Appl. Phys. Lett.* 102(4):041911, 2013.
13. A. Putnis, L. Fernandez-Diaz, and M. Prieto, Experimentally produced oscillatory zoning in the (Ba, Sr)SO4 solid solution. *Nature* 358:743, 1992.

https://doi.org/10.1142/9789811257155_0032

Chapter 32

Chemical Composition

Sarah Sebastian
Northampton U.K.
sarah@shinystudio.co.uk

Chemical Composition. Sarah Sebastian.

This work is created using a molecular chemistry solution. By binding pigments to different groups of liquids, a marbling effect if you will, we are able to see the instant movement of water and gravity. Rather than a traditional water surface tension marbling, this process I have named Chemical Composition, is the marbling of the liquid solution in its entirety. As the water rolls, alongside the other components we are able to see the 3D instant impression of the molecular movement. Once the solution moves with a fluid motion, an instant impression is created. Room for error is a part of the process as the only predictability of Chemical Composition is within the solution, not the outcome. I have personally been humbled and inspired to work with natural elements to facilitate rather than control the impression. This has helped me in my life, to understand and respect movement, harmony, and dissonance all within the same solution. Most astoundingly though, is the impression taken by viewers of the works. It provides the human mind with a Rorschach type test. As it is fluid and organic, each person that sees Chemical Composition has to personally make sense of what they see. Animals, sex, death, and landscapes are some of the typical feedback I have received. It gives a deeper meaning to the science behind free movement and creativity from outside and within.

https://doi.org/10.1142/9789811257155_0033

Chapter 33

How to Face the Complexity of the 21st Century Challenges? The Contribution of Natural Computing

Pier Luigi Gentili

Department of Chemistry, Biology, and Biotechnology, Università degli Studi di Perugia, Italy

pierluigi.gentili@unipg.it

The 21st century challenges are complexity challenges because they take into consideration complex systems and other types of complexities, such as computational, descriptive, and bioethical complexities. This chapter proposes some strategies to tackle the compelling challenges of this century. A promising approach is the interdisciplinary research line of natural computing that includes artificial intelligence.

1. Introduction

A fundamental role of science is solving practical problems and improving the psychophysical well-being of humans. Science succeeds in playing this role when it promotes technological development. Mutual positive feedback action exists between science and technology: science sparks technological development. At the same time, new technologies allow an always more in-depth analysis of natural phenomena. Cutting-edge technologies let us manipulate materials at the molecular and atomic scale, send robots to other planets of our solar system, and engineer living beings. Despite many efforts, there are still compelling challenges that must be won. They are the so-called 21st-Century Challenges included in the 2030 Agenda composed by the United Nations. Examples of these challenges are all those diseases that are still incurable. Some challenges concern human activities. Our manufacturing processes must become circular, minimizing waste. They should not perturb the fragile stability of natural ecosystems and contribute to climate change. Poverty should be eradicated from Earth, and our societies should ensure justice. Whenever we tackle one of these challenges, we need to deal with Complex Systems, such as living beings, ecosystems, climate, and human societies. When we focus on human health, the immune and the nervous systems are other examples. Complex Systems appear so diverse. Currently, they are investigated by distinct disciplines. The burgeoning Complexity Science tries to point out the features shared by all Complex Systems, i.e., the characteristics of Natural Complexity (NaC).[1–4]

2. Complex Systems

All Complex Systems can be described as networks with nodes and links. Different Complex Systems usually have distinct networks' architectures; the nodes and links are often diverse and evolve in time. Complex Systems are constantly out-of-equilibrium in the thermodynamic sense. The behavior of inanimate matter is driven by force fields, whereas that of living beings is information-based. Moreover, Complex Systems exhibit emergent properties. The integration of nodes and links features gives rise to properties that belong to the entire network. The whole is more than the sum of its parts. Finally, there is another universal attribute: Complex Systems cannot be described exhaustively. In other words, science finds many difficulties in predicting the behavior of Complex Systems, especially in the long term. These difficulties are due to three principal reasons.

The first reason has a computational character. Most of the computational problems regarding Complex Systems and their simulation, such as scheduling, machine learning, financial forecasting, solving the Schrödinger equation, and the Traveling Salesman problem, are solvable but intractable. According to the Computational Complexity[5] (CoC) theory, all the solvable problems can be grouped into two sets: Polynomial Problems and Exponential Problems.

A problem is polynomial when the number of computational steps is a polynomial function of the dimension of the problem. Polynomial Problems (P) are problems of recognition. They are tractable because it is possible to achieve the exact solution in a reasonable time interval with the available computing machines. On the other hand, an Exponential Problem, whose number of computational steps is an exponential function of the problem's dimension, is tractable only if it has a small dimension. Exponential Problems with large dimensions are intractable. In these cases, they are transformed into Non-deterministic Polynomial Problems (NP). After fixing an arbitrary criterion of acceptability, solutions are generated through heuristic algorithms, and they are checked if acceptable or not. Meanwhile, some scientists, allured by the amount of money promised by the Clay Mathematics Institute in Cambridge, are trying to rigorously verify if the NP problems are reducible to P problems or this reduction is impossible.

The second reason why we find unsurmountable difficulties in describing Complex Systems is that they show variable patterns. Variable patterns are objects or events whose recognition is made difficult by their multiple features, which vary and are extremely sensitive to the context. Examples of variable patterns are: human faces, voices, and fingerprints; handwritten cursive words and numbers; biological species; patterns and symptoms in medical diagnosis; patterns in apparently uncorrelated scientific data; aperiodic time series; political, economic, and social events. It is necessary to formulate algorithms for recognizing every type of pattern. The steps of pattern recognition are (I) acquisition of instrumental data; (II) selection of the features that are considered as representative of the pattern; and (III) application of an algorithm for the classification step. Despite many attempts, it is still necessary to propose universally valid and effective algorithms for recognizing variable patterns. This difficulty generates a third type of Complexity that it might be named "Descriptive Complexity" (DeC).[4]

The third reason we find difficulties in predicting the behavior of Complex Systems derives from the intrinsically limited predictive power of science. In the description of the microscopic world, it is mandatory to consider the Heisenberg Uncertainty Principle. Such a principle asserts the impossibility of simultaneously and accurately determining every microscopic particle's position and momentum. Therefore, the Uncertainty Principle places concrete limits to the deterministic dream of describing the dynamics of the universe, starting from the description of its microscopic constituents. We might think of describing Complex Systems only at the macroscopic level, neglecting their microscopic "bricks." However, Complex Systems can exhibit chaotic dynamics. Chaotic dynamics are aperiodic and extremely sensitive to the initial conditions. Since the determination of the initial conditions is always affected by unavoidable experimental errors, the chaotic dynamics are unpredictable in the long term by definition.

3. Complexity Challenges

The limited predictive power of science makes many ethical issues related to technological development fiercely arguable. The unstoppable technological development induces humanity to continuously raise a fundamental question: "Is it always fair and safe to do what technology makes doable?" It is a tormenting question that has accompanied humankind since the beginning. Suffice to think about the Greek myth of Prometheus or the most recent novel *Frankenstein*, written by Mary Shelley in the 19th century. Cutting-edge technologies allow for manipulating and re-engineering life. Therefore, bioethical issues arise. There are bioethical issues that concern the beginning of a new life. Examples are: "Is it fair to manipulate human embryonic stem cells?"; "Is it safe to originate genetically modified organisms?" Other bioethical issues concern suffering and the end of human life. Examples are: "Is euthanasia fair?"; "What can we state about the therapeutic obstinacy?". Finally, there are forefront technologies that enhance human intellect and physiology. "Is it fair to exploit such enhancement techniques?". It is tough to find shared solutions to all these queries. They are intrinsically linked to the meaning we give to our lives. Furthermore, from both a biological and a physiological point of view, every living being is a Complex System, and we have already declared the limitations we encounter in predicting the behavior of Complex Systems. The bioethical issues mentioned above generate another type of Complexity, which can be named as Bioethical Complexity (BEC).[4]

From this discussion, it is spontaneous to name the challenges of the 21st century as Complexity Challenges. They involve NaC and BEC, which are interlinked with CoC and DeC. How can we think of winning these Complexity Challenges?

4. Strategies for Tackling Complexity Challenges

First of all, we need an interdisciplinary approach. Natural Complexity must be faced by all the scientific disciplines, including the social and economic ones. Philosophers can help to formulate new epistemological models and new methodologies. When we tackle the Bioethical Complexity, the involvement of scientists, philosophers and jurists, artists, and theologians is appropriate. The artists, guided by their intuitions, could spark new ideas and unconventional ways for interpreting Complexity. The theologians offer an extra dimension for giving meaning to our lives. Universities worldwide should offer Interdisciplinary courses on Complex Systems,[6–8] and the formation of genuinely interdisciplinary research groups should be favored by public and private funding.

The investigation of Complex Systems cannot be performed by relying only on the reductionist approach because Complex Systems exhibit emergent phenomena. A systemic approach is also needed. Furthermore, when we study the behavior of Complex Systems, we cannot trust anymore in one of the cornerstones of the scientific method, which is the reproducibility of the experiments. Experiments on Complex Systems are usually historical events. The philosopher Karl Popper has effectively described this state of affairs by declaring[9] that, in the past, science had been occupied with clocks, which are simple, deterministic systems having reproducible behaviors. Currently, instead, science has to deal with clouds, which are Complex Systems having unique and hardly replicable behaviors.

The investigation of Complex Systems requires monitoring them continuously because their behavior is highly dynamic. Therefore, it is necessary to collect, process, and store massive datasets, i.e., the socalled Big Data. Furthermore, it is becoming evident that an alternative way of doing experiments on Complex Systems is to perform simulations with computers. To deal with the considerable volume and the fast stream of data, their variety, and variability and to extract insights from them, it is necessary to speed up our computational machines, extend their memory space, and always contrive more powerful algorithms.

There are two relevant strategies to succeed. The first strategy consists of improving current electronic computers. Electronic computers are based on Von Neumann's architecture, wherein the memory, storing both data and instructions, is physically separated from the central processing unit. The pace of computers' improvement has been described by Moore's law, stating that the number of transistors (i.e., the ultimate computing elements that are binary switches) per chip doubles every 2 years. There is worldwide competition in devising always faster supercomputers. It is the TOP500 project. According to the last list compiled in November 2021, the fastest supercomputer in the world is the Japanese Fugaku, which reaches the astonishing computational rate of 442 PFlops/s. Meanwhile, Chips' producers invest billions of dollars in contriving computing technologies that can go beyond Moore's law.

4.1. *Natural Computing*

The second strategy is the interdisciplinary research line of Natural Computing. Researchers working on Natural Computing draw inspirations from nature to propose:

- New algorithms
- New materials and architectures to compute and store information
- New methodologies, new models, and possibly a new theory to interpret Natural Complexity

It is based on the rationale that any distinguishable physicochemical state of matter and energy can be used to encode information. Every natural transformation of these distinguishable states is a kind of computation. Within Natural Computing, there are two important research programs. The first one exploits the physicochemical laws to make computations. Every physicochemical law describes a causal event, and any causal event can be conceived as a computation. In fact, the causes are the inputs, the effects are the outputs, and the law governing the transformation is the computation algorithm. The second research program of Natural Computing mimics the features and performances of the natural information systems that belong to living beings. We might mimic living cells (called Biomolecular Information Systems), nervous systems (called Neural Information Systems), immune systems (called Immune Information Systems), and societies (Social Information Systems). All these systems are capable of exploiting matter and energy to encode, collect, store, process, and send information.

With human intelligence as its emergent property, the human nervous system is particularly attractive when facing complexity. It allows us

- To handle accurate and vague information, computing with numbers and words.
- To reason, speak, and make rational decisions in an environment of uncertainty, partiality, and relativity of truth when the Incompatibility Principle holds: "As the complexity of a system increases, accuracy and significance become almost mutually exclusive characteristics of our statements".
- To recognize quite easily variable patterns.

Therefore, it is worthwhile studying human intelligence and trying to reproduce it by developing Artificial Intelligence. Artificial Intelligence is revolutionizing our lives and societies. It is used in basic and applied science, medicine, well-being, economy, and security. There are two strategies to develop AI.[10,11] One strategy consists in writing human-like intelligent programs running in computers or special-purpose hardware. The other is through neuromorphic engineering. In neuromorphic engineering, surrogates of neurons are implemented through non-biological systems either for neuroprosthesis or to design "brain-like" computing machines. Surrogates of neurons can be implemented through specific solid materials in hardware. Such hardware can be rigid if made of solid inorganic compounds or flexible when based on organic films. Alternatively, surrogates of neurons can be implemented through specific nonlinear chemical systems' solutions in wetware (see Figure 1). Finally, specific hybrid electrochemical systems can play as surrogates of neurons. In our research group, we are exploiting molecular, supramolecular, and systems chemistry to mimic some performances of human intelligence and develop Chemical Artificial Intelligence.[12,13]

Specifically, we are devising modules for futuristic chemical robots. A "chemical robot" is thought of as a molecular assembly that reacts autonomously to its environment through molecular sensors; it makes decisions by its intrinsic Artificial Neural Networks and performs actions upon its environment through molecular effectors. The intelligent activities of a chemical robot should be sustained energetically by a metabolic unit. Chemical Robots should be easily miniaturized and implanted in living beings to interplay with cells or

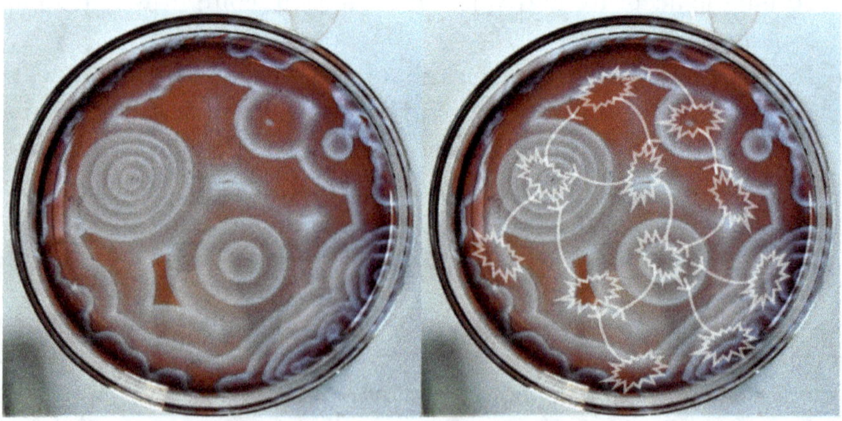

Fig. 1. The phenomenon of chemical waves generated by the Belousov–Zhabotinsky reaction (on the left) and its interpretation according to the theory of Natural Computing that describes the thin film of the solution as a collection of artificial neuron models communicating chemically (on the right).

organelles for biomedical applications. They should become auxiliary elements of the human immune system and help us to defeat the still incurable diseases.

5. Perspectives

The research line of Chemical Artificial Intelligence hopefully will give clues about the origin of life on Earth. The appearance of life on Earth was a phase transition or sudden change in how chemical systems can process and use information. In the beginning, the world was abiotic, and any chemical matter was unable to process information. About 4 billion years ago, the phase transition from a purely abiotic to a biotic world occurred. What happened at that time? The answer to this question might favor a new general theory on Natural Complexity.

References

1. M. Mitchell, *Complexity. A Guided Tour.* Oxford University Press, New York, USA, 2009.
2. P. Charbonneau, *Natural Complexity: A Modeling Handbook.* Princeton University Press, Princeton, NJ, USA, 2017.
3. P. L. Gentili, *Untangling Complex Systems: A Grand Challenge for Science.* CRC Press, Taylor & Francis Group, Boca Raton, FL, USA, 2018.
4. P. L. Gentili, Why is Complexity Science valuable for reaching the goals of the UN 2030 Agenda? Rend. *Fis. Acc. Lincei* 32:117–134, 2021.
5. O. Goldreich, *Computational Complexity. A Conceptual Perspective.* Cambridge University Press, Cambridge, UK, 2008.
6. P. G. Mahaffy, A. Krief, H. Hopf, G. Mehta, and S. A. Matlin, Reorienting chemistry education through systems thinking. *Nat. Rev. Chem.* 2(4):0126, 2018.
7. S. A. Matlin, G. Mehta, H. Hopf, and A. Krief, One-world chemistry and systems thinking. *Nat. Chem.* 8:393–396, 2016.
8. P. L. Gentili, Designing and teaching a novel interdisciplinary course on complex systems to prepare new generations to address 21st-century challenges. *J. Chem. Educ.* 96(12):2704–2709, 2019.
9. K. Popper, Of clouds and clocks. In *Objective Knowledge. An Evolutionary Approach.* Revisited edition, 1972, Oxford University Press, Oxford, K, 206–255.
10. S. J. Russell and P. Norvig, *Artificial Intelligence. A Modern Approach.* Prentice-Hall, USA, NJ, 2009.

11. C. Mead, Analog VLSI and Neural Systems. Addison-Wesley, USA, MA, 1989.
12. P. L. Gentili, The fuzziness of the molecular world and its perspectives. *Molecules*, 23:2074, 2018.
13. P. L. Gentili, Small steps towards the development of chemical artificial intelligent systems. *RSC Adv.* 3:25523–25549, 2013.

https://doi.org/10.1142/9789811257155_0034

Chapter 34

An Unconventional Look at AI: Why Today's Machine Learning Systems are not Intelligent

Nancy Salay

The Department of Philosophy,
John Watson Hall, Queen's University,
Kingston, ON K7L3N6, Canada
salay@queensu.ca

Machine learning systems (MLS) that model low-level processes are the cornerstones of current AI systems. These "indirect" learners are good at classifying kinds that are distinguished solely by their manifest physical properties. But the more a kind is a function of spatio-temporally extended properties — words, situation-types, social norms — the less likely an MLS will be able to track it. Systems that can interact with objects at the individual level, on the other hand, and that can sustain this interaction, can learn responses to increasingly abstract properties, including representational ones. This representational capacity, arguably the mark of intelligence, then, is not available to current MLSs.

1. Introduction

The current rhetoric has it that AI is here, or, at least, around the corner. But if this claim is false, then treating systems as such, that is, relying on them to make judgements, could have grave consequences. Autonomous vehicles are just one example of the many new AI technologies poised to enter the public space. Since stemming this technological tide seems futile, the academy has a responsibility to raise the public's understanding around what constitutes intelligence. Here I begin this effort by arguing that we should stop thinking about current machine learning systems (MLSs) as "intelligences" since they do not model the learning necessary for intelligent behavior.

2. What is Intelligent Behavior?

The very first question we might ask, then, is "What is it to act intelligently?" Examples help orient our thinking so let us begin with a few here. A robot that has been programmed to perform basic household chores, but which freezes when confronted with an unfamiliar task or situation, is not behaving intelligently: it can do only what it has been programmed to do. Similarly, any machine that does only what the limits of its design permit and no more — calculators, toasters, cranes — does so without what we would call intelligence. In contrast, beings who make novel responses to novel situations, who "figure out" what to do next, provide us with paradigmatic intelligent behavior. Think here of the battery of researchers currently working on the problem of developing a Covid-19 vaccine. The task draws on much background knowledge, to be sure, but it also requires a capacity for solving new kinds of problems. If we were not confident that these scientists were capable of such "out-of-the-box" thinking, we'd have no reason for hope.

But we need to be careful — a novel response is not always an indicator of an intelligent response. Many animals are capable of adapting to changing situations in what seem like "pre-programmed" ways: viruses mutate when hosts become resistant; cockroaches change food sources during resource scarcity; and, primates shift their group dynamics when territories diminish. We call the capacity for this sort of behavioral change "adaptability". While there is a relationship between adaptability and intelligence — they are both capacities for learning new responses to new situations — the former is on an evolutionary time-scale, one that extends across individuals to species, and the latter is on a developmental time-scale, one that extends across an individual over its own lifetime. From the perspective of the individual, adaptive responses are canned responses, backup strategies that kick in when first pass responses fail; only from the perspective of the species are such responses novel.

Intelligent responses, in contrast, are local strategies that individuals develop in response to new, locally experienced, situations.

2.1. *Intelligent action is not continual*

Note that having a capacity for intelligent action does not entail that one uses this capacity all the time. That it does was the working assumption in the early days of AI — "Good-Old-Fashioned-AI" (GOFAI). On those completely top-down models, every action (limited to simple screen outputs in most cases since GOFAI systems were not equipped with bodies) was the product of a line of "reasoning", a decision. Today, thanks to the insights of Embodied Cognitive Science and, more broadly, Phenomenology, we understand that even intelligent individuals mostly act in automatic or unconsciously directed ways. Such actions might be skillful, as when someone plays an instrument or adeptly traverses a narrow cliff-side path, but such behavior unfolds according to learned responses to the occurrent features of an ongoing situation. When an obstacle looms, the body swivels to avoid it; when a note is played, the next note in the learned sequence is anticipated. In intelligent behavior, in contrast, factors beyond the occurrent features of the ongoing situation influence our behavior: a note is played, but now the anticipated next note is not played. My musical execution of Beethoven's *Emperor* Concerto might be skillful, but when I intersperse it with the melodic line for Happy Birthday, because I know that today is yours, I have acted intelligently as well. That it is your birthday is not a physical fact in the occurrent situation in which we are participating and yet, as intelligent beings, it is something to which we are both capable of responding.

2.2. *Representation makes intelligent behavior possible*

How do we become responsive to such "offline" factors? We represent them to ourselves and one another. A classic demonstration of the dramatic behavioral effect this capacity for representation can have is the infamous marshmallow test.[1] A subject, usually a child, is presented with a single marshmallow. He is told that he is welcome to eat it but, if he can sit patiently for X minutes without tasting the marshmallow, he will receive another marshmallow in addition to the first. He will receive one marshmallow if it is eaten right away and two marshmallows if he waits until the experimenter returns. Young children find it very difficult to resist the sensory draw of the marshmallow and generally give in and eat it before the experimenter returns. As children mature, however, they are increasingly able to wait for the arrival of the second marshmallow. They are capable of suffering through short-term deprivation — not tasting the marshmallow that is present — for the sake of increased future gain — two marshmallows. To repeatedly not succumb to the marshmallow temptation, to what is sensorily present, children must behave now in a way that takes into account factors that are not spatio-temporally present, namely that there is a potential for future gain. Younger children, perhaps because they have not developed the relevant linguistic representational skills, are capable only of responding to the occurrent, sensory factors of the situation. Being thus completely in the moment, they gobble up the sweet-smelling treat.

Another version of the marshmallow experiment,[2] this time a reverse contingency test with chimpanzee subjects, demonstrates the critical role of representations in intelligent behavior even more clearly. A chimpanzee is presented with two plates of treats, one having visibly more than the other, and is invited to choose one of them. A second chimp sits by, observing. The plate chosen is given to the second chimp and the chooser is awarded the one not selected. Similarly to young children, though they understand the terms of the offer perfectly well, chimpanzees are incapable of resisting the overwhelming sensory draw of the treats — their smell, touch, appearance — and they invariably choose the plate with the greatest amount. Repeatedly they watch, furiously, as the larger pile of candies goes to the lucky bystander. But when a slightly altered version of the experiment was run with a chimpanzee — Sheba — who already knew some rudimentary mathematical symbols, the results were very different. Instead of being asked to choose between two plates with visible piles of candies, Sheba was offered two plates with lids labelled with the number of treats underneath. Since numbers have no sensory draw whatever, Sheba was now able to consistently make the self-maximizing choice of the plate with fewer candies. By using the numbers, representing the possibility of future treats, Sheba was able take into consideration factors that were not spatio-temporally present and thereby make the intelligent choice.

2.3. *Representations have unusual properties*

A word needs to be said here about representations themselves since they are peculiar things. A representation — for example, a sentence utterance, a physical token such as a game playing piece, or an occurrent thought — has, in addition to the usual physical properties that all physical things have representational properties that extend beyond these. My arm is a configuration of cells and impulses, and this is all that it is, nothing more than these physical properties. But the sentence token, "My arm is a configuration of cells and impulses", has the physical properties constituted by the ink on this page, the molecules in the paper, and so on, as well as the property of being about my arm. It represents the world, with respect to my arm at least, as being a certain way. The technical term for this property of representation is "intentionality:" anything that is about/represents something beyond itself in this way is an intentional thing. All signs,[a] then, are intentional objects: words and sentences; numbers; icons. And if we think that thoughts are internal representations of the way the world is or could be, then they are intentional objects as well.

Unlike physical properties, however, representational/intentional properties are not intrinsic to things. A stop sign is a symbol in the context of road traffic systems, but outside of these contexts, to a squirrel, for example, the stop sign is merely a physical thing in the world, representing nothing. Likewise, the sentence "My arm is a configuration of cells and impulses", to someone who does not speak English or to something not capable of speaking at all is just

[a] "Sign" is a technical term for a unit of meaning.

the physical thing that it is, the letter-shaped ink patterns on the page, representing nothing. Physical objects become symbols only in the context of a larger system within which individuals respond to them in ways that point beyond themselves. Such individuals are themselves intentional beings since they can respond to a symbol's representational properties as well as its physical ones. Intentionality, then, is a key aspect of intelligence.

To build an intentional machine, namely one that can use representations to guide its behavior, is one of the central tasks of AI. The foundational assumption that continues to drive research in the field, and which has spawned the current zeitgeist of mainstream cognitive science more generally, is this: intentionality is (exhaustively) reducible to low-level processes. If we want to understand the intentional capacity of human beings, we need to look inside human beings, at their neural circuitry mostly, to see which aspects of it are responsible for intentional behavior. In other words, the capacity to respond to a symbol's representational properties (and not just its physical ones) is nothing more than a (possibly very complex) combination of the low-level processes that constitute the response behavior.

The problem with this idea, however, is that it is wrong: low-level processes and personal-level intentional behavior are not comparable, and hence not identifiable, activities. Yes, of course, the sub-personal processes of Fred taken together are necessary for Fred to learn a new language, but, they are not sufficient: factors that are external to Fred, e.g., the way that symbols are used in his linguistic community, determine whether and how Fred's learned responses are about anything at all, which is to say, whether they are intentional at all. Furthermore, a model of just the sub-personal learning part of Fred's linguistic skill, placed in the appropriate environment, will not yield the appropriate intentional behavior. Without coordinated, personal-level activity, Fred's neural activity is just a series of low-level processes, not linguistic at all. In other words, intentional behavior is an personal-level activity, not a low-level one. This is not a distinction that is often made in cognitive science, but it is a critical one in the context of artificial intelligence research. If intentionality is not exhaustively accounted for by sub-personal activity, then AI systems that only model sub-personal activity, which is precisely what current MLSs do, will not be intelligent. Here I will try, in broad brush strokes, to explain how and why low-level processes cannot constitute intentional behavior.

2.4. *Low-level processing is not intentional*

MLSs model, albeit simplistically, the low-level neural processes that underwrite organism learning. Today's MLSs have achieved much success in classification, in "learning" to distinguish between different kinds of objects, pictures of cats and dogs, for example. I call this "learning" in scare quotes because the relevant classifying behavior is achieved only indirectly, by way of the pixel-level features that consistently correlate with object-level kinds such as dogs and cats. That there really are natural kinds in the world — individuals that share a statistically significant sub-set of features with other individuals — is what makes this kind of indirect learning useful. If the world were not regular in this way, if there were no natural kinds, such an approach would be useless. Imagine a world in which low-level patterns did not correspond to anything useful at the personal level, where a low-level sequence meant CAT one day and then DOG the next! This is an important factor to keep in mind: the learning success of indirect classifiers is partly determined by the high-level homogeneity of environmental conditions.

But given that our world is populated by dependable, statistical regularities, isn't this indirect learning good enough? For simple applications, in the context of web searches for example, it might be. As a model of personal-level behavior, however, it falls short: there is a critical granularity gap between low-level processing and object-level action. At the low-level, interaction is with sensory bits — in the case of our example MLS these are pixels — not with medium-sized objects such as cats and dogs. One low-level processing time-step does not register at the object-level of granularity at all and, conversely, a single action at the object-level corresponds to thousands of low-level processing steps. Relative to the object-level, processing occurs at speeds that do not register as activity at all; while, relative to the processing level, object-level change occurs so slowly that, again, it does not register as change at all during a single, low-level time-step. The two levels of granularity are thus spatio-temporally distinct. Since MLSs cannot interact directly with medium-sized objects, but only indirectly over many low-level time-steps and by way of regular, low-level features, the object "learning" they achieve will always be brittle. Change one lower-level feature of an object that the MLS has not had training on, and it breaks. For this reason, no matter how robust the low-level training is, MLSs will always be subject to adversarial attack. But, more saliently with respect to the question of intelligence, such systems could never learn to respond to an object's representational properties. More on this momentarily. The only sense in which a cat/dog classifier represents a distinction between the concepts CAT and DOG is from our spatio-temporally extended vantage point relative to the network, namely the vantage point of the personal-level, from which we can see that the ongoing activity of the network corresponds to a distinction between cats and dogs. From the classifier's vantage point, however, there are only ever pixels and responses to them; there are no cats and dogs at all.

To help clarify what it is for an individual to interact and learn about an object directly, let us first consider another personal-level activity: locomotion. A cat jumps from the ground to a branch by virtue of a multitude of ongoing, low-level processes that constitute its personal-level activity, but none of the low-level processes are themselves locomotions. Locomotion is something that organism wholes do. A single locomotive step, so to speak, spans a multitude of low-level processes and a single, low-level process does not map to anything at all that would count as a step at the personal-level. Some machines locomote as well. When a car moves down the road, it is the car-whole that is moving, not its parts. The parts, working together, make the car locomotion possible, but there is no part or set of sub-parts that is doing the locomoting.

Perception, which is the method by which organisms interact with their environment, is likewise a personal-level activity. As with locomotion, myriad low-level (sensory) processes underwrite perceptual activity, but perception itself is a personal-level interaction with the objects in an individual's environment. For humans these are mostly medium-sized objects such as cats, dogs, tables, computers, and the like. Of course, organisms that are capable of perception do not always and only navigate their environment with perception; rather, there is an ongoing, dynamic shifting between unconscious/indirect sensory tracking and conscious/direct perception, so rapid that most perceiving individuals are themselves completely unaware of the oscillation between modes of interaction. The frequency of this oscillation, however, is a critical factor in an individual's capacity to learn to respond to the representational properties of objects since the degree to which an individual is capable of sustaining perception — of maintaining ongoing direct interaction with an object — will determine the degree to which that object can become a symbol for an individual. Before we can see how and why this is the case, a few words need to be said about symbols themselves.

2.5. *Symbol abstractness is an indicator of intentionality*

As we have seen, a sign is anything that represents something to an agent. The more removed a sign's representational properties are from its physical ones, the more symbolic it is, while the less distinct a sign's representational properties are from its physical ones, the more natural a sign is. Natural signs and human language lie at opposite ends of this abstractness continuum. Smoke, for example, is a physical by-product of fire, serving as a natural sign of fire to any individual capable of tracking it. Many animals that already have an avoidance response to fire learn, through association, to exhibit the same avoidance response to smoke. Because smoke is sensorily more dispersed than fire, and so can be perceived more quickly, it is a useful natural sign.

As we move along the sign abstraction continuum, however, representational and physical properties increasingly diverge. The physical properties of the *utterance* "Fire!" for example, are entirely distinct from its representational properties. Learning the latter requires not only a series of learning situation experiences in which different utterances are associated with different fire situations, but it also requires an entire system of symbol use within which utterances evoke appropriate fire responses. It requires a community wide practice of linguistic use. Because these representational properties extend to the features of the situations within which sign vehicles and objects occur — a child learns the word "ball" as a result of myriad and various BALL situations — learning a response to them requires sustained interaction with the relevant sign objects, long enough for the situation to unfold. And there's the rub. MLSs track objects only indirectly, they never interact with them at all; consequently, they will not be able to sustain interaction with objects, at least not across the sorts of complex, varied situations in which humans learn. Remember that from the vantage point of the cat/dog classifier, there are no cats and dogs, only pixels. And the more abstract a sign is, the more its representational properties will trade on spatio-temporally diverse situational features rather than on a sign vehicle's physical ones.

In theory, of course, MLSs could be trained to indirectly track situations, just as they do objects. Newer deep learning models attempt to do just that.[3] But success will be elusive: the properties of situations exponentially out-number those of objects, even supposing that there are statistically reliable patterns that underwrite them. In most cases there are not. As Ludwig Wittgenstein famously observed,[4] even seemingly regular situation-types such as games are impossible to pin down: some games such as baseball and soccer involve many players, others such as chess and tennis just two, and some such as solitaire only one; some games have win/lose conditions, while others — many new board games for example — are cooperative; some have fixed rules; others — e.g., "playing house" — do not, and so on. Thus, the possibility for tracking even a simple type of situation such as game playing by tracking the low-level properties of the objects that might figure in them seems very small. What chance, then, is there for indirectly tracking complex, language-learning situations?

Systems such as MLSs, then, that are capable only of indirect object tracking by way of low-level properties, could learn responses to natural signs but will not be capable of more abstract symbolic responses. And it is arguable that since the symbolic properties of such natural signs trade completely on their physical ones, individuals that are capable only of learning to respond to them, and not to more abstract symbols, are not intentional agents. Most new cars today are outfitted with an array of sensors that track the objects in a car's immediate surroundings. When my car is in reverse, for example, it will emit different sequences of "beeps" according to how close a potential obstacle is to the rear fender. In one sense, this seems like an intentional action since it looks like the car is responding to what things in the environment signify — potential obstacle — rather than to them directly as physical objects. But, of course, the car is not beeping because it perceives a potential obstacle; rather, the triggering of one set of sensors triggers another set that in turn triggers the beep mechanism. The car has been designed so that its parts work in concert with one another in such a way that the car signals precisely when a potential obstacle is present. This is clever design to be sure, but the car itself is behaving in the sort of unintelligent, pre-programmed way we began this discussion with. At best, we can say that it has been designed to respond to a natural sign.

3. Conclusion

Because they cannot interact directly at the object-level, current MLS-based systems are not capable of the sustained interaction with objects required for developing responses to abstract symbols. As they are fundamentally non-intentional entities, then, we should not treat them as intelligent systems, no matter how clever their design. Does this mean that there is no possibility of creating an AI system, perhaps out of a complex configuration of these networks? No. But a potential AI will need a capacity analogous to basic perception, namely, a capacity for direct interaction with the objects in its environment. In the case of humans this is sentience, sometimes called "pre-reflective consciousness:" our basic capacity to experience our world, not simply infer it. But we still need a better understanding of what this is before we can start building systems that exhibit it.

References

1. W. Mischel, and E. B. Ebbesen, Attention in delay of gratification. *J. Pers. Soc. Psychol.* 16(2):329–337, 1970.
2. S. T. Boysen, G. Bernston, M. Hannan, and J. Cacioppo, Quantity-based inference and symbolic representation in chimpanzees (Pan troglodytes). *J. Exp. Psychol. Anim. Behav. Process.* 22:76–86, 1996.
3. J. Chen, K. Li, Q. Deng, K. Li, and S. Y. Philip, Distributed deep learning model for intelligent video surveillance systems with edge computing. *IEEE Trans. Industr. Inform.* 1–8, 2019.
4. L. Wittgenstein, *Philosophical Investigations.* Trans. G. E. Anscombe. Blackwell. §83, 1967.

https://doi.org/10.1142/9789811257155_0035

Chapter 35

Composing Music with Bio-Technology: An Intelligent Algorithmic Composition System Using *Physarum polycephalum*-based Memristors

Satvik Venkatesh, Edward Braund*, and Eduardo Reck Miranda
Interdisciplinary Centre for Computer Music Research
University of Plymouth
United Kingdom
* *edward.braund@plymouth.ac.uk*

Computer-assisted and automated composition systems often harness Artificial Intelligence models such as Markov chains, Neural Networks, and Genetic Algorithms to generate musical material. The field of Unconventional Computing (UC) explores non-digital ways of data storage, processing, input, and output. UC paradigms such as Biocomputing and Quantum Computing delve into domains beyond the binary bit to handle complex nonlinear functions. In this chapter, we explore harnessing the biological computing substrate *Physarum polycephalum* as a memristor to process and generate musical material within an algorithmic composition system. It details the journey and process of creating a piece of popular music using this novel technology, which we originally showcased at the New Interfaces for Musical Expression (NIME) conference in 2020.[1] In this piece, entitled *Creep into my Lawn*, the *Physarum polycephalum*-based memristors act as creative collaborators in the composition process. Our work aims to demonstrate the potential of UC paradigms in creative applications and to disseminate this technology to non-experts and musicians so that they can incorporate it into their creative processes.

1. Introduction

Research that explores computational processes to exhibit creative behavior is vast.[2] Arguably, two things often influence a composer's process: music they have heard and may have prominently featured in their education and new and innovative ideas.[3] The former encompasses characteristics of the composer's style and inspirations. The latter has been explored in computational creativity through stochastic processes, which aim to induce randomness in the output of an algorithmic composition system.[4]

Many Artificial Intelligence (AI) models such as Markov chains,[5] Genetic Algorithms (GA), Artificial Neural Networks (ANNs)[6] have been successfully adopted in musical systems. Markov chains compute a conditional probability distribution of future states depending on the current state.[4] The architecture of ANNs is inspired by the human brain's structure and enables us to model complex nonlinear functions. Additionally, other models such as self-similarity and cellular automata,[7] which do not strictly fall under the AI bracket, have been explored as creative tools by composers.[8]

Physical, chemical, and biological systems have long inspired AI algorithms and have also given rise to other forms of computation. The field of Unconventional Computing (UC) looks to the information processing abilities of the natural world to develop new concept computer architectures for data storage and processing.[9] Computational creativity has almost exclusively been implemented on conventional computing technology — i.e., technology based on the Turing Machine and von Neumann architecture. The fundamental processing and storage unit in these devices is the binary bit, which exists as either a 0 or 1. However, technologies being developed in UC explore analog forms of computation and storage. For example, quantum computing uses *qubits* as the fundamental unit of processing. Using the principles of quantum mechanics, these computers perform better than classical computers in some applications of cryptography,[10] quantum search,[11] and in solving linear equations, to cite but three examples.

Physarum polycephalum (Figure 1) is a plasmodial slime mold that has been widely harnessed in UC for biocomputing applications.

Fig. 1. A photograph of a 1-day-old culture of plasmodium of *Physarum polycephalum*.

The organism belongs to the order *Physarales*, subclass *Myxogastromycetidae*, class *Myxomycete*. Research over the past decade suggests that this organism may be a potential gateway for the non-expert to experiment with aspects of UC, a field that comprises laboratory prototypes that are often inaccessible. Uniquely, this biological computing substrate requires comparatively fewer resources than most other substrates. The organism is cheap, openly obtainable, considered safe to use, and has a robustness that allows for ease of application.[9] Moreover, as a result of such unique accessibility for a biological computing substrate, an excellent instructional book on how to use the organism for computational purposes has been published for the UC novice.[9]

Researchers have utilized *Physarum polycephalum's* ability to optimize its shape according to the environment to solve mazes,[12] build logic gates,[13] and find shortest paths.[14] Another interesting property of the organism is that it can act as a memristor.[15] Memristors are the fourth fundamental passive circuit element, the first three being the resistor, capacitor, and inductor. It was theoretically demonstrated by Leon Chua in 1971,[16] but was physically realized in 2008.[17] The word memristor is a contraction of "memory resistor", which describes the element's function: a resistor that remembers its history. Memristors alter their resistance as a function of the previous voltage that has flown through and the length of time that it has been applied. Furthermore, when you stop applying voltage, the memristor retains its most recent resistance state.

In contrast to the other three fundamental elements, memristors are intrinsically nonlinear. We can observe such nonlinearity in their current vs. voltage profile, which takes the form of a pinched hysteresis loop when applied with an AC voltage — a Lissajous figure formed by two perpendicular oscillations creating a high and low resistant state (Figure 2). Hysteresis is where the output of a system is dependent on both its current input and history of previous inputs.

Research into memristor technology is gaining traction. The element is the first to combine storage and processing into one unit, and, therefore, has the potential to revolutionize computing architectures. Moreover, the component's behavior has a likening to the way that synapses in the brain function: researchers have been able to model spike-timing-dependent plasticity (STDP)-like behavior using networks of memristors.[18, 19]

STDP functions by one neuron sending an electrical spike to another neuron. The receiver neuron's synapse evaluates the importance of the incoming signal by contrasting it with its own state that it stores locally and the strength of the connection between the two neurons. The synapse then updates its state accordingly and sends the result of the comparison to the body of neuron two, which may fire an impulse to another neuron. This process propagates across a neuronal network, which gives rise to complex spiking observations and facilitates Hebbian learning. STDP has been adopted in feed-forward ANNs for pattern recognition algorithms.[20]

Physarum polycephalum-based memristors serve as an inexpensive and accessible UC prototype when compared to quantum

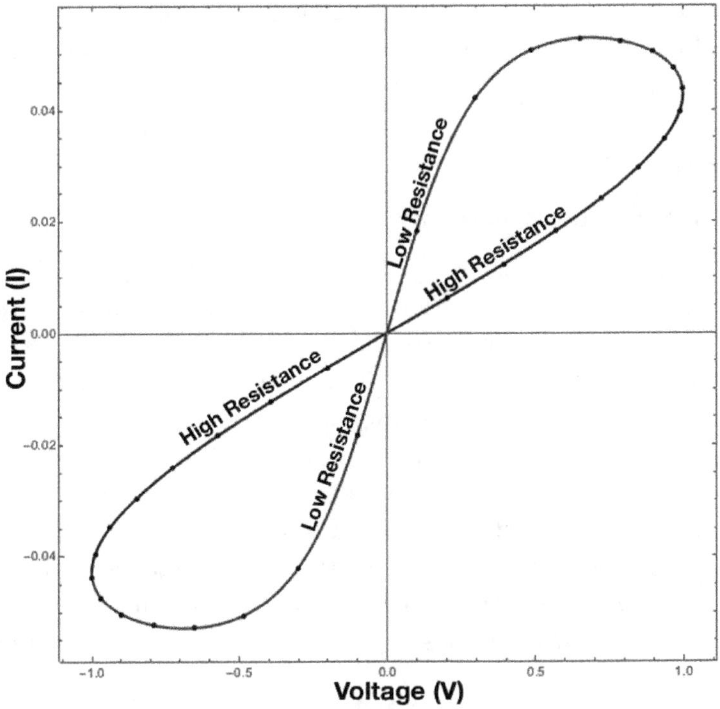

Fig. 2. Example of hysteresis in an ideal memristor (arbitrary values used).

computers and chemically manufactured memristors. To harness these in a computing architecture, we have designed 3D printed receptacles to enclose the organism into a portable electrical component.[21] Furthermore, we have developed a compact kernel — named the Phybox — to interface with biomemristors to allow creative practitioners and D.I.Y. prototyping enthusiasts to begin experimenting with this technology.[22]

In this chapter, we demonstrate how *Physarum polycephalum*-based memristors can be used in composing popular music. Here, we explore using the organism as an intelligent collaborator in the creative process. As we are using a nonlinear component as the processing device, input parameters that govern the composition process require non-digital forms of representation. This study renders a practice-based approach toward understanding nonlinear input parameters, which may not be as obvious as binary logic. It aims to break down the creative process and explores how these parameters influence the output of the PhyBox.

2. System Overview

2.1. *Hardware-wetware framework*

We maintain a *Physarum polycephalum* farm[9] grown in plastic containers as shown in Figure 3. In order to grow *Physarum polycephalum*-based memristors and make them a part of the composition process, we transfer a section of the organism into receptacles.[21] These receptacles allow the organism to act as electrical components and are operated through the PhyBox,[22] a portable biocomputer for creative applications as shown in Figure 4. PhyBox is powered by a Raspberry Pi model B.

Training data of music is converted into voltages with the help of a mapping procedure.[23] In this chapter, training data refers to the music originally composed by the authors. These voltages are

Fig. 3. *Physarum polycephalum* culture grown in a plastic container.

Fig. 4. Setting up the PhyBox for musical composition.

serially applied across the memristors as shown in Figure 5. The current flowing through the memristor depends on its resistance (or memristance). The current is again translated to music through a similar mapping procedure. Therefore, the memristors create new music based on the training data.

Music is fed into the PhyBox through the MIDI protocol. MIDI data are broken down into four parameters — pitch, velocity, time between note-ons (rhythm), and duration. Each memristor processes one of these parameters in parallel — i.e., there are four memristors in the circuit.

2.2. *Mapping procedure*

The framework for generating creative data harnesses the spiking behavior of memristors. Each time there is a voltage change across the memristor, there is a spike in current as shown in Figure 5. This behavior is different from that of linear components like resistors in electronic circuits. The following explains its behavior. When the

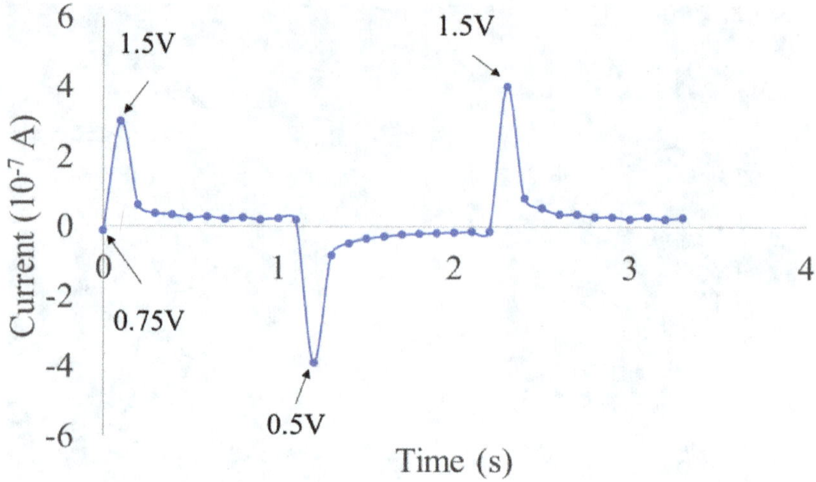

Fig. 5. Spike generated by the memristor on applying four voltage changes — 0.75 V, 1.5 V, 0.5 V, and 1.5 V. Positive spikes are produced for positive voltage changes and vice versa.

positive change in voltage is produced across the memristor, the resistance is initially low. With time, it develops resistance and there is a gradual decrease in the current flowing through the circuit.

2.2.1. *Music to voltages*

In this system, these voltage transitions are attributed to musical note transitions. A musical note is assigned to a specific voltage on a *priority* basis. Musical notes that occur more often are assigned higher priority and vice versa. If there are two events with the same number of occurrences, the more recent event is given a higher priority. This is analogous to the learning rule in STDP, where the new event is good and the old event is bad.

As we are using a Raspberry Pi, the voltage range is 0 V to 3 V. It is assigned by Equation (1).[23]

$$v_{\mathrm{a}} = \frac{3}{N}\left(p - \frac{1}{2}\right)\mathrm{V}, \qquad (1)$$

where v_a is the assigned voltage, N is the number of musical notes and p is the priority of the event. A relationship between popularity and assigned voltages is devised by the following analogy. The rate of change of current (or slope) in the beginning of the graph is greater when compared to the end. Moreover, taller spikes are generated for greater changes in voltage. The stability of resistance is related to the popularity of the event. Therefore, more popular events are mapped to lower voltages and less popular events are mapped to higher voltages.

Note that when the training data are initially fed into the PhyBox, there are no events in the memory. It is trained during the course of the song. Therefore, the number of variations that can possibly occur gradually increases with more data.

2.2.2. *Current to music*

A similar procedure is followed, while converting current readings to music. The voltage range is well-defined to be 0 to 3 V. However, a discrete current range is not defined because the resistance of the memristor is unknown. Empirically, studies have approximated the resistance in the order of mega-ohms.[22] Let I_{max} and I_{min} be the maximum and minimum current sensed during the spike, respectively. The current range $[I_1, I_2]$ is defined by the following equation:

$$[I_1, I_2] = \left[\frac{\min(V_a)}{(v_a/I_{\text{min}})}, \frac{\max(V_a)}{(v_a/I_{\text{max}})}\right], \tag{2}$$

where V_a is the set of all assigned voltages and v_a, the input voltage at that specific moment in time. The current range is divided into equal divisions and converted to music based on *midrise quantization*.[24] Again, the events are ordered according to *priority*, where lower current values stand for more popular events and vice versa.

2.2.3. *Resistors*

One of the advantages of this mapping procedure is that it provides a way to incorporate resistors in the process of musical composition. It defines a resistor's behavior as linear. Therefore, when the training data are fed into a resistor, the output of the system is exactly identical to the input. This offers creative pathways to explore in the process of composition, such as the rhythm can be identical to the input and the pitch can be composed by the memristor. It is important to note that the value of the resistance does not influence the output. In these experiments, we used 330 $k\Omega$ resistors.

2.3. *Parameters*

There are two parameters that monitor the output produced by the system — dwell time and measurement offset. Dwell time specifies how long the voltage is applied across the memristor. This basically determines the processing time of the memristors for each musical event. For example, if the dwell time is 1 s, the memristor takes 20 s to process 20 musical events. The measurement offset determines at what point of time the current is value calculated in the circuit. For instance, on applying 1.5 V at 0.1 s in Figure 5, 0% offset considers the current at 0.1 s and 100% offset considers the current at 1.1 s (Here the dwell time was 1 s).[22,23] have conducted some experiments to understand how these nonlinear parameters effect the output of the memristors. 0% offset produces smaller deviations from the input data when compared to 100% offset.[23] However, there is still no clear delineation of these functions due to their complex nonlinear behavior.

3. Experiments

In this chapter, we use *Physarum polycephalum* as a collaborator in the process of musical composition. As the current interface only accepts MIDI data, we have harnessed it to compose parts for piano, drums, and the vocal melody. The dwell time was constant at 1 s. The measurement offset was altered between 0%, 30%, 50%, and 100%.

3.1. *Structure*

The time and key signature of the song is 4/4 and E minor, respectively. The genre of the song falls under Rock. The song starts with the main riff played only on the piano, which converges into the other instruments joining the main riff. Followed by that, we have one verse, pre-chorus, and chorus, which are sung by a vocalist. The chorus has the main riff in the background. After the chorus, there is a piano solo followed by a drum solo. The song finishes with another chorus at the end. This song follows the structure adopted by most popular music compositions. However, it was slightly altered, for instance omitting the second verse, in order to make things more interesting in view of *Physarum polycephalum*.

In order to explain the structure of the song with clarity, the verse was divided into two parts. While feeding in data for piano and drums, we divided the whole song into three blocks. This was done to avoid overloading the model with too many possibilities and transitions. The first block consisted of the intro and first part of the verse. The second block comprised the second part of the verse and the pre chorus. The third block contained the remainder of the song. These blocks were not divided with regards to time, but were divided with respect to number of MIDI events and structure of the song. For instance, chords in the verse contained more MIDI events than melody lines. The vocal melody was transcribed as a MIDI file. MIDI events like sustain pedal were removed from the training data.

3.2. *Main riff*

The main riff of the song is 4 bars in length. A piano tune is backed up by drums and guitars for the main riff. The first two bars of the piano melody were composed by the author. We wanted the third and the fourth bar to follow a similar pattern as the first two, but with minor variations. The aim is to create a feeling of repetition alongside some anticipation in the listener. We conducted experiments with measurement offset values of 0%, 30%, and 100%. The guitars and drums were composed by keeping the piano track as a reference.

3.3. *Verse*

The instrumentation for the verse consisted of vocals, piano, and drums. The first part of the verse had simple piano chords at the beginning of each bar. We experimented with measurement offset values of 0%, 30%, and 100%. The objective of this experiment was to make *Physarum polycephalum* improve the chord pattern, chord inversions, and the voicing.

The second part of the verse had arpeggios played on the piano. For this section of the song, we felt that there was less flexibility with altering the rhythm. We wanted to maintain the rhythmic structure originally composed by the author. Therefore, we attached memristors for pitch and velocity and resistors for time between note-ons and duration. The measurement offset was set to 100% and this configuration was referred to as 100R.

3.4. *Piano solo*

The harmonic and rhythmic complexity of the piano solo is higher than the other sections of the song. The piano solo was originally composed by the author. Also, there was a scale shift within the solo from E minor to E Phrygian and then C major. At the end of the piano solo, the song returns to E minor. We wanted to explore how *Physarum polycephalum* would respond to scale changes and complex melodic ideas. This is the most interesting section of the song because it provides high degree of freedom with regards to musical options. Additionally, it does not restrict *Physarum polycephalum* with respect to the song's instrumentation because the other instruments do not play during the piano solo. Tests were conducted for measurement offsets of 5%, 30%, 50%, 100%, and 100%R.

3.5. *Drums*

We wanted to harness the organism to create new drum fills based on existing ones. Additionally, we wanted to explore its creativity for the drum solo. We tested it with measurement offset values of 0%, 30%, 100%, and 100%R.

3.6. *Vocal tune*

A vocal melody was composed by the author and transcribed as a MIDI track. Our objective of using *Physarum polycephalum* for the vocal tune was to make it sound more catchy and less monotonous. We explored measurement offsets of 0%, 30%, 100%, and 100%R.

3.7. *Dissimilarity*

In order to objectively calculate the difference between original melody and *Physarum polycephalum*'s rendition, we defined a metric called *dissimilarity*, which is the number of musical events that are different from the original input. Each of the parameters — pitch, velocity, time between note-ons, and duration — would return a value of dissimilarity. The *total dissimilarity* is defined as the sum of individual dissimilarities of the four parameters.

4. Results

All audio files that were composed by the author and *Physarum polycephalum* can be found in https://bit.ly/2U6lI9f. The naming format for the audio files are {block _Instrument _Type _Offset}. The title of the song is *Creep into my Lawn.* The model used to generate musical data only uses note transitions. This gives high flexibility to *Physarum polycephalum*, but opens up possibilities like time signature mismatch and disharmonious chords. Therefore, few modifications were made by the composer to suit the song. As the model is based on musical transitions, in *Physarum polycephalum*'s output, the first note will be missing because no transition has occurred yet. Hence, while evaluating its output, the first note was added from the input.

4.1. *Dissimilarity*

Table 1 contains the *total dissimilarities* for different values of measurement offset. Consistently, 0% was more similar to the input when compared to other offsets. 30% was closer to the input when compared to 100%, but the difference was much smaller.

4.2. *Main riff*

The main riff originally composed by the author is shown in Figure 6. *Physarum polycephalum*'s output for measurement offset of 0% was

Table 1. Table of tests conduced with PhyBox for different values of measurement offset (MO). Numbers within the MO columns show the *total dissimilarity* and '-' indicates no test was conducted. The 'final song' column shows the values of MO chosen for the final song. The values in bold indicate that they were modified to suit the song. There are repetitions of values across rows because the dissimilarities are calculated for the whole block and not only the section.

	Physarum MO (%)					
Section	0%	30%	50%	100%	100%R	Final Song
Main Riff	69	214	—	261	—	100%
Verse Pt.1 Piano	69	214	—	261	—	**0%**
Verse Pt.2 Piano	—	—	—	—	150	**100%R**
Pre Chorus Piano	—	—	—	—	150	**100%R**
Piano Solo	414	738	743	860	313	50% and **100%R**
Verse Pt.1 & Pt.2 Vox	106	229	—	265	81	**100%**
Pre Chorus & Chorus Vox	106	229	—	265	81	—

Fig. 6. The original main riff composed by the authors.

Fig. 7. *Physarum polycephalum*'s rendition of the main riff for measurement offset of 100%.

not interesting. It resembled the original melody except for changing one crotchet to a dotted crotchet. For measurement offsets of 30% and 100%, the generated output sounded more appropriate for the main riff. The authors felt that 30% offset had more resemblance to the original tune when compared to 100%. We chose 100% and made no modifications to *Physarum polycephalum*'s output. It is shown in Figure 7. For the third and fourth bar of the main riff, guitar and drums parts were re-composed to match the new tune composed by the memristors.

4.3. *Verse part 1*

The first part of the verse essentially contained piano chords, drums, and vocals. 0% offset subtly altered the harmonic structure of the verse. 30% output collapsed all the individual notes into one point. This was because the time between the chords was 0, which was obviously not useful for the composition. 100% made too many variations that did not suit the genre of the song and it was difficult to perceive the harmonic structure. Hence, we chose 0% offset for this section.

The original chord pattern of the verse was *Em* — *G* — *Am* — *D*. All chords were first inversions. While processing the output of *Physarum polycephalum*, the root notes of chords were left unchanged. The first chord was the same as the input. The second chord was a triad of the notes *E*4, *G*4, and *A*4 (the numbers stand for the octave), which sounded very close to the chord *Esus4/G*. In order to form a triad that sounds harmonically more stable, *G*4 was changed to *C*5 and thus, obtaining *Esus4/G*. The third chord generated by *Physarum polycephalum* contained *E*2 as the lowest note, which sounded like the second inversion of *Am*. Therefore, the third chord was changed to the second inversion of *Am*. The fourth chord generated by *Physarum polycephalum* contained *F*#4, *G*4, and *A*4, which sounded close to the chord *Gsus2/D*. However, *F*#4 and *G*4 are only a semi-tone apart and therefore, made the chord sound unstable. Hence, *F*#4 was modified to *D*5 and thus obtaining the chord *Gsus2/D*.

The original vocal tune composed by the authors was a simple melody line. 0% produced a melody that was similar to the input, but had irregular rhythms and few variations in pitches. It did not fulfill our goals of making the vocal tune sound more catchy. 30% produced interesting variations in the output. However, the output sounded very continuous to suit a vocal line. Its rhythmic structure did not have enough stops. 100% offset formed a good foundation for the vocal line. Furthermore, it went well with the new chords composed by *Physarum polycephalum*. However, few notes were omitted and altered in order to suit the song.

4.4. *Verse part 2*

The rhythmic structure was unaltered because resistors were connected instead of memristors for time between note-ons and duration. During the second bar, the memristors produced an accidental *Bb* which was not anticipated by the composer. This accidental (or blue note) is pretty popular in jazz and blues genres, which gave the song a new flavor. After performing few modifications, the arpeggio composed by *Physarum polycephalum* was fit into the song.

4.5. *Pre chorus*

The pre chorus of the song mainly consisted strumming and piano patterns in close association with the drum beat and, hence, we wanted to maintain the rhythmic structure. For this part of the song, we used resistors for rhythmic parameters and memristors for pitch and velocity. The memristors changed the voicing of chords and also performed some chord inversions. Few modifications were made to incorporate it in the song.

For the vocal tune of the pre chorus, we felt that the output of memristors did not match the backing chords. Furthermore, the original melody line went well with the subtle changes produced by memristors in the piano line. Therefore, we used the original tune.

4.6. *Chorus*

The chorus of the song had the main riff running in the background alongside vocals. Therefore, the instrumental section was a replica of what was done earlier.

In order to compose the vocal tune for the chorus, we analyzed the output of resistors connected to rhythmic parameters. For the chorus of the song, we felt that the original version was more suitable and *Physarum polycephalum*'s output did not improve it. As we progressed toward the chorus, the song was gaining structure and form. Furthermore, it had the main riff of the song playing in the background. Therefore, there was more rigidity in the number of suitable musical options. Hence, we chose the original tune created by the composer.

4.7. *Piano solo*

Among all values of measurement offsets, 50% was chosen due to the following explanation. In the original solo composed by the authors, there was a mode shift from E minor to E Phrygian and then C major. Among all outputs produced by *Physarum polycephalum*, 50% created a solo that was distinctly perceived as G major. Surprisingly, there was no training data that existed in G major. However, it is the relative major of the E minor scale. This was a very interesting observation and *Physarum polycephalum* sounded like a highly skilled pianist! This demonstrated that the intelligence of the organism can be harnessed to create unique trajectories in creative processes. There were no modifications made by the composer on *Physarum polycephalum*'s output. Furthermore, 30% offset created few bars that were perceived as G major. The output of 30% and 50% were similar, but 50% was preferred by the composer. 100% offset created many stops during the solo which made it unsuitable. Additionally, 100% offset sounded distinctly different from 30% and 50%.

The final piano solo starts with the original one composed by the authors. This was mutated with outputs produced by 0% and 100%R. Just before the scale shift to C major, *Physarum polycephalum*'s solo in 50% offset begins, which lasts for 9 bars. In the final song, this part spans from 2 min 22 s to 2 min 42 s.

4.8. *Drums*

For all values of measurement offsets, the musical files sounded random and unsuitable for the song. This might be due to the four parameters — pitch, velocity, time between note-ons, and duration — not being suitable to analyze percussion tracks. It might need higher level musical structures such as time signatures. Furthermore, in such a song, drums are generally a supporting instrument. *Physarum polycephalum*'s output sounded more like a melody line and not a drum beat. Due to the above observations, we decided to not use *Physarum polycephalum* for the drum solo.

4.9. *Discussion*

This chapter presented the process of composing with *Physarum polycephalum*-based memristors. As mentioned earlier, the model adopted for composition uses note transitions. Therefore, it does not account for higher-level musical structures like time signatures and harmonic functions. Therefore, we had to make modifications in *Physarum polycephalum*'s output to incorporate it in the song. It is important to note that modifications to the output produced by *Physarum polycephalum* were only done in parts where there was a high involvement of multiple tracks and presence of chords. For instance, slightly altering piano and vocal lines during the verse, but not making any modifications in the main riff. Furthermore, individual tracks were passed separately into the PhyBox, which may have contributed to this behavior.

As no changes were made in outputs produced for the main riff and piano solo, it suggests that the model can be improved by further understanding higher-level structures. Examining how instruments interact with each other opens up possibilities for creating networks of memristors that collectively produce a coherent and ready-to-use output.

The authors felt that the memristors have changed the original composition significantly and added their own footprint. It was an interactive process where *Physarum polycephalum* makes suggestions and the composer evaluates them. However, it is important to note that many observations and results were drawn through subjective measures and aesthetic preferences of the composer.

This chapter used different values of measurement offsets for different sections of the song. In Figure 5, it can be observed that 0% captures the value at the top of the spike. The next sample recorded by the PhyBox is after a steep descent. Therefore, in order to have more flexibility in measurement offset, the sampling rate of the system needs to be increased.

We used *dissimilarity* as the metric to compare different values of measurement offsets. This can be improved by borrowing concepts like *entropy* from information theory.[25,26] This will enable us to explore the statistical structure of music in greater detail.

5. Conclusion

This chapter presented a composition in collaboration with *Physarum polycephalum*. It showed how the nonlinear behavior of memristors can be harnessed for the process of popular music production. It aims to make UC technologies reach a wider audience so that non-experts in UC and musicians can adopt them in the process of music production. Harnessing *Physarum polycephalum* has great potential in computational creativity and popular music.

The authors felt that the memristors have changed the original composition considerably and added their own footprint. It was an interactive process where *Physarum polycephalum* makes suggestions and the composer evaluates them.

We showed that resistors can be creatively incorporated alongside memristors during musical composition. We demonstrated that a dwell time of 1 s is suitable for obtaining the memristor's nonlinear behavior. This is faster than an earlier model that used 2 s.[22] *Dissimilarity* values show that the measurement offset considerably influenced the melody produced by the memristor. For each parameter, the output created by *Physarum polycephalum* was unique. This can be verified with sound files provided in the database.

In this chapter, we trained the model with only one song. Future work would explore how the system can be scaled up. This might include creating more complicated learning models such as networks of memristors. Additionally, higher-level data structures such as time signatures and harmonic functions need to be explored.

References

1. S. Venkatesh, E. Braund, and E. Miranda, Composing popular music with physarum polycephalum-based memristors. In *Proceedings of the International Conference on New Interfaces for Musical Expression*, R. Michon and F. Schroeder, editors. Birmingham City University, Birmingham, UK, July,

2020, 514–519. Doi: 10.5281/zenodo.4813507. URL: https://www.nime.org/proceedings/2020/nime2020_paper98.pdf.
2. S. Colton, G. A. Wiggins, *et al.*, Computational creativity: The final frontier? In *Ecai*, vol. 12, 2012, 21–26.
3. D. Cope, *Experiments in Musical Intelligence.* A-R Editions, Inc., 1996.
4. S. M. Ross, *Stochastic Processes*, 2 edn. JohnWiley & Sons, Inc., 1996.
5. K. Jones, Compositional applications of stochastic processes. *Comput. Music J.* 5(2):45–61, 1981.
6. P. M. Todd, *Comput. Music J.* 13(4):27–43, 1989.
7. E. R. Miranda and J. Al Biles, *Evolutionary Computer Music.* Springer, 2007.
8. J. D. Fernández and F. Vico, Ai methods in algorithmic composition: A comprehensive survey. *J. Artif. Intell. Res.* 48:513–582, 2013.
9. A. Adamatzky, *Physarum Machines: Computers from Slime Mould*, vol. 74, *A*, World Scientific, 2010.
10. P. W. Shor, Polynomial-time algorithms for prime factorization and discrete logarithms on a quantum computer. *SIAM Rev.* 41(2):303–332, 1999.
11. L. K. Grover, A fast quantum mechanical algorithm for database search. In *Proceedings of the Twenty-Eighth Annual ACM Symposium on Theory of Computing*, STOC '96, Association for Computing Machinery, New York, NY, USA, 1996, 212–219. ISBN 0897917855. Doi: 10.1145/237814.237866. URL: https://doi.org/10.1145/237814.237866.
12. T. Nakagaki, H. Yamada, and Á. Tóth, Intelligence: Maze-solving by an amoeboid organism. *Nature.* 407(6803):470, 2000.
13. A. Adamatzky, J. Jones, R. Mayne, S. Tsuda, and J. Whiting, Logical gates and circuits implemented in slime mould. In *Advances in Physarum Machines*, Springer, 2016, 37–74.
14. T. Nakagaki, H. Yamada, and A. Toth, Path finding by tube morphogenesis in an amoeboid organism. *Biophys. Chem.* 92(1-2):47–52, 2001.
15. E. Gale, A. Adamatzky, and B. de Lacy Costello, Slime mould memristors, *BioNanoScience.* 5(1):1–8, 2015.
16. L. Chua, Memristor-the missing circuit element. *IEEE Trans. Circuit Theory.* 18(5):507–519, 1971.
17. D. B. Strukov, G. S. Snider, D. R. Stewart, and R. S. Williams, The missing memristor found. *Nature.* 453(7191):80, 2008.
18. G. S. Snider, Spike-timing-dependent learning in memristive nanodevices. In *Proceedings of the 2008 IEEE International Symposium on Nanoscale Architectures*, 2008, 85–92.
19. B. Linares-Barranco and T. Serrano-Gotarredona, Memristance can explain spike-time-dependent-plasticity in neural synapses. *Nat. Prec.* 4(09): 2010.
20. O. Bichler, D. Querlioz, S. J. Thorpe, J.-P. Bourgoin, and C. Gamrat, Extraction of temporally correlated features from dynamic vision sensors with spike-timing-dependent plasticity. *Neural Netw.* 32:339–348, 2012.
21. E. Braund and E. R. Miranda, On building practical biocomputers for real-world applications: Receptacles for culturing slime mould memristors and component standardisation. *J. Bionic Eng.* 14(1):151–162, 2017.

22. E. Braund, S. Venkatesh, and E. R. Miranda, Phybox: A programmable interface for physarum polycephalum-based memristors. *Int. J. Unconv. Comput.* 14(3-4):217–233, 2019.
23. S. Venkatesh, E. Braund, and E. R. Miranda, A nonlinear approach to generate creative data using physarum polycephalum-based memristors. In *Slime Mould in Arts and Architecture*. A. Adamatzky, editor, River Publishers, 2019.
24. M. Bosi and R. E. Goldberg. *Introduction to Digital Audio Coding and Standards*. Kluwer Academic Publishers, 2003.
25. C. E. Shannon, A mathematical theory of communication. *Bell Syst. Tech. J.* 27(3):379–423, 1948.
26. M. T. Pearce and G. A. Wiggins, Auditory expectation: The information dynamics of music perception and cognition. *Top. Cogn. Sci.* 4(4):625–652, 2012.

https://doi.org/10.1142/9789811257155_0036

Chapter 36

The Recent Biotechnology of Tech-tattoo: An Approach from Science, Art, and Philosophy

Catarina Pombo Nabais*

> "*The Panopticon is a discipline machine for dissociating the see/being seen dyad: in the peripheric ring, one is totally seen, without ever seeing; in the central tower, one sees everything without ever being seen*" Michel Foucault, *Discipline and Punish. The Birth of Prison*

More or less archaic in terms of techniques and tools, more or less artistic in terms of the quality of the painting, tattooing is a transformative body practice that has been continuously performed, in all civilizations.[a] Its practice is deeply connected with two of the biggest ambitions of humankind: immortality and enhancement. Alongside mummification practices, the art of marking the skin is the most primitive way of turning the body memorable for eternity.

*Catarina Pombo Nabais holds a PhD in Philosophy from the University of Paris 8 (2007). Researcher at the Faculdade de Ciências da Universidade de Lisboa, Departamento de História e Filosofia das Ciências. Integrated member of the Centro de Filosofia das Ciências, Universidade de Lisboa, Campo Grande 1749-016 Lisboa, Portugal. Email: ccnabais@fc.ul.pt. This work is financed by national funds through FCT — Fundação para a Ciência e a Tecnologia, I.P., within the scope of the Norma Transitória - DL57/2016/CP1479/CT0063.

[a]The most ancient tattooed body dates from 5,000 years ago, but tattoos are known to be practiced since the Upper Paleolithic period (10–30,000 years ago). For further information concerning a historical overview of tattoo, see Taliaferro, Charles and Odden, Mark (2012)[1] and Pombo Nabais, Catarina (2017).[2]

Also, in several cultures, tattooing appears connected to diverse beliefs on body enhancement, being generally practiced with a religious and spiritual determination, such as bringing fertility to women or offering divine, magical strength to warriors and hunters.

Endowed with diverse uses, purposes, and meanings in each culture and throughout history, tattooing has been both socially valued and repressed or even marginalized. That is the case of European societies where, at the end of the 19th century and beginning of the 20th, the tattoo became the symbol of an indigent, clandestine, and underground life, sometimes even criminal. In Europe at the end of the 19th century, some medical academies considered tattooing to be the expression of a psychopathological disorder. In this context, tattooed people were stigmatized as people with an innate predisposition to criminal practice. A little throughout Europe, several collections of tattooed skin were made, almost always accompanied by a detailed documentary record that aimed to establish the typical portrait of the criminal.[b] This documental practice responds to the constitution of a new scientific discipline: criminology. And this new discipline was able to circumscribe a new object of study: the delinquent.

However, at the end of the 20th century, an unexpected phenomenon erupted. The tattoo became fully democratized all over the world, asserting itself as a common expression of individuals of all kinds, ages, and cultures. As the world itself, the tattoo became global. Everywhere there are tattoo studios providing a professional, licensed, and recognized service. Every day the tattooed body is glorified by an increasing number of well-known tattoo artists, with diverse styles and artistic techniques — from figurative to abstract and from watercolors to *pointillisme* — thus reinforcing tattoo's artistic dimension. More and more relevant and widespread,

[b]The French philosopher Michel Foucault is the one who best explains the emergence and the operational mode of the documentary machine that supports this type of collections. In his book *Discipline and Punish. The Birth of Prison* (1975), he explains how the constitution of such collections responds to a program of knowledge and control of individuals by medical, prison, or military institutions.

tattoo became the object of attention of diverse artistic media, from literature to cinema, from photography to music and fashion.[c]

Now, in the midst of this globalization, tattooing has also become invested by science and technology. A new kind of very sophisticated tattoo emerged: the technological tattoo (tech-tattoo). Done with the most up-to-date techniques and tools, tech-tattoo is becoming more, much more, than a simple engraving of the skin. It may only be visible in the dark, overcoming the eye's capacity of seeing in the day light; it may be practiced inside the body; it may be connected to the body's biochemical system, as a transdermal sensor; it may be linked to technological apps and computing services as a remote wireless sensor. In sum, science and technology transformed tattooing into a high-tech functionality, providing it with new skills, new capacities, new possibilities, and new horizons. If the tattoo was traditionally considered to have mystical, magical powers, with science and technology, now it is endowed with new powers that go literally beyond the natural limits of the body. It is thus amazing to realize that one of the most ancient practices of body inscription

[c]In literature, the first great reference would be the classic *Moby Dick* (1851) by Herman Melville. One of the more complex characters, Queequeg, is tattooed. Also, in 1951, Tennessee Williams wrote the play *The Rose Tattoo.* We can also spot *The Illustrated Man* (1951) by Ray Bradbury. In cinema, examples are innumerable. As a 20th century art, cinema could not forget the practice of tattooing, which has become a cultural trait of youth and irreverence. From the classic *The Night of the Hunter* (1955), in which a serial killer priest (Robert Mitchum) tattooes the fingers of each hand with the words "Love" and "Hate" or Robert De Niro in *Taxi Driver* (1976) and *Cape Fear* (1991) whose tattoos are a sign of his violence and perdition. In photography: the case of the Whand-Og portrait, titled *The Last Living Mambabatok* in the Philippines, which shows a tattooed woman who is also the last traditional tattoo artist of the Butbut tribe in the Philippines by artist Rhonson NG. Even in music, there is the case of Chico Buarque with his magnificient song titled *Tatuagem* (Tattoo), written in 1972, as well as the famous song *I've Got You Under My Skin* that Cole Porter created in 1936. Finally, having become a fashion in itself, the tattoo could not fail to appear in fashion. We highlight the irreverent and "politically incorrect" gesture of Issey Miyake when he presented his *Tattoo* fall/winter collection in New York in 1971. A trend that marked several other fashion designers, such as Jean-Paul Gaultier, *Les Tatouages* (Spring 1994).

such as tattooing becomes a recent area of research in science and technology, which take the idea of tattoo into unsuspected levels.

1. Overcoming Visibility

Figure 1 shows how recent tattoo developments constitute a challenge to the usual visibility, both the day-light and the natural body's opacity. The first example concerns tattoos made with ultraviolet reactive inks. The surprising particularity of these tech-tattoos is the fact that they are nearly invisible to the naked eye in daylight and even in the dark. In fact, these tattoos are only visible on a special light, the black light. Besides, they are not only visible; they also literally glow. They glow because they are fluorescent. They only get visible under an ultraviolet black-light bulb. Even if phosphor started to be the ink's main ingredient to achieve the glow-in-the-dark effect due to its capacity of absorbing energy and reemitting it as visible light, it is very toxic and dangerous to the skin, having diverse and serious side-effects.[d] Besides, UV ink is thinner than the regular ink

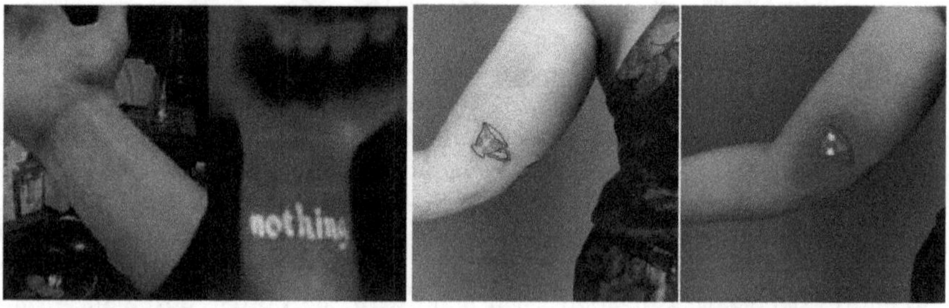

Fig. 1. On the left, a UV tattoo made 100% with UV ink, thus being totally invisible in daily light. Image credit: pinterest.com. On the right, a UV tattoo made by the tattoo artist Kayla Newell that mixes traditional ink with UV ink. Image credit: Kayla Newell.

[d]That is why almost any of the UV inks used for this kind of tattoos is yet to be approved by the U.S. Food and Drug Administration, because there are some restrictions about the neon pigments which are highly toxic chemicals, such as phosphorus, which is suspected of being a carcinogen. According to this institution, only the luminescent zinc sulfide is approved for UV tattoos. Also, many of the ink's pigments are made mostly for printers or automobile painting.

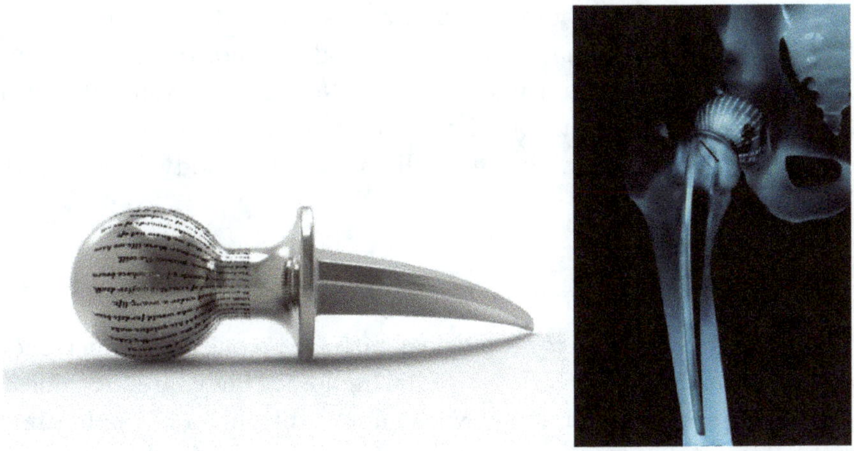

Fig. 2. On the left, the image (photography) of one tattoo engraved prothesis, before the operation. On the right, the image (X-Ray) of one tattoo engraved prothesis after the medical operation. Both images are from Olga Noronha's *Jewelry Collection*. Image credit: Daniel Pires.

for traditional tattoos and very hard to work with, thus requiring a specialized tattoo artist to apply it.

The second example is even more radically invisible to naked eyes, completely unnoticed by sight, even if under black light. This is a tattoo that is not made on the surface of the skin, but on the inside of the body, registered in the applied medical prostheses (Figure 2). That is, a tattoo that is only visible by X-ray. Olga Noronha, a Portuguese artist designer, has created a jewelry, series called *Medical Jewelry*, which consists of medical prostheses where a tattoo can be imprinted.

The creation of this medical jewelry-tattoo articulates, not only science (by the development of new medical prostheses), technology (by the application inside the body of biotechnological implants), but also design (the jewelry) and art (or, in this more precise case, bioart). Noronha's work falls within the proximity of artists like Sterlac or Orlan who conduct radical experiments in transforming their own bodies, thus contributing to the debate on the reconfiguration of the body and its value as a work of art. But, if the art of Orlan or Sterlac is visible to the naked eye, that of Noronha is completely invisible. By inscribing tattoos, not in the superficial skin but into prostheses, that is, by turning tattoos not only private and intimate, but totally invisible to others, Noronha's jewelry collection highlights a new dimension of tattooing. Tattoo thus exists neither on nor under the skin, but literally inside the body, close to the bones.

2. Entering the Biowearable Smart World

The next examples of technological tattoos belong to a much more advanced and profound level of technology. They are not just a practice of engraving an image with an invisible ink or of implanting a technological device inside the body. They go much further. They are data emitters with the capacity of analyzing data and sending it to computer systems they are connected to, that is to say, these tattoos are digital, smart, intelligent devices. Made of electronic components such as electro-conducive ink, biosensors, curvy wires, and thermochromic ink, tech-tattoos are exponentially expanding. Even if still temporary,[e] these tech-tattoos aim at becoming permanent biowearable smart devices. Due to the fast-technological evolution and to their cheap and easy process of fabrication, tech-tattoos are being appropriated for an enormous range of concrete purposes having direct impact in daily life. They may have utilitarian purposes such as providing a payment system, tracking individuals in space, or giving instructions to the Wi-Fi devices to which the tattooed individual may be connected. Or they may have a deeper bio-medical-political dimension when, e.g., they make possible measuring the body's temperature, the heart beats, the level of alcohol or the blood pressure, supervising fitness, computing sleep patterns, in a word, monitoring vital, bio-metric data.

The so-called *Soundwave Tattoo*™ is inscribed in the skin as a traditional tattoo, but is made with a special ink that is linked to Wi-Fi devices (Figure 3). It can reproduce a sound that is previously

[e]For now, this new form of tech-tattoo is still temporary because conductivity is lost through skin's natural resistance.

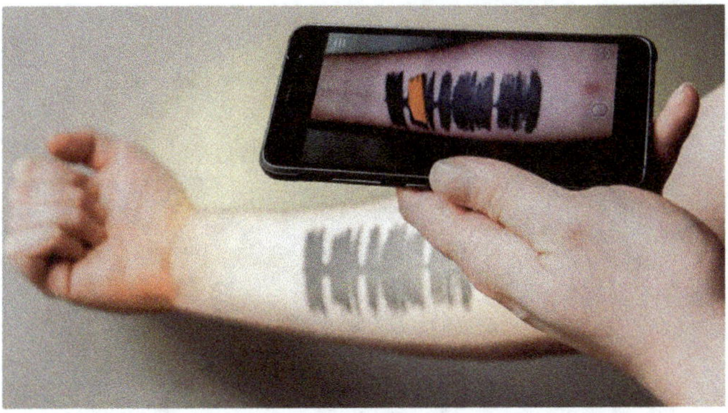

Fig. 3. In the *Skin Motion* website, there are some recommendations about the *Soundwave tattoo*: It should be no larger than 6 inches or 150 mm to be fully captured by the camera on the phone, it should be done in a flat surface such as the inner-forearm and it should be placed where one can easily hold the phone to play the tattoo back with the *Skin Motion* app. Image credit: Sarah Tew/CNET.

memorized in the drawing. First, an audio clip is uploaded on the website of *Skin Motion* company and then a certified artist will grave it in the skin as a traditional tattoo but with a conductive ink. It was invented in April 2017, by the tattoo artist Nate Siggard who shared his invention in a video on Facebook that immediately went viral (over 150 million views in the first month). He then created the *Skin Motion* company, specialized in what he expresses as "personalized Augmented Reality Tattoos".[f]

The next example, *Duo skin*, developed by researchers at the MIT Media Lab (Cindy Hsin-Liu Kao, Asta Roseway, Christian Holz, Paul Johns, Andres Calvo, Chris Schmandt) in collaboration with Microsoft Research, is constituted by an electronic circuit containing a set of biosensors which send data wirelessly when touched and manipulated (Figure 4). The main material used in *DuoSkin* is gold leaf, which is, at the same time, a very strong, resistant, very flexible, and easy to apply material since it adheres directly on the skin. As epidermal electronics, this tech-tattoo functions as a double skin. It is used as a remote control by its wireless communication capacity.

[f] *Skin Motion. Tattoos brought to life*, https://skinmotion.com/soundwave-tattoos/. Assessed May 10, 2021.

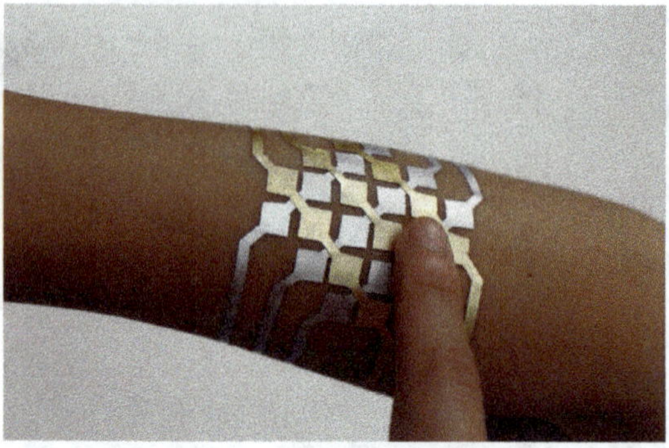

Fig. 4. *Duoskin.* Besides the gold leaf, these tattoos are made of layered adhesive, thermochromic pigment, and a silicone interface. Due to their aesthetical value, they are also provided as jewelry and may be customizable. Image Credit: MS Media/MIT Media Lab.

Made with sensor-based technology it functions as a touch-sensitive interface for controlling daily services such as bank payments, or other smart devices installed at home, at the car, or in the cellphone.

The *DermalAbyss* tech-tattoo is the result of a collaboration between MIT researchers (Katia Vega, Xin Liu, Viirj Kan, and Nick Barry), and Harvard Medical School researchers (Ali Yetisen and Nan Jiang) (Figure 5). It aims at identifying the levels of glucose and sodium in the blood to help in diabetes medical care.[g] This tech-tattoo is made with biosensitive ink which transforms the skin into a color sign emitter of biochemical changes in the body. Depending on blood glucose, sodium, or pH levels, the skin emits various shades

[g]There are a lot of tech-tattoos developed with medical purposes. In 2017, researchers tattooed pigskin, which had been removed from the pig, with molecular biosensors that use color to indicate sodium, glucose, or pH levels in the skin's fluids. In 2019, a team of researchers expanded on that study to include protein sensing and developed smartphone readouts for the tattoos. This year, they also showed that electrolyte levels could be detected with fluorescent tattoo sensors. In 2018, a team of biologists developed a tech-tattoo made of engineered skin cells that darken when they sense an imbalance of calcium caused by certain cancers. They demonstrated the cancer-detecting tattoo in living mice, https://theconversation.com/dynamic-tattoos-promise-to-warn-wearers-of-health-threats-133040. Accessed February 4, 2022.

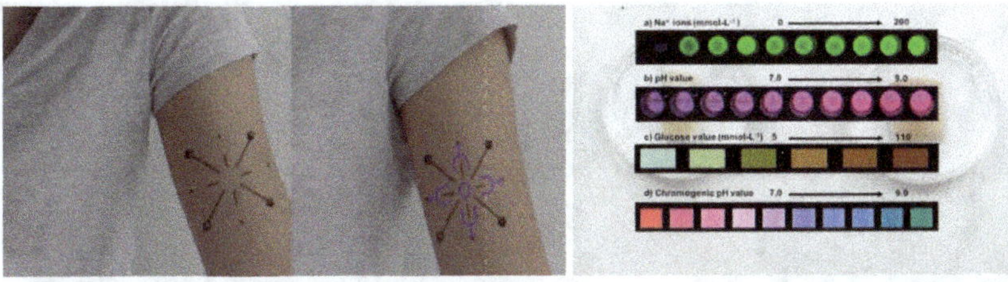

Fig. 5. On the left, *Dermal Abyss* tattoo. On the right, *Dermal Abyss* color palette. Both photos are from Nan Jiang, postdoctoral fellow at the School of Engineering and Applied Sciences at Harvard University who has been working at Harvard Medical School undertaking research on nano biotechnology and biosensors for the *Dermal Abyss* project.

Fig. 6. When UV ray light illuminates the skin (on the right), two blue spots appear on the skin (on the left). Both images are taken from TED talk by Carlson Burns (14/05/2019).[h]

of green. The use of biosensitive ink solves the problem of biomedical monitoring devices, which always depend on the battery life of the sensors.

The last example comes from Carson Burns, a researcher in molecular nanotechnology at the University of Colorado, who is developing a tattoo that goes from invisible to colored spots when the skin is overexposed to sun light and to UV rays (Figure 6). Powered by solar energy, this tech-tattoo has in its microparticles a UV sensitive color changing dye. Working as an alert, appearing only when the skin is overexposed to UV radiation, this tech-tattoo may help in sunburn or cancer prevention.

---

[h] https://www.ted.com/talks/carson_bruns_could_a_tattoo_help_you_stay_healthy/transcript#t-403948 Accessed January 14, 2022.

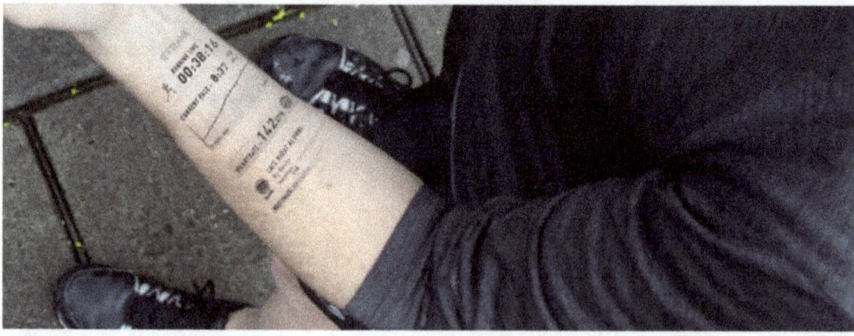

Fig. 7. Software company *Chaotic Moon* has developed a tech-tattoo that is implanted into a person's arm, capable of sending a person's financial and medical information. According to the *Chaotic Moon*'s hardware creative technologist Eric Schneider, "with the tech-tattoo you can carry all your information on your skin and when you want your credit card information or your ID, you can pull that up automatically through the system".[i] Image credit: www.emergeinteractive.com.

Tech-tattoo is incredibly expanding. However, it is important to note that the investment that is being done from international companies allied with research centers and science laboratories makes the biopolitical power of tech-tattoo very clear. With tech-tattoo, data emission on individuals' lives is done not just from the cell phone but from bio-nano-sensors of a body's artificial device (Figure 7). The cell phone — which is already a smart tool — is able to spread lots of information about our movements, tastes, social milieus, number of daily steps, and so on. But with its bio-nano-sensors, tech-tattoo goes deeper, informing about the diseases, the genetics, the very chemistry of each individual. And in providing all this multitude of inner, intimate data, tech-tattoo is in total, instantaneous, and permanent connection with the control technologies around us. Therefore, we may say that "bio-wearable" tech-tattoo turns the body into a touchscreen/touchpad, a digital cyborg exempted from privacy. It transforms the body into a bio-tech surface, a smart and quantifiable canvas which can be easily used by power structures.

[i]https://newyork.cbslocal.com/2016/01/29/tech-tattoos-chaotic-moon/. Accessed October 20, 2021.

3. Tech-tattoo and Control: A Philosophical Approach from Foucault and Deleuze

What if one could transform the skin into the most intelligent, smart, interactive technological device? What if one's skin would be able to exhibit all kinds of internal data about one's body's biochemistry and nervous system? What if one's skin, the most thin and fragile surface of one's body, was the most profound surveillance device? To put it in a straightforward way: what if one's skin was the greatest obstacle to one's freedom?

In his book *Discipline and Punish. The Birth of Prison* (1975),[3] Foucault stresses that the disciplinary society of the 18th and 19th centuries is governed by a regime in which individuals are manipulated, formatted, shaped, by learning how to behave in certain social spaces (spaces of working, of learning, of leisure, of living, etc.), acquiring specific gestures and postures according to specific physical spaces (ex: children learn to be seated in the classroom, soldiers learn to never fix their eyes on the ground at the military service, and so on). Thus, fixed, specific physical enclosed spaces such as schools, industries, hospitals, military barracks, prisons, and so on, are especially projected in order to modulate, to regulate, to discipline all individual's gestures, activities, and even use of time. This disciplinary model goes in parallel with the invention of the Panopticon, which is as an architectonic configuration ruled by two simple symmetrical rules: to observe (the individual) without being observed; to be observed (by power in the figure of the police, the shrink, the judge, and so on) in every single corner of the structure. With the Panoptic regime, all the buildings, from hospital, factories, schools, become organized by the same control model of the prison.

Fifteen years later, in the famous chapter "Postscript on the Societies of Control" from *Negotiations* (1990),[4] Deleuze stresses the need of updating this Foucauldian concept of "Disciplinary Society" into the concept of "Society of Control". Deleuze took the term "Control" from William Burroughs as best benefiting the new power regime put forward since the World War II: "the new monster" (Deleuze, 1990: 178). In this new regime, a much more sophisticated — subtle, diffuse, imperceptible — model of control is put forward. New control devices are adopted. They go in line with the huge technological and computational advancements spread all over the world. Deleuze seems to be terribly prescient: in the early 1990s, prior to the sovereignty of the internet, he understood the continuous, fluid world that was coming. "Disciplinary man produced energy in discrete amounts, while control man undulates, moving among a continuous range of different orbits" (Deleuze, 1990: 180).

Contrary to disciplinary societies in which spatial enclaves were crucial, in control societies, with the advent of internet, space became an endless continuum, characterized by differential potentialities. The internet transformed the relation of the individual with the space. But it has also contributed to a radical change at the level of the language system itself: disciplinary society is analogical; control society is digital. If, in disciplinary societies, the individuals are "numbers or places in a register standing for their position in a *mass*" (Deleuze, 1990: 179), in control societies, they are no longer individuals but "dividuals", that is, virtual entities identifiable by codes and passwords within a generalized database. "The digital language of control is made up of codes indicating whether access to some information should be allowed or denied. We're no longer dealing with a duality of mass and individual. Individuals become '*dividuals*', and masses become samples, data, markets, or '*banks*"' (Deleuze, 1990: 180).

This new kind of power regime, coming from molar architectural space-time to molecular and continuous space-time, has also changed the concept of body which is no longer the mute body shaped by discipline of space and time. In control societies, body is marked by a regime of signs as the inscription of debt inside the capitalist flux which, by digital data, has a perfect and total command of the body's life. The modern body has become the locus of constant social management, the satellite unit, a control station without any difference between interior/exterior. The modern body is permanently connected to power structures (either — or perhaps mostly — in its free time). In a disciplinary society, when I am out of the factory, my time is my own, until I go back tomorrow, for another working day. In a society of control, while freedom seems to be increased on the one hand, the control of our activities expands on the other. While disciplinary societies function based on physical confinement and enclosure, control societies create an ambiguous experience of freedom: beyond the easy, fast, and free communication among individuals that are no longer restricted to space (as space became virtual, fluid), control becomes a permanent, intimate, and diffuse state of affairs.

4. From the Panopticon to the Superpanopticon

Surprisingly, for Foucault, the Panoptic regime was the characteristic of the 18th–19th centuries disciplinary society and, at the same time, the first step to enter the contemporary society of control. Although it is a material, physical, architectonical structure, restricted to space confinement and time limitations, the Panopticon, as a total surveillance model, already contains in itself the possibility of its application in a diffuse, modular, fluid, virtual control society (Figure 8). As Foucault noted, the Panopticon — since its very creation and idealization by Bentham — has in itself the potentiality of becoming a transversal and widely distributed model of control. As Foucault writes: "The celebrated, transparent, circular cage, with its high tower, powerful and knowing, may have been for Bentham

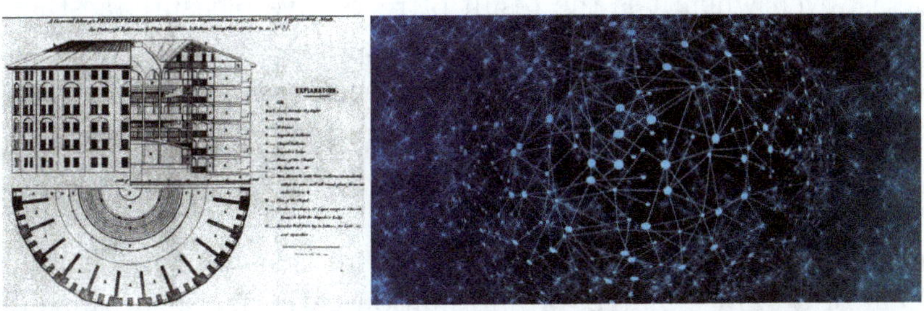

Fig. 8. On the left: image of an American circular prison built in accordance to the Panopticon model idealized by Jeremy Bentham in the 18th century. This surveillance device has in its center a tower with a perfect overview of every single prisoner who only moves in the delimited circular perimeter. Image credit: https://www.abc.net.au/. On the right: iconic representation of the web. Image credit: https://pixabay.com/.

a project of a perfect disciplinary institution; but he also set out to show how one may 'unlock' the disciplines and get them to function in a diffused, multiple, polyvalent way throughout the whole social body (···). Bentham dreamt of transforming it into a network of mechanisms that would be everywhere and always alert, running through society without interruption in space or in time. The panoptic arrangement provides the formula for this generalization" (1995: 208–209). This means that the Panopticon is enhanced when connected to the internet, when upgraded to a virtual, diffused, world wide web, when expanded toward an unlimited space-time. From a restricted physical application, it becomes an unrestricted, a transversal, an immanent device of control anywhere and anytime.

5. Tech-tattoos or the Superpanopticon Body

What I would like to stress is that tech-tattoo may turn Bentham's dream into reality, as it takes the panoptic control regime into another level of accuracy and immanence. Tech-tattoo belongs to this new paradigm of a full-immanent Panopticon, the one that operates from inside the body. Both disciplinary and control regimes, even if operative and efficient, have a limit: the body itself and its insurmountable opacity. Now, with tech-tattoo, control may surpass, for the first time, the body's opacity. Even when it is well-intended, that is, even when it is the result of recreative pleasure, aesthetical experimentation, or medical concern with the individual's life quality, tech-tattoo allows control to become much more intrusive, much more intimate, in a word, immanent and unlimited.

Rather than Jeremy Bentham's physically limited model of the Panopticon, with a rigid centralized focal point from which individual's activities are surveyed, tech-tattoo actualizes Bentham's dream of a diffuse matrix of information gathering algorithms. The "Panopticon" becomes what I suggest naming as "Superpanopticon": we know that we are being watched but we are encouraged not to worry about it. And this normalization of *surveillance* has become constitutive, intimate, immanent to the modern body, inside a world of "universal modulation".

I will finish with an acute quotation from Deleuze: "Felix Guattari has imagined a town where anyone can leave their flat, their street, their neighborhood, using their (dividual) electronic card that opens this or that barrier; but the card may also be rejected on a particular day, or between certain times of the day; it doesn't depend on the barrier but on the computer, that is making sure everyone is in a permissible place, and effecting a universal modulation" (1990: 181–182).

References

1. C. Taliaferro, and M. Odden, Tattoos and the tattooing arts in perspective: *An overview and some preliminary observations.* In Tattoos: Philosophy for everyone: *I Ink, Therefore I Am*, Robert Arp, editor. John Wiley & Sons, Oxford, 2012, 4.
2. Pombo, Nabais, Catarina The most profound is the skin — The power of tattoos. In *Taboo-Transgression-Transcendence in Art & Science*, Honorato, Dalila & Andreas Giannakoulopoulos, editors. Corfu: Department of Audio & Visual Arts — Ionian University, 2017, 128–148.
3. M. Foucault, (1995). *Discipline and Punish. The birth of the Prison*, Eng. Transl. by Alan Sheridan, New York: Vintage Books/Random House [original edition, *Surveiller et Punir. La naissance de la prison.* Paris, 1995], 1995.
4. G. Deleuze, *Negotiations*, Eng. Transl. by Martin Joughin, Columbia University Press, New York, [original edition: *Pourparlers*, Paris: Minuit, 1990], 1995.

https://doi.org/10.1142/9789811257155_bmatter

Index

J

K

L

M

T

U

www.ingramcontent.com/pod-product-compliance
Lightning Source LLC
LaVergne TN
LVHW010537300926
849162LV00002B/158